Practical Neurochemistry

Edited by

HENRY McILWAIN Ph.D., D.Sc.

Professor of Biochemistry in the University of London at the Institute of Psychiatry (British Postgraduate Medical Federation); Honorary Biochemist, The Bethlem Royal Hospital and the Maudsley Hospital

SECOND EDITION

CHURCHILL LIVINGSTONE
Edinburgh London and New York 1975

CHURCHILL LIVINGSTONE
Medical Division of Longman Group Limited

Distributed in the United States of America by Longman Inc., New York and by associated companies, branches and representatives throughout the world

© Longman Group Limited 1975

All rights reserved. No part of this publication may be reproduced, stored in a retrieval system, or transmitted in any form or by any means, electronic, mechanical, photocopying, recording or otherwise, without the prior permission of the publishers (Churchill Livingstone, 23 Ravelston Terrace, Edinburgh).

First edition 1962
Second edition 1975

ISBN 0-443-01258-X

Library of Congress Cataloging in Publication Data
Main entry under title:

Practical neurochemistry

    Includes bibliographies and index.
    1. Neurochemistry. I. McIlwain, Henry.
[DNLM: 1. Neurochemistry. WL104 P895]
QP356.3.M3 1975   612'.8'042   74-21555

Printed in Great Britain

Preface

Study of neural systems has often been in the forefront of biochemical discovery, and yet it features little in the available accounts of practical biochemistry. Writing the present book was prompted by this deficiency and by the needs of neurochemistry itself. The book has developed from practical teaching and demonstrating to postgraduate and undergraduate students, in faculties of medicine and science; and from research and research training based primarily on mammalian neural systems. The needs of these categories of students and research workers have been borne in mind throughout, and it is intended that the book should be of service to all who are concerned with neural systems in chemical, biochemical, physiological, pharmacological or pathological study.

Annual courses in practical neurochemistry have been run at this Institute since 1960. Lecturers and demonstrators who have contributed to the courses now total twenty or more, and students over two hundred. All have contributed to the knowledge and experience displayed in this book, and it has been appropriate to enlarge its authorship, so that it now becomes an edited volume to which seven authors contribute the specialities which they teach and practise in London University. Neurochemical teaching in this University has greatly increased since the inauguration, in 1970, of an M.Sc. degree in neurochemistry. Teaching for the degree given in three Institutes of the British Postgraduate Medical Federation, involves a large element of practical laboratory work: partly with detailed instruction and partly as research projects. This book is intended to assist both types of work.

For the abbreviations and conventions involved in chemical description we have followed the practice of *The Biochemical Journal* and *The Journal of Biological Chemistry*; temperatures are in degrees Centigrade.

Institute of Psychiatry,
London, 1975.   H. MCILWAIN

Contributors

H. S. BACHELARD, D.Sc., Ph.D. Reader in Neurochemistry,
Dept. of Biochemistry, Institute of Psychiatry (British Postgraduate Medical Federation, London University), De Crespigny Park, London SE5 8AF.

M. L. CUZNER, B.Sc., B.A., Ph.D. Honorary Lecturer,
Dept. of Neurochemistry, Institute of Neurology (British Postgraduate Medical Federation, London University), Queen Square, London WC1 3BG.

H. McILWAIN, D.Sc., Ph.D. Professor of Biochemistry,
Dept. of Biochemistry, Institute of Psychiatry (British Postgraduate Medical Federation, London University), De Crespigny Park, London SE5 8AF.

R. M. MARCHBANKS, B.A., Ph.D., M.A. Lecturer in Biochemistry,
Dept. of Biochemistry, Institute of Psychiatry (British Postgraduate Medical Federation, London University), De Crespigny Park, London SE5 8AF.

R. RODNIGHT, D.Sc., Ph.D. Reader in Biochemistry,
Dept. of Biochemistry, Institute of Psychiatry (British Postgraduate Medical Federation, London University), De Crespigny Park, London SE5 8AF.

J. M. TURNBULL, B.Sc., Ph.D. Lecturer in Biochemistry,
Dept. of Biochemistry, Charing Cross Hospital Medical School (London University), Fulham Palace Road, London W6 8RF.

M. J. VOADEN, B.Sc., Ph.D. Senior Lecturer in Biochemistry,
Institute of Ophthalmology (British Postgraduate Medical Federation, London University), Judd Street, London WC1H 9QS.

Contents

	Page
Preface	v
Contributors	vii

CHAPTER

1.	Obtaining, Fixing and Extracting Neural Tissues. *R. Rodnight*	1
2.	Constituents of Neural Tissues: Fluids, Some Electrolytes, Amino Acids and Ammonia. *R. Rodnight*	17
3.	Constituents of Neural Tissues: Intermediates in Carbohydrate and Energy Metabolism. *H. S. Bachelard*	33
4.	Lipids: Extraction, Thin-layer Chromatography and Gas-Liquid Chromatography. *M. L. Cuzner and J. M. Turnbull*	60
5.	Individual Lipids: Preparative and Analytical Methods. *M. L. Cuzner*	88
6.	Preparing Neural Tissues for Metabolic Study in Isolation. *H. McIlwain*	105
7.	Metabolic Experiments with Neural Tissues. *H. McIlwain*	133
8.	Electrical Stimulation of the Metabolism of Isolated Neural Tissues. *H. McIlwain*	159
9.	Maintenance of Isolated Parts of the Brain for Electrical Measurements. *H. McIlwain*	191
10.	Cell-Free Preparations and Subcellular Particles from Neural Tissues. *R. M. Marchbanks*	208
11.	Isolation and Study of Enzymes from Neural Systems. *H. S. Bachelard*	243
12.	Subsystems and Regions of the Mammalian Brain; The Retina. *H. McIlwain and M. J. Voaden*	261
13.	Bodily Metabolites and Drugs. *R. Rodnight*	293

Index   323

1. Obtaining, Fixing and Extracting Neural Tissues

R. RODNIGHT

Obtaining the brain from laboratory animals	1
Fixation of the brain *in situ*	4
Methods available	5
Illustrative experiment using liquid nitrogen	7
Comment on the validity of fixation methods	9
Extracting neural tissues for analysis	10
Apparatus	10
Extracting agents	11
References	15

Sensitivity and speed of response are characteristics of neural systems; they have their counterparts in the lability of many of their chemical constituents, and changes ensuing in brief periods while tissues are being sampled should always be taken into account in planning neurochemical investigation. The brain is the neural organ which yields most material and is most accessible for chemical study, and this account therefore begins by describing procedures for sampling the brain of laboratory animals: in the fresh state as required for metabolic experiments or analysis of stable constituents, and also by methods which fix the tissue *in situ* and permit determination of labile constituents.

OBTAINING THE BRAIN FROM LABORATORY ANIMALS

The chemical composition of the brain is influenced by the physiological and behavioural state of the animal at the time of death. As far as possible, therefore, the experimental animal should be kept in its normal environment and subjected to only the minimum of restraint during the period immediately preceding an experiment. The following procedure permits removal of the cerebral hemispheres and midbrain from small animals such as rats and guinea pigs within 2 min of death, but in the process the cerebellum and brain stem are damaged. If the latter structures are required the procedure given below for rabbits is applicable to these animals, although it is more lengthy.

Procedure: rats and guinea pigs
   The only instruments required are a pair of sharp-pointed scissors (12–15 cm) and a narrow, stainless-steel spatula (about 5–7 mm wide). The animal is stunned by a light blow accurately delivered on the back

of the neck about 1·5 cm behind the posterior tip of the skull. A direct blow on the head or too heavy a blow on the neck will cause cerebral haemorrhage and damage the brain; a steel bar weighing 0·5 kg is a suitable instrument for this purpose. Immediately after the blow the neck is cut so as to sever the major blood vessels; exsanguination minimizes the likelihood of bleeding into the skull while the brain is being removed. The skin above the skull is now quickly dissected away with one or more cuts of the scissors, exposing the unshaded portion of Fig. 1.1. With the scissors held vertically a cut is made through the muscles, bone and spinal cord just posterior to the occipital bones (Fig. 1.1, a–a_1). With the scissors still held vertically, but turned at right angles, a second cut is made through the occipital bone (Fig. 1.1, b–b_1). Then with the scissors still in the bone and with their points held a little apart, the scissors are twisted so as to cause the cut to extend as a crack along the sagittal suture of the cranium. The scissors are then withdrawn and used to prise away the parietal bones. To do this one point of the scissors is inserted below the bone near to the midline and the other point rests on the edge of the skull outside it; a twist of the scissors then lifts the bone. The other parietal bone is removed in a similar fashion. This procedure may or may not remove the frontal bones; if not, these must be removed in the same manner by prising them upwards.

The complete removal of the parietal and frontal bones is important if the cerebral cortex is afterwards to be sliced, for projections readily cut the cortex when the brain is being removed from the cranium. Care must also be taken to ensure that the points of the scissors do not damage the brain while the skull is being opened; the point should therefore be in contact with the skull (and not the brain) throughout the procedure. It is to avoid such damage that the cranium is opened as has been described by cracking it, rather than by a cut along the sagittal suture of the bone.

During the second vertical cut with the scissors, or in the process of prising the skull away, the dura is likely to be cut at the base of the hemispheres. If not a small incision in it is made at this point. The flat blade of the spatula is now slipped between the cortex and the dura, and, keeping the surface of the blade parallel to the cortical surface, the dura is cut so as to free almost entirely the cerebral hemispheres. The blade of the spatula is next used in a vertical position to cut through the brain stem by inserting it between the cerebral hemispheres and the cerebellum and pressing through to the bone; having reached the bone the spatula is moved from side to side as well as forward across all the base of the skull until it reaches the front of the cavity. In so doing the spatula severs the cranial nerves and optic tracts, leaving the cerebral hemispheres and associated structures free. These are then scooped up with the spatula, steadied on it with a light touch of the forefinger, and placed in a Petri dish containing moist filter paper.

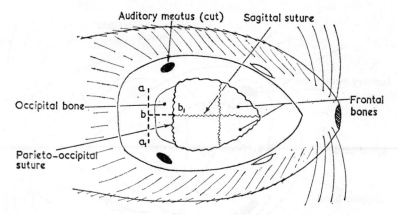

Fig. 1.1 Obtaining the brain from a guinea pig. Cuts made by scissors through the neck muscles and vertebral column: $a-a_1$; and through the occipital bone, $b-b_1$.

To extend the opening in the skull of the guinea pig, 'toe-nail cutters' (Scholl, London) are useful. These are scissor-like, 11 cm long and carry a single cutting blade 9 mm long, which operates perpendicularly to a plate which is easily inserted between the skull and dura without damage to tissue below. In this way the skull opening can be extended forward to expose the olfactory bulbs.

Procedure: rabbits

In addition to scissors and spatula, bone forceps and a scalpel are required. The animal is suspended by the hind legs with one hand and a blow sufficient to stun is delivered on the back of the neck. A correctly aimed blow need not be forceful and will stun the animal sufficiently to enable it to be exsanguinated by cutting the carotid vessels. The muscles covering the back of the skull and neck are dissected away with a scalpel. This should expose the foramen magnum, but if this is not immediately evident the head is tilted forward in relation to the neck, when the membrane covering the spinal cord as it leaves the cranium will be seen between the atlas and the foramen. The membrane is cut with pointed scissors. One point of the bone forceps is carefully inserted into the foramen in a direction parallel to the spinal cord. The occipital bone is then broken with a series of short nibbling cuts as far forward as the parieto-occipital suture and the same process is repeated on the other side (Fig. 1.2, $a-a_1$, $-a_2$). The occipital bones may now be removed by prising them upwards and forwards, when they will break along the line of the parieto-occipital suture (Fig. 1.2). If the floccular lobes are required the lateral cuts on either side are made through the temporal bones below the auditory meatus (Fig. 1.2, $a-c_1$, c_2); care is then needed when the bone cap is removed to avoid tearing the floccular

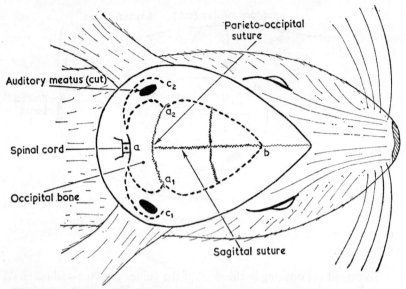

Fig. 1.2. Obtaining the brain from a rabbit. Dotted lines indicate the routes taken by bone forceps through the cranium.

lobes from the cerebellum and it will probably be necessary to break the temporal bones in a second direction. The parietal and frontal bones covering the cerebral hemispheres are removed by continuing to cut the bone along the outer edges of the cranium on each side in the direction indicated in Fig. 1.2 (a_1, a_2–b). The bone cap may be removed in two pieces by prising it upwards from each side, when it will break along the line of the sagittal suture; alternatively it may be removed in one piece provided that care is taken to prevent any membrane adhering to the bone from tearing the brain. After removing the bone the remaining membrane is cut with scissors. To remove the brain a curved spatula is used to lift it gently off the base of the cavity, starting at the rostral end. The optic tracts and cranial nerves may be cut with the spatula or with fine scissors.

FIXATION OF THE BRAIN *IN SITU*

The labile substances of the brain cannot be determined by analysis of tissue removed after death in the manner just described. For these constituents, methods of fixing the brain *in situ* have been developed. The most widely used method utilizes some form of rapid freezing of the brain by liquid nitrogen at a temperature of $-180°C$. Fixation by microwave irradiation is a recent innovation of great promise, requiring special equipment which is described subsequently.

Methods available

(1) *Methods applicable to anaesthetized animals.* Fixation with liquid nitrogen was first applied to relatively large laboratory animals such as the cat, dog or monkey and involved exposing the brain under anaesthesia and then freezing the cortex with copious quantities of liquid nitrogen (Kerr, 1935). The frozen brain is then chiselled out with chilled instruments and ground in suitable fixative. A major disadvantage of this method is that it introduces the use of an anaesthetic, which in itself alters the concentrations of labile constituents in the brain. For animals such as the cat or dog, however, it remains the major method available.

(2) *Total immersion methods.* In an alternative procedure originally due to Stone (1938) and which has since been extensively applied, a small animal such as a mouse or young rat is dropped into a vessel of liquid nitrogen and frozen whole. Brain tissue is removed from the frozen head with aid of a sharp chisel. Total immersion methods have been criticized in recent years on the grounds that the rate of cooling of cerebral tissue which occurs on immersion of the intact animal in liquid nitrogen is not sufficiently fast to ensure metabolic arrest before the outset of significant degrees of anoxia and ischaemia, particularly in the deeper parts of the brain. Some of these criticisms are considered below, but it remains a fact that for many purposes total immersion methods properly applied give acceptable results, although they appear to fall short of those attainable with the freeze-blowing technique mentioned under (4) below. The illustrative experiment described later in this chapter is an example of a total immersion method applied to rats of up to 100 g body weight, which should give values for creatine phosphate in the cerebral hemispheres in the range 3–4 μmol/g of frozen tissue.

Lowry and Passoneau (1971) considered immersion of the animals in Freon–12 (CCl_2F_2) cooled to $-150°C$ with liquid nitrogen to be superior to direct immersion in liquid nitrogen on the grounds that the extensive gasification which occurs with the latter procedure to some extent insulates the head. However, Swaab (1971) and Ferrendelli *et al.* (1972) found that the rate of cooling in liquid nitrogen alone to be as good as or superior to that obtained by immersion in cooled Freon-12.

(3) *Regional analysis.* A major disadvantage of total immersion methods in which the brain is sampled immediately on removal from liquid nitrogen is that satisfactory regional dissection of the brain is impossible. One approach to this difficulty was developed by Takahashi and Aprison (1964) in a procedure which avoids the actual freezing of the brain and therefore permits dissection. Animals were immersed in liquid nitrogen for intervals of time just sufficient to bring the temperature of the brain to 0°C; they were then removed, decapitated and the brains dissected in a cold room. The optimum time needed to cool the

brain to 0°C without freezing was predetermined in control animals with the aid of thermocouples inserted in the brain at various depths. The method gave satisfactory results for acetylcholine, but does not appear to have been applied to other labile constituents.

An alternative approach to the problem of regional analysis was developed by Lowry and his colleagues (Folbergrova et al., 1969). It has been extensively applied in studies of energy metabolism in mouse brain. The animal is frozen in liquid nitrogen, but before sampling the brain is allowed to warm up to −20°C. At this temperature dissection into the major regions is possible, while there is no loss of labile constituents. The dissection may be done in a cold room or kryostat held at −20°C.

(4) *Freeze-blowing.* To surmount the problems of inadequate heat transfer associated with total immersion methods Veech et al. (1973) have developed a radically different method, which requires, however, special equipment. The method is essentially a modification of the freeze clamping technique of Wollenberger et al. (1960) in which the tissue is pressed into a thin wafer between two aluminium blocks precooled in liquid nitrogen. In the freeze-blowing apparatus, brain tissue is very rapidly expelled from the cranial vault into a cold metal chamber by positive air pressure. The method is applicable to conscious rats (200–250 g) which are held in a restraining cage; two metal probes operated by solenoids are driven into opposite sides of the cranium and air under pressure enters one probe and expels the supratentorial portion of the brain through the other probe into the chamber precooled in liquid nitrogen. The tissue is deposited as a thin wafer on one surface of the chamber. Judging by the values obtained for $[NAD^+]/[NADH_2]$ and $[ATP]/[ADP]$ ratios, the method is superior to total immersion methods; it also gives higher values for intermediates of carbohydrate metabolism and for creatine phosphate and lower values for lactate, than do total immersion methods. It possesses the advantage of being applicable to mature rats. However, regional analysis is excluded and the actual area of the brain represented in the sample is poorly defined and not necessarily the same as that sampled for ATP in other investigations.

(5) *Microwave irradiation.* This method was originally used in studies of cyclic nucleotide metabolism in rat brain (Schmidt et al., 1971, 1972), the whole animal being exposed to microwave irradiation for 20–30s in a commercial oven of low power. Although giving better results for cyclic AMP than achieved without rapid fixation, this procedure was critized by Lust et al. (1973) on the grounds that the relatively long exposure could result in uneven inactivation of enzymes and diffusion of metabolites through damaged cell membranes. The development of more powerful ovens in which the radiation is focussed on the animal's head appears to have overcome these initial problems. Guidotti et al. (1974) used specially designed equipment producing

2 kW at 2·45 MHz in which all the radiation is directed to the head. For rats of 175–200 g exposure times of 2 s resulted in values for cyclic nucleotides, intermediates of carbohydrate metabolism and acetylcholine very close to those obtained by freeze-blowing; ATP and phosphocreatine, however, were about 30–40 per cent lower. Satisfactory results for the energy-rich phosphates in mouse brain were obtained by exposure for only 0·5 s. Medina *et al.* (1975) used a 6 kW oven (Amperex Corp. New York, U.S.A.), operating at a frequency of 2·45 MHz and delivering 5·5 kW directly to the animal's head. Exposure of mice for 0·4 s in this equipment gave satisfactory results for the energy-rich phosphates and intermediates of carbohydrate metabolism in the brain.

An important advantage of fixation by microwave irradiation is that it enables detailed dissection of the brain to be carried out after fixation has been achieved.

Illustrative experiment: fixation and removal of rat brain using liquid nitrogen

A pair of tongs and a sharp woodworking chisel (about 6 mm wide) are placed in a small Dewar vacuum flask lined with a cloth and about one-third filled with liquid nitrogen. Another Dewar vessel, of capacity 1–2 litres, is half filled with liquid nitrogen and covered with a cloth. A stout wooden board and several clean sheets of paper are required. Small vessels for containing liquid nitrogen can be made by insulating 25 ml or 50 ml beakers with expanded polystyrene; they are useful for keeping pieces of brain frozen pending further procedures. Porcelain mortars used for crushing frozen brain may also be insulated with polystyrene. It is advisable to have a pair of leather gloves available for handling the frozen animal immediately on removal from liquid nitrogen. The animals should not weigh more than 100 g. If rats or guinea pigs are to be used this limits the choice to immature animals. Mice, however, afford a source of adult mammalian brain for this method.

With one hand the cloth covering the large Dewar vessel is removed and with the other hand the animal is dropped head-first into the liquid nitrogen. The cloth is immediately replaced. After 15 s the cloth is removed and the position of the animal is noted to confirm that the head is completely immersed. One minute later, or when bubbling has largely ceased the frozen body is removed with the chilled tongs and wrapped in a cloth so that only the head remains uncovered. Holding the animal firmly in one hand (a glove may be used in place of the cloth) all the hair on the top of the head is rapidly scraped away with the blade of a scalpel. The body is now held firmly on its side on a clean sheet of paper placed over a stout wooden board. By means of a series of downwardly directed thrusts with a corner of the chisel, transverse and longitudinal grooves are cut in the occipital and parietal bones respectively. (Some workers prefer to use a narrow gouge, instead of a chisel, for this operation.) The top of the skull is then removed by means of a glancing blow of the chisel. Usually this detaches the bone from the head in one or two pieces; if not, further glancing blows with the chisel are required, although care must be taken not to remove cerebral tissue at this point. The latter is easily recognised by its slightly pink colour and also by its superficial blood vessels, provided these have not been accidentally sliced away.

At this point the animal is replaced in liquid nitrogen for about 15 s in order to maintain the brain in a fully frozen state during the subsequent stages. The brain is removed by inserting the chisel between the bone and the frozen tissue and levering upwards; the whole of the cerebral hemispheres and underlying structures will then usually come away in one or two large pieces. It is important, before removing the brain, to make sure that no bone fragments lie on its surface.

The following alternative procedure for obtaining cerebral tissue from frozen animals is preferred by some workers. The frozen body is preferred by some workers. The frozen body is held firmly on the wooden block and the head severed with a sharp blow from the chisel. The crown of the head is now placed on the block and the head split longitudinally with a blow from the chisel. The two halves of the brain are then readily separated from the cranium with the chilled chisel.

Several procedures may be used to transfer the frozen tissue to fixative. The tissue may be placed in a chilled stainless steel crusher (Minard and Davis, 1961) or in a Wedgwood-porcelain mortar containing a little liquid nitrogen, and rapidly crushed to a powder with a precooled steel crusher or pestle. If necessary more liquid nitrogen is added during the crushing to maintain the low temperature. Any excess nitrogen is then allowed to boil off and immediately this has occurred the powdered tissue is rapidly transferred with a cooled spatula, without allowing it to thaw, to a tared test-tube homogenizer containing the extracting agent. It is ground in the extractant and the homogenizer reweighed, giving the tissue weight by difference. A simple apparatus for the pulverization and rapid quantitative transfer of tissue frozen in liquid nitrogen has been described in Ladinsky *et al.* (1972).

An alternative procedure is useful for obtaining several small samples. After removal the brain is returned to the liquid nitrogen for a few seconds and then placed on a clean sheet of paper. Using the chisel it is quickly broken into several fragments of about 100 mg and the fragments are immediately placed in a dry test-tube immersed in liquid nitrogen or in a container of solid carbon dioxide and acetone. The fragments are removed one at a time and rapidly weighed on a torsion balance with the aid of a light alloy rider bent to an appropriate shape. Immediately after weighing, each fragment is dropped into extractant at 0° in a test-tube homogenizer and ground in the usual way. In this procedure care must be taken to grind frozen tissue immediately on immersion in extractant; otherwise there is a danger that part of the tissue will reach 0°C before being adequately fixed, at which temperature appreciable breakdown of labile phosphates may occur. To avoid this problem Minard and Davis (1961) extracted frozen tissue in a stainless steel mortar and pestle with 10% (w/v) trichloroacetic acid in acetone at the temperature of solid carbon dioxide. The slurry was transferred to an evaporating basin resting in crushed ice and an equal volume of cold aqueous 10% trichloroacetic added. The acetone was then evaporated by a stream of nitrogen. In an alternative approach used by Folbergrova et al. (1969) the frozen samples (about 50 mg) were weighed at $-20°C$ and then placed in tubes containing 0·1 ml of 0·1 M-HCl in 99% methanol at the temperature of solid carbon dioxide in ethanol. By crushing and stirring with a glass rod the tissue was dispersed in the solvent at this temperature before addition of the acid fixative at 0°C.

Comment on the validity of fixation methods

Using mice, Stone (1938) found 2–3 s were needed for the brain to freeze when the animals were dropped into liquid air; results for lactic acid comparable to those of Kerr were found. The applicability of Stone's method to larger animals was investigated by Richter & Dawson (1948b). Young rats of about 35 g were dropped into liquid air and thermocouples, previously implanted in the skull to different depths, were used to determine the rate at which the temperature of various parts of the brain fell to below zero. The surface of the brain was observed to fall to 0° within 4–5 s of immersion, but in the deeper parts the temperature remained near 37° for some 10–20 s and only then dropped below 0°. Nevertheless values for creatine phosphate and lactic acid were obtained comparable to those reported earlier by Kerr and by Stone.

More recent attempts to assess the validity of total immersion methods have also determined the rate of cooling in different parts of the brain with indwelling thermocouples. Swaab (1971) recorded rates of cooling in decapitated rat or mouse heads on immersion in Freon–12 at $-150°C$. Unacceptably slow rates of freezing were observed with rats weighing more than 100 g: more than 30 s were required for the cerebral cortex and 150 s for the hypothalamus respectively to reach 0°C. Cooling rate in the cortex decreased strikingly if a piece of skin was removed from the skull before immersion and if animals of 50 g were used; in the latter case less than 3 s was required to reach 0°C. However, cooling of the hypothalamus was still comparatively slow and occurred gradually, in contrast to the observations of Richter & Dawson (1948b). In mice Swaab found that the cerebral cortex of decapitated heads reached 0°C in 6·4 s using either Freon-12 or liquid nitrogen as coolant. In the study of Ferrendelli *et al.* (1972) rates of cooling were carefully measured in several areas of the mouse brain on immersion of the intact animals in liquid nitrogen. The time for the tissue temperature to fall to 0°C varied from 2–4 s for the parietal or occipital cortex, from 14–19 s for the hypothalamus and from 10–25 s for various regions of the brain stem. These two studies confirm that total immersion methods are applicable only to immature rats or to mice. Many workers accept an upper limit of 100 g body weight, but it is evident that even in animals of this weight analysis should be restricted to superficial parts of the brain if precise results are required.

Procedures which involve decapitation of an animal so that the severed head drops straight into liquid coolant are definitely unsatisfactory. Richter & Dawson (1948a) found the modification gave higher values for cerebral ammonia than did total immersion of intact animals, a result which they ascribed to the shock of decapitation, which occurred, of course, a fraction of a second before the animal entered the liquid air. A similar discrepancy was observed for lactic acid by Richter & Dawson (1948b) and for cerebral phosphates by McIlwain *et al.* (1951), and Ferrendelli *et al.* (1972).

EXTRACTING NEURAL TISSUES FOR ANALYSIS

The following sections concern the extraction of relatively simple water-soluble components, especially in instances in which neural tissues present peculiarities which distinguish them from other tissues. Extraction of neural systems for lipid constituents in described in Chapter 4 and for enzymes in Chapter 11.

Apparatus

Test-tube homogenizers are very satisfactory for extracting neural tissues when the combined volume of tissue and extracting agent does not exceed about 15 ml, and they can be used for volumes up to 40 ml. They may be constructed entirely in glass, or from glass tubes with stainless steel-Teflon pestles. The pestles are usually driven by an electric motor, although for many purposes operation by hand is satisfactory. Tissue may also be ground with extracting agent in an ordinary glass or porcelain pestle and mortar, but quantitative work is less satisfactory using this technique.

Often, in treating soft tissues such as those of the brain, the pestles of test-tube homogenizers are more satisfactory with their ends ground in the tube to a smooth fit and without glass teeth. Homogenizers with smooth pestles are generally suitable for extracting the grey and white matter from the brain of small animals with aqueous solutions at temperatures below 80°. When extracting with organic solvents or with hot aqueous solutions it may be advantageous to use pestles fitted with shallow cutting teeth. These should also be used for peripheral nerve, which is in any case a particularly difficult tissue to extract satisfactorily in glass homogenizers.

Some workers have found it useful to freeze neural tissues after removal from the animal, using liquid air or nitrogen, and before extraction to crush them to a powder in a chilled mortar by the procedure used for tissue fixed *in situ*. This variant is a valuable preliminary to the extraction of tough neural tissues as peripheral nerves which are more readily broken up in the frozen state, but it has no applicability to the analysis of labile constituents which must be determined on tissue fixed *in situ*. Comminution in this way was used by Whittaker *et al.* (1972) in a study of cholinergic vesicles of the electric organ of various species of torpedo fish. Homogenization in conventional equipment proved unsatisfactory owing to the high collagen content of the tissue. The tissue was therefore frozen in liquid Freon-12 in a ceramic mortar and crushed to a coarse powder with a pestle before extraction with medium at 0°C. Electron microscopy showed that freezing did not destroy the vesicles.

Extracting agents

Tissue extractants have two functions: first of precipitating major constituents of the tissue, usually proteins and lipids, so 'fixing' enzymes and preventing further metabolism; and secondly, of bringing into solution the constituents under investigation. It is usually assumed that a reagent which satisfactorily precipitates the tissue proteins so as to give, on centrifugation or filtration, a clear extract, fixes the tissue from the moment homogenization is complete. In certain instances, however, some chemical change may occur after this stage (cf. Schurr *et al.*, 1950) and for this reason it is advisable as a general rule to carry out extractions at low temperatures and to separate the precipitate as soon as possible. In some circumstances however, it may be feasible to ensure complete fixation by heating the tissue and extracting agent.

In devising an analytical procedure for a particular constituent the choice of the most suitable extractant for the tissue requires careful consideration. Besides fixation of enzymes and solution of the constituent, a satisfactory reagent should not encourage loss of the constituent through chemical change or through adsorption on protein precipitates or glass surfaces. It is often an added advantage, particularly if chromatographic analysis is to follow, if procedures are available for removing the reagent from the extract without loss of the constituent being analysed. There is no reagent which is entirely satisfactory in all respects for all compounds; for newly investigated constituents, therefore, it is good practice to apply more than one extracting agent and to adopt the most suitable.

A major problem encountered in extracting neural tissues, particularly white matter, arises from their high lipid content. Because of this factor reagents giving clear extracts with other tissues often fail to do so when applied to neural tissue. In the following discussion this and other general aspects of a variety of extracting reagents are considered.

INORGANIC ACIDS

Some tissue proteins are soluble, to a greater or lesser extent, in most inorganic acids but they are quantitatively precipitated by certain acids, notably tungstic, metaphosphoric and perchloric acids. Tungstic and metaphosphoric acids have been used to extract brain in the past (e.g. Damron et al., 1952; Blass, 1960), but are rarely applied now.

Perchloric acid, on the other hand, is widely used for extracting neural tissues. It is an efficient agent, but as it has oxidizing properties it should not be used in the analysis of easily oxidized substances; this gives added reason for carrying out extractions with it under cold conditions. By neutralizing the extract with potassium hydroxide solution and allowing the neutral extract to stand at 0° for 10 min, the major part of the excess acid precipitates as its potassium salt and may be removed by filtration; perchlorate at about 0·06 M is left in the filtrate. Addition of potassium chloride to the base aids precipitation of the insoluble perchlorate. For example, Folbergrova et al. (1969) neutralized perchloric acid extracts with 1·6 M-KOH containing 0·4 M–imidazole base and 0·4 M-KCl; this mixture was designed to neutralize 90% of the acid with KOH and the rest with imidazole, leaving half of the imidazole base in excess to buffer the solution at about pH 7·0.

Hydrochloric acid. Although extracts made with this acid contain appreciable amounts of protein, hydrochloric acid has proved useful in extracting brain tissue for analysis of cyclic 3′,5′-adenosine monophosphate. The following procedure, derived from one of Butcher et al. (1965), was used by Weller et al. (1972). Brain tissue was homogenized in 0·1 M-HCl at a protein concentration of 100 mg/ml and then placed in a boiling water bath for 2 min. After cooling and centrifuging the clear extracts were adjusted to pH 4·0. Some workers (Kakiuchi et al., 1969) have used successfully 0·03 M-HCl at 0°C to extract the nucleotide from incubated slices of cerebral cortex, but see Weller et al. (1972).

ORGANIC ACIDS

Trichloroacetic acid. This substance is extensively used as an extractant for neural tissues and is satisfactory for most amino acids, phosporrylated intermediates (see Chapter 3) and many other acid soluble metabolites. It is less satisfactory for extracting easily oxidized substances such as thiols and indoles, and for readily hydrolysed compounds unless used at 0°. It also interferes with the estimation of reducing substances by copper-reduction methods. To extract brain, the tissue is homogenized in a 5–20% (w/v) solution of the acid at 0°; 0·5–1 g of acid should be used for 1 g of tissue. A large proportion of the trichloroacetic acid in the protein-free filtrate may be removed by extracting the solution several times with equal volumes of ethyl ether or carbon tetrachloride. Alternatively, if acidic constituents are not sought, the extract may be passed through a column of the ion-exchange resin Dowex-2 in the chloride form.

Picric acid. A solution of picric acid for deproteinizing blood was used by Hamilton & Van Slyke (1943) and applied to tissues by Tallan *et al.* (1954). The reagent is a very satisfactory extractant for amino acids in brain (Gaitonde, 1974). The tissue (1 g) is homogenized in 10 ml of a 1% (w/v) solution of the acid and the mixture is centrifuged for 1 h at 2800 g. The extracts are usually cloudy due to the presence of suspended lipids and lipoproteins; some of the lipid may be removed by extraction with petroleum or benzene, but the extract remains cloudy.

The picric acid remaining in the extract may be removed by passing the extract through a column of an anion exchange resin such as Dowex-2 in the chloride form. These resins have a high affinity for picric acid, making it possible to use small columns and thus reducing the absorption of tissue anions; most of the acidic constituents which are absorbed are readily removed by washing the column with dilute hydrochloric acid or with more picric acid. Thus a column 0·9 × 2 cm of Dowex-2 resin of 100–200 mesh is adequate for 25 ml of extract, and three washes with 2 N-HCl will then displace virtually all the anions except the picrate ion. After use the resin cannot be satisfactorily regenerated and is therefore discarded.

Sulphosalicylic acid. A solution of this acid is sometimes used to extract brain for amino-acid analysis (Saifer, 1971; Gaitonde, 1974). For 1 g of tissue 10–15 ml of 3% (w/v) sulphosalicylic acid are used; the tissue is homogenized in the acid and the mixture left at 0° for 30 min, after which it is centrifuged. The extracts are cloudy and contain lipoprotein material.

Sulphosalicylic acid may be removed from the extract by passage through a column of Dowex-2 resin in the chloride or hydroxyl form. However, it is not easy to recover acidic substances from the columns without displacing absorbed sulphosalicylic acid, although separation from basic substances is quite feasible by this technique. Sulphosalicylic acid cannot be removed from aqueous solution by extraction with organic solvents. For amino-acid analysis on columns of cation exchange resins, separation of the acid before chromatography is unnecessary (see for example Enwonwu & Worthington, 1973).

ORGANIC SOLVENTS

The particular advantage of solvents as tissue extractants lies in the fact they are generally mild neutral reagents which do not encourage the breakdown of unstable constituents, leaving intact, for example, the catechol and indole amines. Excess solvent in the extract, moreover, may be simply removed by *in vacuo* evaporation. Their major disadvantage is that they extract a proportion of the tissue lipids, which in the case of neural tissues results in extracts of high lipid content. Also, many of the more polar tissue constituents are only partially soluble in organic solvents.

Apart from the organic solvents which precipitate protein, certain solvents immiscible with water may be used to extract soluble constituents from aqueous or saline homogenates of tissue with considerable specificity. For example, benzene or toluene have been used to extract tryptamines from homogenates of brain at pH 11 (Hess & Udenfriend, 1959; Saavedra & Axelrod, 1972).

Acetone. Extraction of 1 g of brain with 4 ml of acetone was used by Correale (1958) in the determination of serotonin by bioassay; acetone at 20 ml/g of brain gave lower recoveries of the amine. This procedure is no longer used for serotonin determination, but acetone extraction remains a treatment that merits trial, in obtaining still-unidentified constituents. If bioassay is to be conducted on the acetone extract, the usual procedure consists of removing the acetone by *in vacuo* evaporation at a low temperature, followed by repeated extraction of the aqueous residue with light petroleum, to remove the lipids. After this treatment the extracts are still cloudy and contain lipids associated with protein. Acetone is employed in a different fashion in extracting enzymes (q.v. Chapter 11), when 'acetone powders' are first prepared.

Ethanol. Extracts of brain suitable for amino-acid chromatography may be prepared by homogenizing 1 g of tissue in 3–4 ml of ethanol, but higher recoveries are obtained if 10–13 ml of 75% (w/v) ethanol are used (Roberts & Frankel, 1950; Porcellati & Thompson, 1957; Evans, 1973). After separating the precipitated protein by filtration or centrifugation, the ethanol is removed by evaporation *in vacuo*, leaving a cloudy aqueous extract which may be cleared by centrifuging at 20,000 g for 0·5 h.

In estimating tissue lactate under special circumstances (McIlwain & Tresize, 1956) ethanol extracts from slices of 0·1 g were evaporated to dryness in a desiccator at 0–5° and the residue extracted with the reagents of the subsequent determination. In some circumstances it is possible to isolate a desired constituent by direct ion-exchange chromatography of an ethanol extract. This method was used by Irreverre & Evans (1958) to isolate γ-guanidinobutyric acid from calf brain.

ZINC HYDROXIDE

The ordinary Somogyi method of protein precipitation has been applied to cerebral tissues by several workers. The zinc-tissue precipitate is, however, a strong adsorbent and this should always be borne in mind when using the method for substances whose properties with respect to adsorption on the precipitate are unknown.

In the procedure adopted by Blass (1960) brain was homogenized in 0·3 N-$ZnSO_4$ in the proportions of 0·6 g of tissue to 1 ml of reagent. Barium hydroxide (1 ml of 0·3 N) was then added and the mixture was homogenized again and centrifuged. The use of barium instead of sodium hydroxide to neutralize the zinc sulphate gives an extract free of sulphate ions, but may be attended by additional losses of some constituents by adsorption on the barium sulphate precipitate.

HEAT PRECIPITATION

In principle, heat precipitation should be used only in the analysis of heat-stable constituents; but in practice surprisingly good recoveries of some quite labile substances may be obtained. Suitable conditions for applying deproteinization by heat at a specified pH to blood have been studied by Hunter (1957). The optimum conditions for extracting brain by heat were developed from this study by Blass (1960).

For quantities of brain up to 4 g a glass homogenizer tube and pestle containing 20 ml of water are placed in a boiling water bath for 3 min. To the water in the tube 0·3 m equiv. of acetic acid is added for each 1 g of brain to be extracted. The tube is removed from the water bath and the tissue is added and rapidly homogenized. The tube is then returned to the water bath for 10 min. On centrifuging or filtering a clear or sometimes slightly opalescent extract is obtained.

For larger scale extraction, for example of ox brain, the tissue is homogenized in a mechanical blender with 3–5 vol. of hot dilute acetic acid (80–90°, 0·3 m equiv. of acetic acid/g tissue). The suspension so obtained is transferred to another vessel and kept at 100° for 10 min. The extract which results from centrifuging or filtering is cloudy and contains about 3 mg of protein/100 ml. The method is useful as a preliminary to the application of other methods to a large quantity of tissue; for example the small amount of protein remaining may be removed by tungstic acid or trichloracetic acid, preferably after freeze-drying the preliminary extract.

REFERENCES

Blass, J. P. (1960) *Biochem. J.* **77**, 484; Ph.D. Thesis, University of London.
Butcher, R. W., Ho, R. J., Meng, H. C. & Sutherland, E. W. (1965) *J. biol. Chem.* **240**, 4515.
Correale, P. (1958) *J. Neurochem.* **2**, 201.
Damron, C. M., Monier, M. M. & Roe, J. H. (1952) *J. biol. Chem.* **195**, 599.
Enwonwu, C. O. & Worthington, B. S. (1973) *J. Neurochem.* **21**, 799.
Evans, P. D. (1973) *J. Neurochem.* **21**, 11.
Ferrendelli, J. A., Gay, M. H., Sedgwick, W. G. & Chang, M. M. (1972) *J. Neurochem.* **19**, 979.
Folbergrova, J., Passoneau, J. V., Lowry, O. H. & Schulz, D. W. (1969) *J. Neurochem.* **16**, 191.
Gaitonde, M. K. (1974) *Research Methods in Neurochemistry*, **2**, 321.
Guidotti, A., Cheney, M., Trabucchi, M., Doteuchi, M., Wang, C. & Hawkins, R. A. (1974) *Neuropharmacology*, **13**, 1115.
Hamilton, P. B. & van Slyke, D. D. (1943) *J. biol. Chem.* **150**, 231.

Hess, J. M. & Udenfriend, S. (1959) *J. Pharmacol.* **127,** 175.
Hunter, G. (1957) *J. clin. Path.* **10,** 161.
Irreverre, F. & Evans, R. L. (1959) *J. biol. Chem.* **234,** 1438.
Kakiuchi, S., Rall, T. W. & McIlwain, H. (1969) *J. Neurochem.* **16,** 485.
Kerr, S. E. (1935) *J. biol. Chem.* **110,** 625.
Ladinsky, H., Consolo, S. & Sanvito, A. (1972) *Analyt. Biochem.* **49,** 294.
Lowry, O. H. & Passoneau, J. V. (1971) In *Recent Advances in Quantitative Histo- and Cytochemistry*, ed. Dubach, U. C. & Schmidt, U. p. 63. Berne: Huber.
Lust, W. D., Passoneau, J. V. & Veech, R. L. (1973) *Science*, **181,** 280.
McIlwain, H., Buchel, L. & Cheshire, J. D. (1951) *Biochem. J.* **48,** 12.
McIlwain, H. & Tresize, M. (1956) *Biochem. J.* **63,** 250.
Medina, M. A., Jones, D. J., Stavinhoa, W. B. & Ross, D. H. (1975) *J. Neurochem.*, **24,** 223.
Minard, F. N. & Davis, R. V. (1961) *J. biol. Chem.* **237,** 1283.
Porcellati, G. & Thompson, R. H. S. (1957) *J. Neurochem.* **1,** 340.
Richter, D. & Dawson, R. M. C. (1948a) *J. biol. Chem.* **176,** 1199.
Richter, D. & Dawson, R. M. C. (1948b) *Am. J. Physiol.* **154,** 73.
Roberts, E. & Frankel, S. (1950) *J. biol. Chem.* **187,** 55.
Saavedra, J. M. & Axelrod, J. (1972) *J. Pharmac. exp. Ther.* **182,** 363.
Saifer, A. (1971) *Analyt. Biochem.* **40,** 412.
Schmidt, M. J., Schmidt, D. E. & Robinson, G. A. (1971) *Science, N.Y.* **173,** 1142.
Schmidt, M. J., Hopkins, J. T., Schmidt, D. E. & Robison, G. A. (1972) *Brain Res.* **42,** 465.
Schurr, P. E., Thompson, H. T., Henderson, L. M. & Elvehjem, C. A. (1950) *J. biol. Chem.* **182,** 29.
Stone, W. E. (1938) *Biochem. J.* **32,** 1908.
Swaab, D. F. (1971) *J. Neurochem.* **18,** 2085.
Takahashi, R. & Aprison, M. H. (1964) *J. Neurochem.* **11,** 887.
Tallan, H. H., Moore, S. & Stein, W. H. (1954). *J. biol. Chem.* **211,** 927.
Veech, R. L., Harris, R. L., Veloso, D. & Veech, E. H. (1973)*J. Neurochem.* **20,** 183.
Weller, M., Rodnight, R. & Carrera, D. (1972) *Biochem. J.* **129,** 113.
Whittaker, V. P., Essman, W. B. & Dowe, G. H. C. (1972) *Biochem. J.* **128,** 833.
Wollenberger, A., Ristau, O. & Schoffa, G. (1960) *Pflügers Arch. ges. Physiol* **270,** 399.

2. Constituents of Neural Tissues: Fluids, Some Electrolytes, Amino Acids and Ammonia

R. RODNIGHT

Tissue fluid	17
Inulin space	19
Inorganic constituents; Cations	20
Sodium and potassium	21
Magnesium and calcium	22
Copper	22
Chlorides and other ions	23
Chloride	23
Multiple ion determinations	24
Amino acids	25
Chromatographic methods	26
Glutamic acid, glutamine and γ aminobutyric acid	27
N-Acetylaspartic acid and related peptides	28
Glutathione	28
Ammonia	30
References	30

The present Chapter is the first of four (Chapters 2 to 5) which are concerned with neural tissues as the source of chemical compounds or the subject of analysis. Chemical change in neural systems is described in subsequent chapters, for such metabolic studies depend greatly on the methods of characterization, isolation and analysis now to be described.

TISSUE FLUID

The water content of neural tissues is readily determined by drying the tissue under suitable conditions. Fragments or slices of tissue, weighing 50 mg or more, are drained (Chapter 6) if they have been immersed in fluid, and are then placed in a small tared vessel such as 1 ml porcelain crucible or a glass microbeaker. Drying is generally carried out in an oven at 105° for 2-3 h or overnight. The vessels are cooled in a desiccator before reweighing. Alternatively the water may be removed from thin sections or slices of tissue by freeze-drying. This method takes longer but has the advantage of leaving many of the organic constituents of the tissue unchanged and available for analysis in the dried sample. The same type of vessel is suitable. The samples are prefrozen in the vessels with the aid of solid CO_2 and immediately placed in a vacuum desiccator containing $CaCl_2$, and already connected to an oil pump giving a pressure of 0·02 mmHg or less. The complete freeze-drying of neural tissue takes some 18 h depending on the weight of the sample and the efficiency of the equipment, but values to within 3% of the weight of oven dried slices can be obtained by freeze-drying for 5 h.

Tissue fluids have been described as occupying extra- or intracellular spaces (see McIlwain, 1963; McIlwain & Bachelard, 1971). Attempts to measure the magnitude of these, both *in vivo* and *in vitro*, have involved study of the penetration into tissue of substances for which the cell membrane is believed to be impermeable. It may be noted, however, that in the nervous system, as elsewhere, organized structures and material (e.g. synaptic structures, mucopolysaccharides) may exist outside the cell membrane and limit the access of added substances. Moreover, the degree to which a substance passes into the cell can rarely be precisely determined, and while reasonable assumptions can be made for some substances, cell permeability may change considerably during the course of an experiment, as for example permeability to Na^+ and K^+ during excitation. Thus with neural tissues it would appear unwarranted to interpret the results of penetration studies in terms of clearly defined tissue spaces and more appropriate to relate them to the substance concerned and the experimental conditions employed. With these reservations, however, such studies are of great value; for example in observing changes in fluid distribution occurring in tissue in response to applied agents or varying experimental conditions (e.g. Varon & McIlwain, 1961; Bourke & Tower, 1966).

The typical fashion of expressing results of tissue composition in terms of fluid spaces is:

$$\text{Space occupied by substance } (\%) = \frac{\text{Quantity of substance/g wet wt of tissue}}{\text{Quantity of substance/ml of external fluid}} \times 100 \quad (1)$$

In this expression the density of the tissue and fluids is assumed to be unity. The percentage of the tissue spaces not occupied by the substance is given by the total water content of the tissue minus the above value (1).

In vivo, the external fluid of the brain is usually assumed to approximate in composition to that of the cerebrospinal fluid (McIlwain & Bachelard, 1971). In *in vitro* experiments the substance concerned is either incorporated in the appropriate salines from the beginning of the experiment or, if its presence is likely to interfere with other observations, added to the saline at the end of the experiment. In the latter case the tissue is allowed to remain in the saline for a further period of time just adequate for the added substance to diffuse into the accessible tissue space. To assess the optimum period for this, control experiments are required. It should be borne in mind that further changes in fluid distribution in the tissue may occur during the additional period of immersion.

Inulin is the substance most commonly used for space measurements in isolated preparations of neural tissue. Of substances native to the tissue the negative chloride ion has been used for space measurements since it is distributed according to the Nernst equation (which determines the membrane potential) in a ratio of about 1:20 between the intra- and extracellular space (McIlwain, 1963). However, chloride space is difficult to interpret in *in vitro* experiments with cerebral tissues. Thus chloride space of slices of guinea pig, rat or cat cerebral cortex increases considerably during incubation in salines containing Cl^- as the principal anion (Cummins & McIlwain, 1961; Bourke & Tower, 1966). Under aerobic conditions the major part of these additional chlorides appear to remain extracellular, but if metabolism is inhibited there is evidence that chlorides enter the cells.

Inulin space

In vitro studies with mammalian cerebral cortex slices have shown that inulin occupies less of the tissue spaces than do chlorides. During the incubation of slices under conditions of typical metabolic experiments inulin space increases, an increase apparently related to the concomitant uptake of fluid by the slices (Pappius & Elliott, 1956; Varon & McIlwain, 1961; Bourke & Tower, 1966). Part of the inulin taken up by cortical slices is washed out only slowly by immersion of the tissue in inulin-free saline; the inulin in the slice must therefore be determined by tissue analysis.

Inulin is relatively inert metabolically and may be normally incorporated at 1% (w/v) in a metabolic saline from the start of an experiment. Pappius & Elliott (1956), for example, found that this concentration had no appreciable effect on the oxygen uptake of rat cerebral cortex slices. Inulin space may then be determined at the end of the experiment by the following procedure (Varon & McIlwain, 1961). The slice is removed from the vessel, drained and ground with 4 ml of 6% (w/v) trichloroacetic acid in a test-tube homogenizer. The tube is centrifuged and the supernatant decanted into a 25 ml flask. The precipitate is washed and the extract made up to 25 ml. For inulin estimation the following reagent is required.

Resorcinol-thiourea reagent. Resorcinol (0·1 g) and thiourea (0·25 g) are dissolved in 100 ml of glacial acetic acid.

Procedure. The tissue extracts, an aliquot of the medium and inulin standards (2 ml of each) are placed in test tubes immersed in cold running tap water. The standards should contain 5–30 μg of inulin. Resorcinol reagent (1 ml) and 0·46 mM-$FeCl_3$ in 3·5 N-HCl (7 ml) are added and the solutions mixed. The tubes are placed in a water bath at 80° for 15 min and then cooled in running tap water for 5 min. Optical densities are read at 514 nm within 1 h of removing the tubes from the bath. The colours should not be unduly exposed to light during the interval before reading. Inulin space is calculated from the expression (1), given above.

Isotopic methods. Penetration of inulin into tissue slices may be measured with commercially available radioactive inulin, labelled with [*carboxyl*^{14}C] inulin (e.g. Bourke & Tower, 1966; Levi & Lattes, 1970); [*methoxy*-^{3}H] inulin has also been used (Schousboe & Hertz, 1971). An advantage of this procedure is that it enables concentrations of inulin to be used in the medium much lower than the 1 % needed for the chemical method. The labelled inulin should be radioactively pure; this may be checked by the chromatographic procedure of Levi (1969). For space measurements typically 0·4 to 0·5 μCi of the [^{14}C] isotope is added to the medium for one slice. After incubation slices are homogenized in trichloroacetic or perchloric acid and centrifuged. Radioactivity is determined in portions of the supernatant and medium by liquid scintillation spectrometry. Inulin space is then given by: c.p.m. per g of tissue × 100/c.p.m. per ml of medium.

A comparison of the colorimetric and isotopic procedures using [^{14}C] inulin was made by Bourke & Tower (1966); the methods gave identical results.

Space measurements with other sugars. Raffinose gave comparable results to inulin in cerebral cortex slices from guinea pig (Gilbert, 1966; Bachelard, 1971) and cat (Bourke & Tower, 1966). Sucrose space also compared well with inulin space in slices from cat cortex provided the concentration in the medium was less than 0·5 % (w/v) (Bourke & Tower, 1966).

INORGANIC CONSTITUENTS: CATIONS

Flame spectrophotometry, in the emission or atomic absorption mode, is generally the method of choice for cation determinations in tissues, although chemical methods are available for the polyvalent ions. For general information regarding flame spectrophotometry, papers by Margoshes & Vallee (1956), Alcock & MacIntyre (1966) and Sanui & Pace (1968) may be consulted. Technical points requiring attention in the application of this technique to neural tissue are choice of suitable extraction procedures and necessary precautions to avoid cross-interference between cations. A number of procedures are available; some typical values are given in Table 2.1. The illustrative procedure for Na^+ and K^+ given below is suitable for determining these ions in a simple flame photometer using filters (the instrument supplied by Evans Electroselenium Ltd, Halstead, Essex, is satisfactory) in extracts of fresh or incubated cerebral tissues.

Sodium and potassium: illustrative procedure

The tissue (50–100 mg) is homogenized by hand in 5 ml of 6% (w/v) trichloroacetic acid and the pestle withdrawn. After standing for 15 min the tubes are centrifuged at 800g for 10 min, the supernatants are decanted and a sample (0·5 ml) diluted appropriately with deionized water; usually a five- to tenfold dilution is satisfactory. The flame photometer is now calibrated, the curves being prepared from dilutions of a stock standard solution containing 100 mM-$NaNO_3$ and 50 mM-KCl using the appropriate filters. The concentrations sprayed should range from zero to 320 μ equiv. of Na^+/l and from zero to 160 μ equiv. of K^+/l; the curves are not necessarily linear and should be established for each series of determinations. Once the instrument is calibrated the unknown samples are sprayed and their cation content determined. When slices of mammalian cerebral cortex are immersed in salines for metabolic experiments (Chapter 7) they gain Na^+ and lose K^+. These changes are most marked under anaerobic conditions and in the absence of substrate. On incubation with substrate in oxygen slices regain K^+; ion movements on excitation *in vitro* can be observed as indicated in Chapter 7.

Table 2.1 Ion content of cerebral cortex tissue

Ion	Content (μmol/g fresh tissue)	
	Rabbit[a]	Guinea pig[b]
Sodium	59	60
Potassium	103	112
Chloride	47	40
Magnesium	6·31	5·3 0
Calcium	1·31	2·17[d]
Copper	0·05	—

[a] Data from decapitated animals (Hanig & Aprison, 1967).
[b] Data from tissue removed within 2 min of exsanguination (McIlwain, 1963).
[c] Value quoted by Bachelard & Goldfarb (1969) for fresh tissue slices.
[d] Value quoted by Lolley (1963).

Comment. Concentrated HNO_3 may be used to extract the tissue; in this case homogenization is not required and the tissue is left in the acid at room temperature for 3 to 24 h to allow time for digestion. There is no advantage in using hot HNO_3 for digestion and cold acids permit the determination of chloride in the same extract without the necessity of adding $AgNO_3$ before digestion.

In determining K in neural tissues by flame photometry interference from Na present in the extracts must be considered. The degree to which this occurs will depend upon the particular instrument used.

Usually it is necessary to incorporate Na in the K standard in approximately the same molar ratio as is present in the tissue samples, but with some instruments a much higher ratio of Na/K may be required.

Magnesium and calcium

Procedures for determining these cations by atomic absorption spectrophotometry in neural tissues have been described by Chang et al. (1966), Hanig & Aprison (1967), Bradbury et al. (1968), Bachelard & Goldfarb (1969) and Goldfarb & Rodnight (1970). In analysing extract of whole tissue for Ca^{2+} inclusion of 2% $LaCl_3$ in the extract is needed to correct for interference from phosphates (Hanig & Aprison, 1967). In determining bound Ca^{2+} and Mg^{2+} in preparations of membrane fragments from ox brain by atomic absorption spectrophotometry, extraction with acid or ashing was found to be unnecessary, provided the protein concentration of the extract was less than 0·5 mg/ml (Goldfarb & Rodnight, 1970).

Calcium may also be conveniently determined in neural tissues by a fluorometric method using the reagent calcein (fluorescein-bis-methyleneimino-diacetic acid). Lolley (1963) employed this reagent to determine calcium in cerebral tissues by measuring the change in fluorescence of a mixture of the tissue extract and reagent on addition of EDTA solution. Weller & Rodnight (1974) used an adaptation of the procedure of Kepner & Hercules (1963) to determine bound calcium in preparations of membrane fragments from brain. The reagent was prepared by dissolving 12 mg of calcein in 2 ml of 4 M-KOH and making up to 500 ml with water. To decrease the blank fluorescence 1 mM-EGTA was slowly added to the reagent until samples (0·2 ml) mixed with 1 ml of 0·4 M-NaOH showed virtually no fluorescence at 5 nm with an activation wavelength of 405 nm. Membrane preparations were ashed in a furnace at 600° C for 3 h, the ash dissolved in 0·1 M-HCl and the solution neutralized with 1M-KOH. Suitable samples were made up to 1 ml in fluorometer tubes, 0·1 ml of 4 M-NaOH added, followed by 0·1 ml of calcein reagent. Fluorescence was measured at the wavelength given above and calcium content determined by reference to a standard curve, which was linear in the range 1 to 10 nmol of Ca^{2+}. Internal standards were unnecessary.

Copper

Copper determinations are of interest in the condition of hepatolenticular degeneration (Wilson's disease) when tissue levels may be over ten times their normal value (Walshe, 1972). Several methods are available, most of them derived from the procedure of Eden & Green (1940). This depends on the formation of a yellow complex of copper with diethyldithiocarbamate; interference by iron, which if present in a free form gives a colour with the reagent, is avoided by adding pyrophosphate and carrying out the reaction at pH 8 to 9. Other complexing agents giving higher colour yields are available. Curzon & Vallett (1960) used bicyclo-hexanone oxalyldihydrazone, first introduced by Peterson & Bollier (1955), to determine copper in purified human caeruloplasmin. Neocuproin (2,9-dimethyl-1,10-phenanthroline; Smith & McCurdy, 1952; Parsons & Basford, 1967) and bathocuproin (4,7-diphenyl-1,10-phenanthroline; Van de Bogart & Beinert, 1967; Wharton & Rader, 1970) have also been employed for copper determination in tissue proteins. A spectrophotometric method using N,N,N^1,N^1-tetraethylthiuram disulphide was applied to the determination of copper in subcellular fractions prepared from brain tissue by Matsuba & Takahashi (1970). Histochemical methods for copper and other heavy metals in brain have been described by Howell (1959) and Barden (1971).

For the quantitative determination of copper in neural tissues complete oxidation of organic matter is an essential preliminary step. Dry ashing in a furnace or wet ashing with hot acid are both satisfactory (but, see Hanig & Aprison, 1967). Whichever procedure is adopted the usual precautions necessary in trace metal analyses must be rigorously followed. Thus contact of the tissue with metal should be avoided and for dissection, glass spatulas with ground edges are used; these may be fashioned in the laboratory. Water is twice distilled from glass or deionized with suitable resins. Glassware must be exhaustively rinsed before use. Reagents should be of the highest purity and traces of copper present in them controlled by adequate blank procedures or removed.

Copper may also be readily determined by atomic absorption spectrophotometry. For example, Gerstl et al. (1967) wet-ashed brain tissue with a nitric-perchloric acid mixture (7:3, v/v) and determined copper in a Perkins-Elmer instrument in the concentration range 0·1–0·8 µg of copper/ml. Neutron activation has also been used for simultaneous determination of copper, manganese and zinc in neural and other tissues (Parr & Taylor, 1964; Wong & Fritze, 1969).

CHLORIDES AND OTHER IONS

Chlorides

In the following procedure, adapted by Cummins & McIlwain (1961) from Lowry et al. (1954), chloride may be determined in a sample of the same tissue extract as used for the determination of Na^+ and K^+.

Rhodanine reagent. A few hours before use $N-H_2SO_4$, 10% (w/v) gum arabic and 0·05% (w/v) 5-(p-dimethylaminobenzylidene)rhodanine (Eastman Organic Chemicals, Rochester, N.Y., U.S.A.) in 2-methoxyethanol (methylcellosolve) are mixed in the proportions 4:1:0·25 (by vol.). The stock rhodanine solution is kept at 4° and is discarded when it fails to give an optical density of 2 when the final reagent is mixed with excess silver solution.

Procedure. The tissue is ground in 0·1 N-HNO$_3$ or in trichloroacetic acid (2·5 ml of 10% (w/v)/100 mg tissue). Of the trichloroacetic acid extract 0·1 ml is transferred to a 2 ml centrifuge tube and excess mM-AgNO$_3$ added. After standing at room temperature for 30 min the tube is centrifuged. To 0·1 ml of the supernatant 3 ml of the rhodanine reagent are added and the solution mixed. Optical densities are read after 30–60 min at 470 nm. Appropriate internal standards are run in parallel. The chloride content of fresh mammalian cerebral cortex is about 40 μmole/g.

Comment on other methods. Alternative chemical methods for determining chloride which have been applied to neural tissues include titration with mercuric nitrate using diphenylcarbazone as indicator (Schales & Schales, 1941), which was found by Thomas & McIlwain (1956) to give erroneously high values for chloride of cerebral tissue incubated in chloride-free media; and microdiffusion (Conway, 1947), which Thomas & McIlwain (1956) found gave comparable results with tungstic acid extracts to a method based on precipitation of the silver salt. Extraction of the neural tissue with warm HNO$_3$ is sometimes recommended (Valcana & Timiras, 1969), but is not essential. If wet ashing with hot acid is used it is important to include excess silver nitrate in the digestion mixture; if digestion with nitric acid is carried out in the absence of AgNO$_3$, loss of chloride by volatalization occurs. The excess silver is usually determined by titration with ammonium thiocyanate, or, as in the method of Lowry *et al.* (1954) with rhodanine dye.

The amperometric method of Cotlove *et al.* (1958; Cotlove, 1963) for chloride ion determination has also been applied frequently to neural tissues (Bourke *et al.*, 1965; Bourke & Tower, 1966; Valcana & Timiras, 1969). Oxidation of the sample with perborate (Cotlove, 1963) was found essential when analysing cerebral tissue by this method, but was unnecessary for analysis of cerebrospinal fluid (Bourke *et al.*, 1965). Isotopically labelled chloride (^{36}Cl) has been used to measure chloride penetration into the brain in perfusion studies (Bourke & Nelson, 1972) and chloride exchange in synaptosomes (Marchbanks, 1974).

Multiple ion determinations

An excellent study of methods for multiple ion determinations in brain tissue is due to Hanig & Aprison (1967). Three different ashing procedures were compared: dry ashing in platinum crucibles (16 h at 550°C plus 4 h at 650°C); wet ashing with (a) 0·75 M-HNO$_3$ and (b) concentrated HNO$_3$. Dry ashing gave excellent recoveries of all ions except copper (88%) and not unexpectedly, chloride: both were fully recovered by wet ashing with dilute HNO$_3$. The cations were determined in a Perkins Elmer atomic absorption spectrophotometer; 2% LaCl$_3$

was added to the extracts for calcium determination to prevent interference from phosphate. Chloride was determined by the Aminco-Cotlove Chloride titrator. A dual channel integrating microflame photometer for simultaneous analysis of Na^+ and K^+ in nanogram samples of freeze-dried neural tissue was evaluated by Keesey (1968). The method gave results within 3–5% of those obtained on larger samples with conventional flame photometers.

AMINO ACIDS

Some 18 amino acids occur in neural tissues in the free form (Table 2.2). Two of these, γ-aminobutyric acid and N-acetylaspartic

Table 2.2 Typical values for free amino acids in brain

Amino acid	Content (μmol/g of fresh tissue)		
	Mouse[a]	Guinea pig[b]	Human[c]
N-Acetylaspartic acid	8·25	—	5·48
Alanine	0·66	0·65	0·25
γ-Aminobutyric acid	2·52	1·88	0·42
Arginine	0·13	—	0·10
Aspartic acid	3·38	2·36	0·96
Cystine	0·02	—	0·01
Cystathionine	0·06	0·07	2·02
Glutamic acid	11·50	9·51	5·96
Glutamine	4·83	3·88	5·80
Glycine	1·61	0·98	0·40
Histidine	0·05	—	0·09
Isoleucine	0·04	0·04	0·03
Leucine	0·08	0·07	0·07
Lysine	0·18	0·07	0·12
Methionine	0·03	0·10	0·03
Phenylalanine	0·04	0·04	0·05
Proline	0·06	0·08	0·04
Serine	0·78	0·68	0·44
Taurine	9·13	1·61	0·93
Trytophan	0·05[d]	—	0·01
Tyrosine	0·05	0·07	0·06
Valine	0·08	0·06	0·13

[a] Data for whole brain frozen within 30 s (Himwich & Agrawal, 1969).
[b] Data for cerebral cortex obtained at biopsy and frozen within 15 s of removal (Perry et al., 1971).
[c] Value for cerebral cortex obtained at autopsy (Tallan, 1957).
[d] Value for 7-day-old mice (Hunt & Johnson, 1972).

acid, are typical of neural tissue since their concentration in other tissues is either negligible or very low; for the remainder the qualitative pattern is in general similar to that of other tissues. Of special interest to neural systems are the aromatic amino acids, tryptophan, 5-hydroxytryptophan, tyrosine and dihydroxyphenylalanine (DOPA), precursors of the cerebral indole and catechol amines. The acidic amino acids, glutamate and aspartate, and taurine may also have important roles in synaptic transmission, as well as participating in intermediary metabolism (McIlwain & Bachelard, 1971). Marked species differences are shown.

Analytical methods for amino acids include those based entirely on chromatography and capable of determining all or most of the acids in a single sample; and methods designed for individual acids of special interest, which also often utilize chromatography.

Chromatographic methods

Methods employing chromatography on ion-exchange resins, paper and thin layers have all been applied to neural tissues. Automatic analysis on long columns of cation-exchange resins is widely used. Single column procedures using lithium citrate buffers for elution are capable of resolving all the amino acids in brain (Perry *et al.*, 1968; Perry *et al.*, 1972), but are technically demanding; complete resolution of glutamine from glutamate is not always readily achieved (Gaitonde, 1974).

Chromatography on thin layers of silica or cellulose powder is often more convenient than column analysis and has the advantage of greater sensitivity; however, specificity and accuracy, particularly in the case of minor amino acids, is generally inferior to the ion exchange methods. The following procedure is based on thin-layer chromatography on cellulose powder and is suitable for demonstrating and determining to within $\pm$ 10% the five main amino acids in cerebral tissue: glutamic acid, aspartic acid, γ-aminobutyric acid, glutamine and taurine. Less prominent spots of other amino acids are also seen, but are less readily identified.

Procedure. The tissue (200–300 mg wet weight) is homogenized in a test-tube homogenizer in 10 ml of ice-cold 75% (v/v) ethanol. Peripheral nerve usually proves too tough to homogenize satisfactorily in this way and is more conveniently frozen with liquid nitrogen and ground to a powder in a pestle and mortar under liquid nitrogen before extracting with ethanol. The homogenate is centrifuged for 10 min at 2000 g and a measured volume of the clear supernatant evaporated to dryness on a rotary film evaporator with minimum application of heat. The residue is taken up in water to a concentration of 500μg of tissue /ml and centrifuged at 40 000 g for 30 min. From the clear supernatant spots of 10–20 μl are applied to thin-layer plates of cellulose powder (MN300, Camlab Ltd, Cambridge, is satisfactory) about 2 cm from one corner. The plates are developed with 80% phenol in a tank saturated with phenol vapour for 4 h and then dried in a strong current of air overnight or until they no longer smell of phenol. For the second direction isopropyl alcohol+formic acid (90%)+water (40:2:10, by vol.) is used. The tanks need not be equilibrated with solvent vapour for this run: the solvent should reach the top edge of the plates in 3 h. A standard mixture of amino acids (20 mM each, 5–10 µl spots) should be run in parallel. Some workers prefer to reverse the order of the solvents, but in this case particular care must be taken to ensure complete removal of phenol before development with ninhydrin reagent.

The amino acids may be detected by spraying the plates with 0·2% ninhydrin in acetone, followed by gentle heating in an oven (60%). For quantitative work the plates should be stained with the cadmium–ninhydrin reagent of Atfield & Morris (1961) and developed in an ammonia-free atmosphere as recommended by these authors. (The reagent is prepared by dissolving 0·2 g of cadium acetate in 20 ml of water and 4 ml of glacial acetic acid. Acetone (20 ml) is then added and any precipitate that forms is redissolved by shaking. Finally 2g of ninhydrin is added. The solution must be prepared daily.) The amino acids appear as pink spots, which may be scraped off the plates, the colour is eluted with methanol and determined spectrophotometrically at 500 nm.

Comment on other thin-layer procedures. In an elegant and sensitive procedure (Shank & Aprison, 1970a,b) for determining glutamate, aspartate, glutamine, taurine, γ-aminobutyrate, serine, alanine, threonine and glutathione in 10 mg (wet wt) samples of neural tissue the amino acids are first converted to their dinitrophenyl derivatives before chromatography on silica gel. The amino acid derivatives are eluted from the plates with 0·01 M-$NaHCO_3$ and their extinction at 360 nm determined. The lower limit of sensitivity is 0·4 µmol of amino acid/g of tissue.

Glutamic acid, glutamine and γ-aminobutyric acid

Comprehensive procedures using ion-exchange resins for separating and determining these amino acids by the ninhydrin reaction are described by Gaitonde (1974). These are manual methods that permit the isolation of samples of the three amino acids, as well as their quantitative analysis; in isotopic experiments therefore the radioactivity of the individual labelled amino acids may be readily determined. The total amino-acid fraction of the tissue is first isolated from suitable extracts by absorption on a strongly acidic cation-exchange resin and elution with ammonia solution. After taking to dryness and solution of the residue in water the amino acid fraction is run through a column of a weakly acidic cationic resin in the H^+ form; the column retains γ-aminobutyrate which can be selectively eluted with water. Glutamate and aspartate in the effluent from this column are isolated by retention on a further column of an anionic-exchange resin in the acetate form, followed by elution with acetic acid. The amino acids unabsorbed by the basic column include glutamine and the neutral amino acids. If glutamine alone is required, this is determined by hydrolyzing it with acid to glutamate and isolating the latter on another column of anionic resin. The only proviso is that other compounds yielding glutamate on hydrolysis must be absent; this is normally the case since glutathione is retained on the first basic column. Alternatively glutamine and the neutral amino acids may be fractionated on a long column of a cationic-exchange resin using lithium citrate buffers. No problem in the separation of glutamine and glutamate arises, since the latter is absent.

For the selective determination of these amino acids in cases where isolation of the acid is not required, enzymic methods are available (Bergmeyer, 1963). Specific applications to neural tissue include a micro procedure for glutamate and alanine described by Young & Lowry (1966; see also Folbergrova et al., 1969) applicable to samples as small as 1 μg dry wt. For γ-aminobutyrate a micro adaptation of the method of Jakoby & Scott (1959) is due to Kravitz & Potter (1965), who used it to determine the content of the amino acid in inhibitory and excitatory nerves of the lobster. Glutamine may be determined using glutamic dehydrogenase and glutaminase by the procedure of Buttery & Rowsell (1971).

N-Acetylaspartic acid and related peptides

Methods for the isolation and determination of N-acetylaspartic acid, N-acetylglutamic acid and N-acetylaspartylglutamic acid in neural tissues are reviewed by Benuck (1974). Chromatographic procedures using cationic exchange resins are available; after separation the conjugates are determined after hydrolysis by the ninhydrin reaction. N-Acetylaspartate, which is quantitatively the most important member of this group in mammalian brain, and N-acetylglutamate, whose concentration is much lower, require hydrolysis with 3 M-HCl for 1 h at 100°C for complete release of the respective free amino acids. In the case of the dipeptide, N-acetylaspartylglutamate, boiling in 6 M-HCl for 24 h is necessary.

Enzymic methods for determining N-acetylaspartate on a microscale have been described by McIntosh & Cooper (1965) and Fleming & Lowry (1966). These use aspartoacylase (N-acylaspartate amidohydrolase, E.C. 3.5.1.15) from bovine kidney to hydrolyse the conjugate to acetate and aspartate, followed by determination of the latter with aspartate: 2-oxoglutarate aminotransferase, malate dehydrogenase and NADH. A procedure is available (Marcucci & Mussini, 1966) for the separation of N-acetylaspartate by gas chromatography, after esterification of the carboxyl group.

Glutathione

Glutathione occurs in brain almost entirely in the reduced form and oxidation of a proportion of it to the disulphide form takes place within a minute or so of death (Martin & McIlwain, 1959). Thus in order to determine the true *in vivo* levels of the reduced form it is necessary to analyse tissue fixed *in situ*. A very satisfactory extractant for glutathione in neural tissues is sulphosalicylic acid. The tissue (300–400 mg) is ground with 5 ml of a 3% (w/v) solution of the acid at 4° in a test-tube homogenizer. The homogenate is allowed to stand for 30 min at 4° before separation of the precipitate by centrifuging at 0° for 20 min. These conditions result in virtually no oxidation of added glutathione, which remains stable in the extract at 0° for up to 6 h. The extracts may be used for glyoxalase assay without further treatment except neutralization.

Glutathione (GSH) has been determined by amperometric titration or by enzymic methods. Total glutathione (GSH + GSSG) can be determined by ion exchange chromatography at room temperature by the procedures for glutamic acid and related compounds described by Gaitonde (1974). Gluthathione cannot be satisfactorily separated on columns of strongly acidic cation exchange resins operated at raised temperature because ill-defined oxidation leads to the appearance of broad peaks which overlap with peaks of other amino acids (Cohen & Lin, 1972).

The usual enzymic method employs glyoxalase for which reduced glutathione is the specific co-enzyme. The glyoxalase assay is less accurate and more time-consuming than the amperometric method, but it is rather more sensitive (requiring only 0·03 μmole of thiol) and considerably more specific for glutathione. Thomson & Martin (1959) found that of 11 thiol derivatives only glutathione and *S*-acetylglutathione were determined in the glyoxalase system, whereas the amperometric method determined to some degree all compounds possessing a free SH group with the exception of ergothioneine and thiolhistidine. The enzymic method, however, is susceptible to interference by inhibition from a number of glutathione analogues (Kermack & Matheson, 1957). Both methods can be adapted to the determination of oxidized glutathione through a preliminary reduction. In the glyoxalase assay this is achieved by a pre-incubation of the tissue with the yeast apoglyoxalase preparation which contains a glutathione reductase; methylglyoxal is then added and an assay for total glutathione conducted (Martin & McIlwain, 1959). For the amperometric method electrolytic reduction in a commercial desalter (Smith, 1960) was used by Thomson & Martin (1959). Compared with the enzymic method recovery of added glutathione was less satisfactory with this procedure and correspondingly lower values for the disulphide in brain were obtained.

A very sensitive enzymic cycling method for glutathione has been described by Grassetti & Murray (1967), but does not appear to have been applied to neural tissues. The method utilizes the reaction between 2,2'-dithiodipyridine and reduced glutathione catalyzed by glutathione reductase in the presence of NADPH, in which 2-thiopyridone is formed. The latter is determined spectrophotometrically at 343 nm.

AMMONIA

To determine ammonia in neural tissues the colorimetric method of Russell (1944) adapted to a suitable form of microdiffusion analysis (Conway, 1947; Seligson & Seligson, 1951) continues to be widely used (e.g. Godin et al., 1967; Weil-Malherbe, 1969). The method depends on a reaction between ammonia, phenol and the hypochlorite ion to give a stable blue colour. The addition of small amounts of nitroprusside markedly accelerates colour production and obviates the need to introduce a correction for glutamine hydrolysis (Brown et al., 1957).

Enzymic procedures are also available using glutamic dehydrogenase and measuring the oxidation of NADH spectrophotometrically or fluorometrically (Folbergrova et al., 1969; Buttery & Rowsell, 1971). The latter method incorporates a procedure to reduce the ammonia content of the reagents to zero; this is a considerable advantage since the sensitivity of the enzymic method is greatly limited by the occurrence of traces of ammonia in reagents and in commercially available preparations of glutamic dehydrogenase (see Folbergrova et al., 1969).

The ammonia content of the brain increases very rapidly following death and *in vivo* amounts can be determined only on tissue fixed *in situ* by immersion in liquid nitrogen. Cerebral ammonia levels are also affected by the procedure adopted for killing the animal: for example decapitation immediately prior to freezing the head results in higher values than does the preferred technique of immersing the whole animal in coolant.

For the chemical method extraction of the tissue with 10% (w/v) trichloroacetic acid at 0°C is recommended. At least 8 ml of acid/g of tissue should be used and the protein precipitate separated immediately by centrifuging at 0°C. As appreciable hydrolysis of glutamine occurs in trichloroacetic acid solutions kept at room temperature it is important to work throughout at 0°C and to keep the extracts at a low temperature pending analysis. In the case of ammonia determination by the enzymic procedure it is usual to extract the tissue with perchloric acid and remove excess perchlorate ion with KOH. Buttery & Rowsell (1971) used 10 ml of 12M-$HClO_4$ to extract 1–2 g of tissue.

REFERENCES

Alcock, N. W. & MacIntyre, I. (1966) *Meth. biochem. Analysis* **14**, 1.
Atfield, G. N. & Morris, C. J. O. R. (1961) *Biochem. J.* **81**, 606.
Bachelard, H. S. (1971) *J. Neurochem.* **18**, 213.
Bachelard, H. S. & Goldfarb, P. S. G. (1969) *Biochem. J.* **112**, 579.
Barden, H. (1971) *J. Neuropath. exp. Neurol.* **30**, 650.
Benuck, M. (1974) *Research Methods in Neurochemistry* **2**, 361.
Bergmeyer, H.-U. (ed.) (1963) *Methods of Enzymatic Analysis*, pp. 381, 384. Academic Press: New York.

Bourke, R. S., Greenberg, E. G. & Tower, D. B. (1965) *Am. J. Physiol.* **208**, 682.
Bourke, R. S. & Nelson, K. M. (1972) *J. Neurochem.* **19**, 663.
Bourke, R. S. & Tower, D. B. (1966) *J. Neurochem.* **13**, 1071.
Bradbury, M. W. B., Kleeman, C. R., Bagdoyan, H. & Berberian, A. (1968) *J. Lab. clin. Med.* **71**, 884.
Brown, R. H., Duda, G. D., Korkes, S. & Handler, P. (1957) *Arch. Biochem. Biophys.* **66**, 301.
Buttery, P. J. & Rowsell, E. V. (1971) *Analyt. Biochem.* **39**, 297.
Chang, T. L., Gover, T. A. & Harrison, W. W. (1966) *Analytica Chim. Acta* **34**, 17.
Cohen, H. P. & Lin, S. (1972) *J. Neurochem.* **19**, 513.
Conway, E. J. (1947) *Microdiffusion Analysis and Volumetric Error.* London: Crosby Lockwood.
Cotlove, E. (1963) *Analyt. Chem.* **35**, 101.
Cotlove, E., Trantham, H. V. & Bowman, R. L. (1958) *J. Lab. clin. Med.* **57**, 461.
Cummins, J. T. & McIlwain, H. (1961) *Biochem. J.* **79**, 330.
Curzon, G. & Vallet, C. (1960) *Biochem. J.* **74**, 386.
Eden, A. & Green, H. H. (1940) *Biochem. J.* **34**, 1202.
Fleming, M. C. & Lowry, O. H. (1966) *J. Neurochem.* **13**, 779.
Folbergrova, J., Passoneau, J. V., Lowry, O. H. & Schulz, D. W. (1969) *J. Neurochem.* **16**, 191.
Gaitonde, M. K. (1974) *Research Methods in Neurochemistry* **2**, 321.
Gallagher, B. B. (1969) *J. Neurochem.* **16**, 701.
Gerstl, B., Eng, L. F., Hayman, R. B., Tavaststjerna, M. G. & Bond, P. R. (1967) *J. Neurochem.* **14**, 661.
Gilbert, J. C. (1966) *J. Neurochem.* **13**, 729.
Goldfarb, P. S. G. & Rodnight, R. (1970) *Biochem. J.* **120**, 15.
Godin, Y., Mark, J., Kayser, Ch. & Mandel, P. (1967) *J. Neurochem.* **14**, 142.
Grassetti, D. R. & Murray, J. F. Jun. (1967) *Analyt. Biochem.* **21**, 427.
Hanig, R. C. & Aprison, M. H. (1967) *Analyt. Biochem.* **21**, 169.
Himwich, W. A. & Agrawal, H. C. (1969) *Handbook of Neurochem.* **1**, 33.
Howell, J. S. (1959) *J. Path. Bact.* **77**, 473.
Hunt, D. M. & Johnson, D. R. (1972) *J. Neurochem.* **19**, 2811.
Jakoby, W. B. & Scott, E. M. (1959) *J. biol. Chem.* **234**, 937.
Keesey, J. C. (1968) *J. Neurochem.* **15**, 547.
Kepner, B. C. & Hercules, D. M. (1963) *Analyt. Chem.* **35**, 1238.
Kermack, W. O. & Matheson, N. A. (1957) *Biochem. J.* **65**, 48.
Kravitz, E. A. & Potter, D. D. (1965) *J. Neurochem.* **12**, 323.
Levi, G. (1969) *Analyt. Biochem.* **32**, 348.
Levi, G. & Lattes, M. G. (1970) *J. Neurochem.* **17**, 587.
Lolley, R. N. (1963) *J. Neurochem.* **10**, 665.
Lowry, O. H., Roberts, N. R., Leiner, K. Y., Wu, M. L. & Farr, A. L. (1954) *J. biol. Chem.* **207**, 1.
McIlwain, H. (1963) *Chemical Exploration of the Brain. A Study of Cerebral Excitability and Ion Movement.* Amsterdam: Elsevier.
McIlwain, H. & Bachelard, H. S. (1971) *Biochemistry and the Central Nervous System*, 4th edn. Edinburgh: Churchill Livingstone.
McIntosh, J. C. & Cooper, J. R. (1965) *J. Neurochem.* **12**, 825.
Marchbanks, R. M. (1974) *Trans. biochem. Soc.*, **2**, 664.
Marcucci, F. & Mussini, E. (1966) *J. Chromatog.* **25**, 11.
Matsuba, Y. & Takahashi, Y. (1970) *Analyt. Biochem.* **36**, 182.
Margoshes, M. & Vallee, B. L. (1956) *Meth. biochem. Analysis* **3**, 353.
Martin, H. & McIlwain, H. (1959) *Biochem. J.* **71**, 275.
Pappius, H. & Elliott, K. A. C. (1956) *Can. J. Biochem. Physiol.* **34**, 1007.
Pappius, H. M., Klatzo, I. & Elliott, K. A. C. (1962) *Can. J. Biochem. Physiol.* **40**, 885.

Parr, R. M. & Taylor, D. M. (1964) *Biochem. J.* **91**, 424.
Parsons, P. & Basford, R. E. (1967) *J. Neurochem.* **14**, 823.
Perry, T. L., Hansen, S., Berry, K., Mok, C. & Lesk, D. (1971) *J. Neurochem.* **18**, 521.
Perry, T. L., Sanders, H. D., Hansen, S., Lask, D., Kloster, M. & Gravlin, L. (1972) *J. Neurochem.* **19**, 2651.
Perry, T. L., Stedman, D. & Hansen, S. (1968) *J. Chromat.* **38**, 460.
Peterson, R. E. & Bollier, M. D. (1955) *Analyt. Chem.* **27**, 1195.
Russell, J. A. (1944) *J. biol. Chem.* **156**, 457.
Sanui, H. & Pace, N. (1966) *Appl. Spectrosc.* **20**, 135.
Schales, O. & Schales, S. S. (1941) *J. biol. Chem.* **140**, 879.
Schousboe, A. & Hertz, L. (1971) *J. Neurochem.* **18**, 67.
Schroeder, J. R. & Stahman, M. A. (1970) *Analyt. Biochem.* **34**, 331.
Seligson, D. & Seligson, H. (1951) *J. Lab. clin. Med.* **38**, 324.
Shank, R. P. & Aprison, M. H. (1970a) *J. Neurochem.* **10**, 1461.
Shank, R. P. & Aprison, M. H. (1970b) *Analyt. Biochem.* **35**, 136.
Smith, G. F. & McCurdy, W. H. (1952) *Analyt. Chem.* **24**, 371.
Tallan, H. H. (1957) *J. biol. Chem.* **224**, 41.
Thomas, J. & McIlwain, H. (1956) *J. Neurochem.* **1**, 1.
Thomson, C. G. & Martin, H. (1959) Glutathione. *Biochem. Soc. Symp.* **17**, 17.
Valcana, T. & Timiras, P. S. (1969) *J. Neurochem.* **16**, 935.
Van De Bogart, M. & Beinert, H. (1967) *Analyt. Biochem.* **20**, 325.
Varon, S. & McIlwain, H. (1961) *J. Neurochem.* **8**, 262.
Walshe, J. M. (1972) In *Biochemical Aspects of Nervous Diseases*, ed. Cumings, J. N., p. 111. London & New York: Plenum Press.
Weil-Malherbe, H. (1969) *J. Neurochem.* **16**, 855.
Weller, M. & Rodnight, R. (1974) *Biochem. J.*, **142**, 605.
Wharton, D. C. & Rader, M. (1970) *Analyt. Biochem.* **33**, 226.
Wong, P. Y. & Fritze, K. (1969) *J. Neurochem.* **16**, 1231.
Young, R. L. & Lowry, O. H. (1966) *J. Neurochem.* **13**, 785.

3. Constituents of Neural Tissues: Intermediates in Carbohydrate and Energy Metabolism

H. S. BACHELARD

Preparation of tissue for analysis	33
Analysis for carbohydrates and their derivatives	35
Glucose	35
Glycogen	37
Glycosaminoglycans	39
Lactate and pyruvate	40
Pentoses	42
Intermediates in the citric acid cycle	43
Acid-soluble phosphates	43
Fixation and extraction of tissue	44
Fractionation of acid-soluble tissue phosphate by precipitation	45
Fractionation by chromatography	45
Inorganic phosphate	47
Hexose phosphates	49
Cyclic AMP	50
Creatine phosphate	50
Nucleotides	51
Other phosphate esters	53
Acid-insoluble phosphates	54
Nucleic acids	54
Protein–phosphorus in residual fraction	56
References	56

The intermediates of carbohydrate and energy metabolism of importance in the brain are treated separately in different groups in this chapter, although it must be remembered that they are metabolically interrelated. Carbohydrates and derivatives are described first. For convenience, because they are often present together in tissue extracts and may have to be carefully separated from each other, glycolytic intermediates as hexose phosphates are included with other phosphates (inorganic phosphate, phosphocreatine, nucleotides and phosphate esters) under the common heading of acid-soluble phosphates. Intermediates of the tricarboxylic acid cycle and glycosaminoglycans (including mucopolysaccharides) are described under separate headings.

PREPARATION OF TISSUE FOR ANALYSIS

Many of these constituents are labile in the brain after death and thus the tissue in which they are to be determined must be rapidly fixed *in situ*. This applies especially to many of the acid-soluble phosphates, and also to glucose, glycogen, lactate and pyruvate. Rapid freezing techniques for fixing the tissue *in situ* were described in Chapter 1; the most recently developed of these have been devised specifically for the analysis of these constituents in neural tissues. Especial attention has been given to the possibility of avoiding the delays of many seconds which occur before freezing of the whole brain tissue of a small experimental animal is complete, when the whole animal is immersed in a freezing environment. Thus, in one laboratory, the original surgical exposure technique (Kerr, 1936) has been re-examined: liquid N_2 is poured into a plastic funnel, previously inserted directly over the skull through a skin incision, and requires the use of a general anaesthetic (Granholm *et al.*, 1968; Siesjö *et al.*, 1972). This technique results in higher values for glucose, glycogen and labile phosphates, with lower values for lactate and pyruvate (Table 3.1). These are considered to result from the more rapid freezing but the possibility that it is in part

Table 3.1 Labile intermediates in mammalian brain, fixed by different method.[1]

Constituent	Concentration (μmol/g fresh tissue) Method of fixation			
	In situ[2]	Funnel freezing[3]	Freeze blowing[4,5]	Microwave[5,6]
Glucose	1·0 to 1·5	3·3	1·0 to 1·8	1·2 to 1·7
Lactate	2·0 to 2·4	1·2	1·2	1·2
ATP	1·8 to 2·4	3·0	2·45	1·5 to 2·4
ADP	0·3 to 0·4	0·26	0·56	0·65
Phosphocreatine	2·4 to 3·3	4·9	3·5 to 4·1	2·4 to 3·7

[1]In general, higher values for glucose, ATP and phosphocreatine, and lower values for lactate, pyruvate and ADP, are taken to indicate closer approximation to *in situ* values.
[2]Kerr & Ghantus, 1936; Lowry *et al.*, 1964; Strang & Bachelard, 1971a, 1973.
[3]Folbergrova *et al.*, 1972; Siesjö *et al.*, 1972 (with light general anaesthesia)
[4]Veech *et al.*, 1973.
[5]Guidotti *et al.*, 1974.
[6]Medina *et al.*, 1975.

due to cerebral changes caused by anaesthesia itself cannot be excluded (Mayman *et al.*, 1964; Strang & Bachelard, 1973). Also the technique involves immobilisation of the animals with tubocurarine chloride, and the animals prepared in this way have shown elevated blood glucose concentrations (Lewis *et al.*, 1974). The higher values for brain glucose could result partly from its increased availability in the circulation (Bachelard *et al.*, 1973).

More recently, use of a novel technique, 'freeze blowing', has been reported (Veech *et al.*, 1973). This involves the expulsion by air pressure of the brains of conscious rats directly into a cooled chamber. Some but not all of the results indicated that improved values had been obtained (Table 3.1). The use of heat inactivation in a micro-wave oven was originally shown to be of value in fixing tissues for heat stable constituents such as cyclic AMP (Schmidt et al., 1971). The technique has been improved with the development of more intense micro-wave irradiation; its use for very short time periods of less than 1 s has given values for the intermediates of Table 3·1 which are comparable with those obtained using the 'freeze-blowing technique' (Guidotti et al., 1974; Medina et al., 1975).

Specific inhibitors of glycolysis have also been used, especially in research on larger animals; thus for lactate (Kinnersley & Peters, 1930) and for glycogen (Chesler & Himwich, 1943; McIlwain & Tresize, 1956) injecting animals with sodium iodoacetate a few minutes before death gave results similar to those obtained by immersion in liquid N_2.

ANALYSIS FOR CARBOHYDRATES AND THEIR DERIVATIVES

The majority of the analytical techniques now exploit the availability of specific enzymes, usually coupled to reactions which lead to oxidation or reduction of nicotinamide nucleotides, which are readily estimated by ultraviolet spectrophotometry or fluorometry (Lowry & Passonneau, 1972).

Glucose

The usual methods consist of extracting the tissue with a suitable reagent and determining the glucose in the filtrate or in the supernatant after centrifugation. The determinations were originally performed by measuring the reducing properties in the extract but then had the disadvantage of interference by non-glucose reducing substances. The paper chromatographic study of Gey (1956) suggested that glucose constituted only some 40% of the total reducing substance in brain extracts after deproteinization with the $ZnSO_4$, $Ba(OH)_2$ reagent of Somogyi (1952). Current methods utilize specific enzymic analysis of the glucose in tissue extracts prepared by homogenizing in perchloric or trichloroacetic acid.

Tissue extraction. Powdered brain samples, fixed *in situ*, are homogenized in 0·6N-perchloric acid at 0° and the mixture centrifuged to give a clear supernatant, which is neutralized with the calculated amount of $KHCO_3$ at 0°. After storage for 1 or 2 h at 0° the precipitated $KClO_4$ is removed by centrifugation and the neutral supernatant is used without further purification. This 'acid-soluble' extract can be used also for the determination of many of the labile soluble constituents described below.

Determination

(a) *Spectrophotometric.* The following reagents are pipetted into a 1 ml cuvette of 1 cm light-path: 0·1 M-tris HCl buffer, pH 7·6, containing 10 mM-MgSO$_4$ (0·5 ml), 30 mM-ATP (0·1 ml), 4 mM-NADP$^+$ (0·1 ml), hexokinase (3 international units/ml, 0·1 ml), and glucose 6-phosphate dehydrogenase (1 international unit/ml, 0·1 ml). The extinction at 340 nm is measured over 5 min to allow any endogenous glucose 6-phosphate to react. The neutralized tissue extract (0·1 ml), or glucose standard (0·1 to 1 mM, 0·1 ml), or water (0·1 ml) is then added and the increase in extinction at 340 nm is followed at 1 min intervals until it remains constant (usually within 5 min). Since each μmol of glucose in the cuvette will give an increase in extinction of 6·22, the change in extinction is divided by 6·22 to calculate the amount of glucose present, in μmol/sample.

(b) *Spectrophotofluorometric method.* The reagents are similar to those of method (a) with slight differences in concentration, and are placed in a 1 ml fluorometer cell: 0·1 M-tris HCl buffer, pH 7·6, containing 10 mM-MgSO$_4$ (0·5 ml); 10 mM-ATP (0·1 ml); 0·4 mM-NADP$^+$ (0·1 ml); hexokinase (1 international unit/ml), 5 mM-dithiothreitol (0·1 ml) and tissue sample or glucose standard (0·1 ml). The instrument is set at 340 nm (excitant) and 440 nm (emission). For a discussion of instruments and settings see below, and Lowry & Passonneau (1972). A suitable sensitivity range is selected using quinine standards and the fluorescence is read. Glucose 6-phosphate dehydrogenase (10 μl of a solution of 3 international units/ml) is added and the fluorescence read at 1 min intervals until it is constant (usually within 5 min). A suitable range of glucose standards is from 0·01 to 0·2 mM.

Specific activity measurements of radioactive glucose

Most methods are based on separation of the glucose from other soluble constituents by a variety of chromatographic procedures. Charged constituents, as carboxylic acids, amines and amino acids, are removed by passage through mixed bed ion-exchange resins and the glucose separated finally by chromatography on paper or thin layers of cellulose (Lindsay & Bachelard, 1966). An enzymic method is now available (Strang & Bachelard, 1971b) for the determination of uniformly labelled [^{14}C-]glucose in unpurified tissue extracts. The brain is fixed *in situ* and a neutralized perchloric acid extract is prepared as described above. The method involves the enzymic conversion of glucose to 6-phosphogluconate which is then decarboxylated; the evolved ^{14}CO$_2$ is trapped and counted. Simultaneous determination of the glucose and its radioactivity can be performed.

To apply the method the following reagents are placed in a 3 ml cuvette of 1 cm light-path: 0·1 M-tris HCl buffer, pH 7·6 (2 ml); 50 mM-MgSO$_4$ (0·1 ml); 6 mM-ATP (0·2 ml); 10 mM-NADP$^+$ (0·2 ml); hexokinase (28 international units/ml, 10 μl); glucose 6-phosphate dehydrogenase (24 international units/ml, 10 μl) and the sample containing the radioactive glucose (0·1 ml). The reaction is allowed to proceed until there is no further increase in extinction at 340 nm, from which the total amount of glucose present can be calculated. The reaction mixture is then transferred to a small bottle with a rubber seal and heated briefly at 100° to ensure full conversion of 6-phosphogluconolactone into 6-phosphogluconate. 6-Phosphogluconate dehyrogenase (24 international units/ml, 10 μl) is added. After 1 h, the solution is acidified by injecting H_2SO_4 and the $^{14}CO_2$ is expelled by a stream of air into a 10% solution of Hyamine 10-X hydroxide. Since 1 equivalent of CO_2 is driven off from each molecule of glucose, the radioactivity of the hyamine solution must be multiplied by 6 to give the total radioactivity present in the original glucose.

Glycogen

Glycogen was originally extracted from frozen brain samples into hot alcoholic KOH, from which crude glycogen separated on cooling and was recovered by centrifugation. The residue was then washed with organic solvents to remove cerebrosides, dried, suspended in water, hydrolyzed in acid and the glucose in the hydrolysate determined chemically (anthrone) or from its reducing power (Kerr, 1936; Le Baron, 1955). These procedures gave artificially high values since carbohydrate other than glycogen was included in the determination, and accounted for up to 75% of the total measured (Strang & Bachelard, 1971c). Even the use of specific enzymes (glucose oxidase or the hexokinase, glucose 6-phosphate dehydrogenase system) does not yield true values because the glucose in acid hydrolysates is derived also from non-glycogen carbohydrate. Current methods involve the use of enzymes which degrade glycogen specifically.

Tissue extraction

Two alternative methods of extraction are recommended, the choice depending on whether or not other metabolites are to be estimated in the same tissue samples: method (1) if glycogen is the only metabolite to be measured, and method (2) if other metabolites are to be estimated.

Method (1). The frozen powdered brain sample (about 100 to 200 mg) is placed directly into 1 ml of 60% aqueous KOH (w/v) in a calibrated centrifuge tube and heated at 80°C for 10 min. Ethanol (95%, 2 ml) is added, mixed thoroughly, and the tube held in an ice bath for at least 30 min. After centrifugation at 2000 g for 10 min the supernatant is removed and the precipitate is washed free from contaminating cerebrosides by suspension in 2 ml of chloroform : methanol (1:4 v/v), heating briefly at 100° and then centrifuging. This washing procedure is repeated once and the residue is washed once with diethyl ether, and air-dried. The residue is suspended in 1 ml water or buffer by heating at 60° for 5 min. On cooling to room temperature any insoluble material can be removed by centrifugation since it is not glycogen. The aqueous solution is then used directly for the enzymic determination of glycogen.

Method (2). The frozen powdered brain sample is dispersed gently at 0° in 0·6 N-perchloric acid. Crude cerebral glycogen is insoluble under these conditions, provided the tissue dispersion is gentle. The more vigorous the dispersion, the higher the proportion of glycogen which becomes soluble. However if the dispersion is performed by trituration or by gentle hand-homogenization in a tube with a loose pestle, all of the glycogen remains precipitated (Bachelard & Strang, 1973). After centrifugation the pellet is used for glycogen estimation and the supernatant can be used for the determination of acid-soluble constituents. To the pellet is added 1 ml of 60% aqueous KOH, the tube is heated at 80° for 10 min, and 2 ml 95% ethanol are added, with thorough mixing. After cooling at 0° for 30 min, the precipitated glycogen is treated as described in method (1).

Determination

Two enzymic methods are available. The use of glycogen phosphorylase to produce glucose 1-phosphate and glucose originally depended on the contamination of the commercially available enzyme with debranching activity (Lowry & Passonneau, 1972) but it has recently been found necessary to add this activity to the commercial phosphorylase to ensure complete degradation of the glycogen. At the present time (1974) no commercial source of debranching complex is known, so it has to be prepared by the investigator. A commercial source of amylo-1,4-1,6-glucosidase has now appeared (Boehringer u. Soehne, Mannheim, Germany). This enzyme catalyzes the complete degradation to glucose in one experimental stage and the following procedure is recommended (Bachelard & Strang, 1973).

To a sample (0·1 ml) of the aqueous solution of glycogen in 0·1 M-acetate buffer, pH 4·75, is added 10 μl of amyloglucosidase (140 international units/ml) and the mixture is allowed to stand at room temperature for 2 h. To this is added 0·1 M-tris HCl buffer, pH 7·6, containing 10 mM-$MgSO_4$ (0·75 ml); 50 mM-ATP (0·05 ml); 10 mM-$NADP^+$ (0·05 ml) and hexokinase (0·2 international unit). The fluorescence is read before, and 30 min after, the addition of glucose 6-phosphate dehydrogenase (0·03 international unit) as described above for the glucose estimation. Purified liver glycogen is used as the standard.

Specific activity of ^{14}C glycogen. This can be measured after enzymic degradation of the glycogen to 6-phosphogluconate (above) by the method described for glucose, in which the evolved $^{14}CO_2$ is trapped and counted. Alternatively the glucose produced from the amyloglucosidase hydrolysis of glycogen in acetate buffer can be separated by thin-layer chromatography on cellulose in ethylacetate-pyridine-water (10:4:3, by vol.), eluted and counted (Strang & Bachelard, 1971c).

Glycosaminoglycans

These carbohydrate polymers are often found associated with proteins in the brain and may contain neutral sugars, hexosamines, hexuronic acids and sialic acid. They can be divided into two main groups: the mucopolysaccharides which contain no sialic acid or neutral sugars and the sialoaminoglycans which do.

Mucopolysaccharides

These occur in normal mammalian brain in concentrations up to 1·7 mg/g fresh tissue, mainly as hyaluronic acid and chondroitin sulphates A and B. Small amounts of dermatan sulphate and heparan sulphate also occur (Margolis, 1969). This group of macromolecules is important clinically and they accumulate in the brain and body fluids of patients suffering certain metabolic disorders, including *Hurler's syndrome* (gargoylism) and *metachromatic leucodystrophy*. The hexosamine constituents of cerebral mucopolysaccharides (glucosamine and galactosamine) may occur as the N-acetyl derivatives. The hexuronic acid most commonly found is glucuronic acid. N-Acetylgalactosamine, with sialic acid, is a major constituent also of gangliosides (Chapter 5; see also McIlwain & Bachelard, 1971).

The mucopolysaccharides can be extracted from brain samples by homogenization in 5 vol. of 0·2 M-K_2HPO_4 followed by proteolytic digestion with trypsin. The polymers are then precipitated with cetylpyridinium chloride and separated by means of NaCl fractionation. Low concentrations of NaCl (0·4 M) extract the non-sulphated hyaluronic acid and higher concentrations (1·2 M) yield a mixture of chondroitin sulphates, dermatan sulphate and heparan sulphate (Singh & Bachhawat, 1965). These sulphated mucopolysaccharides have been partially separated by chromatography on DEAE-Sephadex columns (Singh *et al.*, 1969).

Sialoaminoglycans

The sialic acid-containing aminoglycans ('sialofucohexosaminoglycans') have been purified according to the following procedure (Brunngraber *et al.*, 1969; Di Benedetta *et al.*, 1969). The fresh brain is extracted by homogenization in chloroform-methanol (2:1, v/v), then chloroform-methanol (1:2, v/v), to remove gangliosides (Chapter 5). The dried residue is suspended in 0·1 M-sodium acetate buffer, pH 5·5 (4·5 ml/g original tissue), containing 5 mM-cysteine and 5 mM-Na_2EDTA. Proteolytic digestion (twice) using papain, and centrifugation, yields an extract containing all the glycosaminoglycans. These have been separated into three groups as follows. Dialysis removes the dialysable sialoaminoglycans and precipitation with cetylpyridinium chloride removes the mucopolysaccharides described above. The centrifuged supernatant is extracted with *n*-pentanol to remove excess cetylpyridinium chloride and the resultant extract contains nondialysable sialoaminoglycans. Sugar constituents detected include sialic acid, fucose and mannose, in addition to glucosamine and galactosamine; no hexuronic acids appear to be present.

Analysis of constituent sugar derivatives
The glycosaminoglycans are hydrolyzed in sealed tubes at 100° in HCl. The concentration of HCl and the time of hydrolysis required varies with the type of macromolecule being studied. *Hexuronic acids* (usually glucuronic acid) are estimated by the carbazole reagent method: 5 ml of 25 mM-sodium tetraborate in concentrated H_2SO_4 is carefully added at 0° to 1 ml of neutralized sample. The mixture is heated at 100° for 10 min and cooled. Carbazole (0·125% in ethanol, 0·2 ml) is added and the mixture heated at 100° for 15 min. After cooling, the extinction is read at 530 nm (Davidson, 1966). *Total hexosamines* are determined by a variation of the Elson-Morgan reaction. The following was used by Ludoweig & Benmaman (1967): 0·5 ml samples in 0·3 N-HCl (after removal of neutral sugars on ion exchange columns) are treated with 1 ml acetylacetone (3% in 0·7 M-Na_2CO_3) and heated at 80° for 1 h. After cooling, 10 ml 95% ethanol and 1 ml Ehrlich reagent (2·67% *p*-dimethylaminobenzaldehyde in ethanolic-HCl) are added. The extinction at 535 nm is used, in comparison with standards, to calculate the total hexosamine present. Glucosamine can be distinguished from galactosamine by a low-temperature modification of the procedure: an 0·5 ml sample in 0·3N-HCl is cooled to 0°, the acetylacetone reagent is added at 0°, and the mixture is stored for 18 h at 0°. Additions of 10 ml 90% ethanol and 1 ml Ehrlich reagent are also made at 0°. The mixture is incubated at 60° for 3 min, cooled to 20–25° for 1 h and the extinction read at 525 nm. Galactosamine, but not glucosamine, reacts (Ludoweig & Benmaman, 1967). Alternative methods for distinguishing between the hexosamines are given by paper chromatography on ethyl acetate–pyridine–*n*-butanol–butyric acid–water (10:10:5:1:5) (Mukerjee & Ram, 1964) or by taking advantage of the specificity of yeast hexokinase. Glucosamine is phosphorylated but galactosamine is not; the phosphorylated product is separated by ion exchange chromatography (Davidson, 1966). Methods for the separation and analysis of mucopolysaccharides and glycoproteins have been reviewed recently by Margolis & Margolis (1972).

Lactate and Pyruvate

Lactate is commonly extracted from cerebral tissues with trichloroacetic acid or perchloric acid; the neutralized acid-soluble extract described previously is suitable. Methods of determination were earlier based on specific colorimetric reactions (Barker & Summerson, 1941) but have been superseded by more sensitive enzymic reactions. These require care due to possibilities of contamination of laboratory glassware: lactic acid can be difficult to remove. The methods of determination must take cognizance of the fact that the equilibrium of the lactate dehydrogenase reaction favours lactate formation rather than its oxidation, and so require the addition of reagents such as hydrazine to trap the pyruvate formed and to drive the reaction in the desired direction. Since the normal concentration of lactate in the brain is high relative to the sensitivity of the method, spectrophotometry is usually sufficiently sensitive. A spectrophotometric cuvette is prepared with the following reagents (final vol., 1 ml), based on the hydrazine-trapping method of Hohorst (1963): 0·6 ml glycine-hydrazine buffer, pH 9·3 (0·5 M-glycine, 0·4 M-hydrazine sulphate), 1 mM-NAD^+ (0·2 ml) and the sample containing lactate in 0·1 to 0·2 ml. The extinction is read at 340 nm, lactate dehydrogenase (20 international units) is added, and the change with time in extinction at 340 nm is followed until constant.

Lowry & Passonneau (1972) have used a coupled enzyme system rather than hydrazine to trap the pyruvate formed; glutamate and glutamate-pyruvate transaminase convert the pyruvate to alanine.

Pyruvate. The concentrations of pyruvic acid which normally occur in the mammalian brain (Table 3.2) are less than one-tenth of those of

Table 3.2 Concentrations of citric acid cycle intermediates in normal mammalain brain

Intermediate	Enzymes used in assay	Concentration (nmoles/g frozen tissue) found by freezing-method:		
		1	2	3
Pyruvate	Lactate dehydrogenase	91	89	91
Citrate	Aconitase, isocitrate dehydrogenase	327	—	—
	Citrate lyase, malate dehydrogenase, lactate dehydrogenase	—	337	263
Isocitrate	Isocitrate dehydrogenase	16	—	—
α-Oxoglutarate	Glutamate dehydrogenase	127	121	213
Succinate	Lactate dehydrogenase, pyruvate kinase, succinate thiokinase	686	—	—
Fumarate	Fumarase	73	—	—
Malate	Malate dehydrogenase	438	370	277
Oxaloacetate	Malate dehydrogenase	4	5	—

Freezing methods:
[1] Mouse brain, frozen by immersion (Goldberg *et al.*, 1966).
[2] Rat brain, 'funnel frozen', see introduction (Bachelard *et al.*, 1974).
[3] Rat brain, 'freeze-blown', see introduction (Veech *et al.*, 1973).

lactate, but the method of estimation is more straightforward: as mentioned above, the equilibrium of lactate dehydrogenase favours lactate formation. Methods for extraction of pyruvate from tissue samples are identical to those for lactate. A convenient spectrophotometric assay for pyruvate contains the following reagents: 0·1 M-tris HCl buffer, pH 7·6 (0·6 ml), 1·0 mM-NADH (0·2 ml) and the sample containing pyruvate in 0·2 ml. The extinction at 340 nm is read, lactate dehydrogenase (10 international units/ml, 10 μl) is added, and the reaction followed until the decrease in extinction at 340 nm ceases (usually within 5 min). Greater sensitivity, for small samples containing less pyruvate, can be obtained using a fluorometer. The only modification is that the concentration of NADH is reduced to one-twentieth of that quoted.

For microanalysis of pyruvate in very small samples, Lowry & Passonneau (1972) have described a cycling method.

Radioactivity measurements. The above enzymic method for conversion of lactate to pyruvate has been adapted to measurement of the specific activity of radioactive lactate in cerebral extracts (Strang & Bachelard, 1971b). As described above, hydrazine is used to trap the pyruvate produced and the pyruvate hydrazone is converted to the less soluble dinitrophenylhydrazone, which can be estimated directly by colorimetry and counted. The radioactivity present in the cerebral pyruvate can be measured in the same way, but with the enzymic stage omitted, as follows. The pyruvate can be converted directly to the hydrazone, partially purified and counted (Bachelard, 1965). This should always be performed when the radioactivity of lactate is determined, as a control for the endogenous pyruvate.

Pentoses

Determination of acid-soluble pentoses in extracts of neural tissue by established methods presents no special problems. Portions of an extract prepared by homogenization of the tissue in perchloric or trichloroacetic acid are shaken with acid-washed charcoal to remove nucleotides. To avoid adding an unnecessary excess of charcoal the efficacy of the removal was monitored (Tower, 1958) by determining the extinction of the filtrate at 260 nm. Aldopentoses (ribose, deoxyribose and arabinose) can be determined in the presence of ketopentoses and hexoses by the phloroglucinol reaction of Dische & Borenfreund (1957).

It should be noted that the pentose 5-phosphates yield a higher extinction than the free pentoses (Ashwell, 1966). The following procedure is used: the sample in 0·1 ml (containing 0·01 to 0·1 μmol) is treated with 1 ml of the reagent (glacial acetic acid, 110 ml; concentrated HCl, 2 ml; 0·8% glucose, 1 ml and 5% phloroglucinol in ethanol, 5 ml). The mixture is heated at 100° for 15 min and cooled. The extinction at 510 nm is subtracted from that at 552 nm; the difference in extinction is linear with a concentration range of up to 1 mM pentose in the sample (Ashwell, 1966). A coupled enzymic method for determination specifically of ribose 5-phosphate has been described (Racker, 1963); the sensitivity is similar to that of the phloroglucinol method.

Intermediates of the citric acid cycle

As with the other intermediates described in this chapter, earlier methods of estimation of tricarboxylic acids and related metabolites were based on chromatographic methods, and these are still important techniques, especially if radioactivity measurements are to be made. For analysis of the levels of the intermediates, a range of enzymic techniques is now available, and these avoid the need for prior separation of the individual metabolites. Silicic acid column-chromatographic methods permit simultaneous determination of most of the acids of the citric acid cycle (Frohman et al., 1951; Busch, 1953); only 10 mg of tissue is required for analysis. Gas–liquid chromatographic methods are also available, and require prior methylation of the intermediates to render them sufficiently volatile for the separation (Barnett et al., 1968). The latter methods are also directly applicable to specific activity measurements of the labelled intermediates. Enzymic techniques were applied by Goldberg et al. (1966) to analysis of the citric acid cycle intermediates in the brain. Tissues frozen in situ were homogenized in perchloric acid and the extract neutralized with $KHCO_3$ as described previously in this chapter. The enzymic assays were performed on this neutralized acid-soluble extract. Details are given of the methods for citrate, isocitrate, α-oxo acids, succinate, malate and fumarate, and are all based on coupled enzyme systems which result in oxidation or reduction of nicotinamide nucleotides, measured fluorometrically. Table 3.2 gives the concentrations of the intermediates normally found in the brain. The amino acids which metabolically are closely related through transamination to citric acid cycle intermediates (glutamate, glutamine, alanine, aspartate and γ-aminobutyrate) were described in Chapter 2.

ACID-SOLUBLE PHOSPHATES

Following the discovery of phosphocreatine in muscle, much of the subsequent exploratory work on tissue phosphates included studies with the brain (Kerr, 1935, 1936), and many of the methods in current use, described below, have developed from studies on neural tissues. The concentrations of the more important acid-soluble phosphates which occur in the mammalian brain in vivo are listed in Table 3.3; the values quoted take no account of possible species differences. Further comments on certain of the values are made later in relation to the determination of individual phosphates.

Table 3.3 Acid-soluble phosphates in the mammalian brain

Compounds	Approximate in vivo concentration (μmol/g fresh or frozen tissue)
Inorganic phosphate	3 –5
Phosphocreatine	2·5 –4·6[1]
Nucleotides	
Adenosine triphosphate	2·5 –3·0
Adenosine diphosphate	0·25 –0·5
Adenylic acid	0·05 –0·2
3′, 5′-Cyclic AMP	0·001–0·002
Guanine nucleotides	0·5 –1·0
Uridine nucleotides	approx. 1
Cytidine nucleotides	Present
NAD+	0·33
NADH	0·03
NADP+	0·01[2]
NADPH	0·02[2]
Hexose phosphates	
Glucose 6-phosphate	0·08
Fructose 6-phosphate	0·016
Glucose 1,6-diphosphate	0·03 –0·08
Fructose 1,6-diphosphate	0·10 –0·20
Triose phosphates	
Dihydroxyacetone phosphate	0·02 –0·05
Glyceraldehyde 3-phosphate	0·005
3-Phosphoglycerate	0·04
2-Phosphoglycerate	0·01

Data from Lowry et al. (1964), McIlwain & Bachelard (1971), Bachelard et al. (1974). For phosphate derivatives of lipids, see Chapter 5.
[1]With light anaesthesia.   [2]Tissues not frozen in situ.

Fixation and extraction of tissue

The importance of fixing the brain *in situ* when cerebral phosphates are to be determined was indicated in Chapter 1 and in the introduction to this chapter, with Table 3.1.

Extraction. The frozen tissue must be rapidly ground with the extracting agent, for, if it is allowed to thaw before grinding is complete, considerable changes in concentrations of labile phosphates may occur. Preliminary grinding of the frozen tissue is therefore essential, as emphasized in Chapter 1. As extracting agents, 5–10% (w/v) trichloroacetic acid or 0·3–1 N-perchloric acid are commonly used at 0° and the resulting extracts must be neutralized and maintained at low temperature, pending analysis. Perchloric acid is usually preferred: much of the perchlorate anion can be removed as the K^+ salt (described previously) and it has been judged to be superior to trichloroacetic acid, which precipitates phospholipids less efficiently (Heald, 1960; see also Wilson

& Thomson, 1969). The method of extraction with perchloric acid is identical to that described above for glucose.

Fractionation of acid-soluble tissue phosphates by precipitation

Earlier attempts to fractionate the phosphates depended on the relative solubilities of their calcium or barium salts. These methods are still applicable, although the recently developed chromatographic and enzymic techniques render them less so. The general scheme of barium salt fractionation which follows (Fig. 3.1) is based on the description of Le Page (1957; see also Stone, 1948). Current techniques

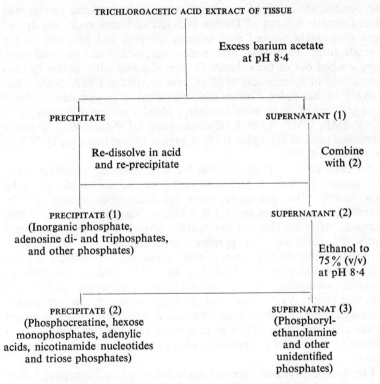

Fig. 3.1 Fractionation of cerebral phosphates as their barium salts.

for further separation are usually based on a combination of chromatographic methods and estimation is by specific enzymic methods.

Fractionation by chromatography

Chromatography using ion exchange resins, gel filtration supports, paper and thin layers, especially in various combinations, are proving of great value in fractionating tissue phosphates and related intermediates.

Purine and pyrimidine bases, nucleosides and nucleotides

A variety of well-tried and reliable methods has been applied to separations from neural tissues of purines and purine nucleotides, but there has been less emphasis on development of techniques for separating pyrimidines and their nucleotides. Piccoli *et al.* (1969) applied modifications of the chromatographic techniques of Cohn (1950) to the separation of cerebral purine and pyrimidine derivatives as follows. The frozen brain samples were extracted into 50% (v/v) aqueous ethanol which was then made 0·3 M in perchloric acid. After removal of most of the perchlorate as the K^+ salt (above), the neutralized extract was placed onto a column of Dowex 1-x8 ion exchange resin, which had been thoroughly washed with organic solvents and prepared in the formate form. Free bases and nucleosides, which did not exchange, were washed out of the column. Elution stepwise with increasing concentrations of formic acid (0·02 to 0·08 N) yielded CMP, NAD^+ and 5'-AMP in that order. Elution with a gradient of formic acid, followed by a gradient of ammonium formate, yielded a progression of peaks of GMP, IMP, UMP, ADP, UDP-coenzymes, UDP-glucose, ATP and a combined peak of GTP plus UTP. A similar system was used by Wilson & Thomson (1969).

An alternative method, suitable for separation on a smaller scale, was applied by Pull & McIlwain (1972) to the nucleotides in tissue superfusates. The derivatives were removed from samples of the superfusate from tissues incubated *in vitro* by absorption onto activated charcoal. After washing the charcoal the absorbed material was eluted with 10% (v/v) aqueous pyridine, concentrated and subjected to chromatography on thin layers of silica gel in butan-1-ol–ethyl acetate–methanol–0·880 ammonia (7:4:3:4, by vol.). Clean separations of adenine, adenosine, inosine, hypoxanthine and cyclic AMP were achieved, with ATP, ADP and 5'-AMP remaining at the origin. Although some loss of material occurred, due to irreversible adsorption to the charcoal, the method gives a rapid reproducible separation of minute amounts and is applicable also to analysis of labelled components.

For microanalysis of cerebral nucleotides, microelectrophetic techniques have been devised. The separated material is not analysed as phosphate but by ultraviolet spectrometry after electrophoresis on a thread of specially-treated cellulose fibre (Edström, 1956). The technique has been applied to base-ratio analysis of single nerve cells and of the membranes and axoplasm of their axons (Edström, 1964).

A more recent technique, high pressure liquid chromatography, has been applied to separation of rat brain nucleotides: 14 different compounds were separated from 1 mg of tissue (Shmukler, 1972; see also Horvath & Lipsky, 1969).

Thin-layer chromatographic methods for the separation of tissue nucleosides and nucleotides have recently come into use, although little has been published on separations from cerebral tissues at the time of writing. These generally give better separations than are obtained on paper and are more rapid than column chromatogaphy. Purine and pyrimidine bases and their nucleosides are separated on cellulose and the nucleotides on various ion-exchange celluloses (Verachtert et al., 1965; Randerath & Randerath, 1967). Recommended solvent systems have been reviewed by Mangold (1969).

An example of separations of those constituents from brain extracts is given by Wong & Henderson (1972; see also Crabtree & Henderson, 1971). These workers used two-dimensional chromatography on cellulose plates to separate bases and nucleosides. Use of acetonitrile–0·1 M-ammonium acetate (pH 7·0)–0·88 ammonia (60:30:10) as first solvent and butan-1-ol–methanol–water–0·88 ammonia (60:20:20:1) as second solvent gave good separations of adenine, adenosine, hypoxanthine, guanine, inosine, guanosine, xanthine and xanthosine; the nucleotides did not run. The nucleotides in the same brain extracts were separated by one-dimensional chromatography on polyethyleneimine cellulose (see Randerath, 1962) by stepwise application of increasing concentrations of sodium formate. The nucleotides ran in increasing order of mobility: GTP, ATP, GDP, ADP, GMP, XMP, IMP, AMP and NAD^+.

Sugar phosphates

Chromatographic procedures for these compounds usually follow the barium fractionation procedure of Fig. 3.1, although the current enzymic techniques described below obviate the need for prior separation, if tissue concentrations only are required. The separation techniques are still advantageous if specific activities of the labelled phosphates are to be determined. The barium fraction containing the sugar phosphates (Fig. 3.1) can be treated with ion exchange resins to remove Ba^{2+} and the constituents of the resultant solution separated by paper or thin-layer chromatography (see, for example, Tower, 1958).

Inorganic phosphate

The concentrations of acid-soluble labile phosphates in neural tissues are relatively high and care has to be exercised in ensuring that estimations of inorganic phosphate do not include the various forms of esterified phosphate present in the extracts. The most commonly used methods are all variants of the method of Fiske & Subbarow (1925), based on measurement of the blue colour formed on reduction of phosphomolybdic acid. The problem of contamination by phosphate esters is overcome by extraction of the phosphomolybdic acid from the aqueous solution in which it has been formed into an organic solvent, e.g. a mixture of benzene and isobutanol (Berenblum & Chain, 1938) before reduction to give the blue-coloured product. Provided the extraction into the solvent is prompt, interference through hydrolysis of esters such as phosphocreatine is virtually absent.

The method described below is derived from that of Martin & Doty (1949) and is suitable for determining inorganic phosphate in the presence of labile phosphate (except acetyl phosphate, but see the comment below). When it is applied to determinations in the acid tissue extracts described above, the addition of 2·5 N-H_2SO_4 is usually omitted and the procedure can be carried out at room temperature.

Procedure. The tissue extract at 0° (1–10 μg of phosphate in 3 ml) is transferred to a glass-stoppered test tube in an ice bath and mixed with 5 ml of 1:1 (v/v) isobutanol–benzene and 0·5 ml of 2·5N-H_2SO_4. Ammonium molybdate (0·5 ml of 10%, w/v) is added and the tube immediately stoppered and shaken for 15 s. After separation of the layers (if necessary by brief centrifugation) a sample (usually 2 ml) is mixed with an equal volume of N-H_2SO_4 in ethanol. Stannous chloride as reducing agent (0·5 ml) is added and immediately and thoroughly mixed: the extinction at 730 nm is read after 15 min. The reducing agent is kept as a 10% (w/v) stock solution of $SnCl_2.H_2O$ in concentrated HCl and is diluted 1:200 with 0·25 N-H_2SO_4 before use.

Comment. Extraction methods are susceptible to variation in the ionic composition of the aqueous phase: a high concentration of ions, including H^+, diminishes the transfer of phosphomolybdate to the organic phase. Initially, therefore, it is advisable to determine the recovery of added inorganic phosphate from a tissue extract. If this is found to be less than obtainable with pure solutions, internal standards, or standards prepared in solutions of similar ionic composition to that of the unknown solutions, should be employed. For instance, concentrations of trichloroacetic acid greater than 5% (w/v) interfere and must be removed by solvent extraction (Ernster *et al.*, 1952). With tissue extracts made in perchloric acid (0·3 N) the addition of further acid before extraction is not necessary, but here also higher concentrations diminish recovery. Martin & Doty (1949) describe a silicotungstate reagent that permits the determination to be made in the presence of a small quantity of protein. The use of this reagent was found advisable in determining inorganic phosphate released by alkali from cerebral phosphoproteins (see below).

If the extract contains acid-labile phosphates, it is important that the time of shaking is not prolonged beyond the recommended 15 s. Once the two phases have been separated, transfer of the organic phase can safely be delayed for a few minutes as Ernster *et al.* (1952) found that without shaking negligible migration of phosphomolybdic acid into the organic phase occurred.

Enzymic determination of inorganic phosphate. This method is potentially more sensitive than the colorimetric method described above, since fluorometric techniques can be applied. It is based on the use of glycogen phosphorylase, which requires inorganic orthophosphate to degrade glycogen to form glucose 1-phosphate. The determination depends on the addition of coupling enzymes to convert glucose 1-phosphate to glucose 6-phosphate which in turn is converted to 6-phosphogluconolactone with reduction of $NADP^+$ to NADPH. Details of the method of Fawaz et al. (1966) have been described by Lowry & Passonneau (1972).

TOTAL PHOSPHORUS

To convert organically-bound phosphorus to inorganic orthophosphate tissue extracts are usually digested by heating with mixtures of sulphuric and perchloric acids; these procedures have been discussed by Heald (1960). A commonly used mixture is 3:2 (v/v) concentrated H_2SO_4—10 M-$HClO_4$ (Hanes & Isherwood, 1949). Oxidation can be hastened by addition of H_2O_2 or HNO_3 without loss of phosphorus; the excess must be removed by prolonging the time of digestion. The addition of H_2O_2 or HNO_3 must be made when the digests are warm, rather than after cooling to room temperature (Le Page, 1948). With sulphuric acid in the mixture some pyrophosphate may be formed: this can be hydrolyzed to orthophosphate by diluting the sample with water and heating at 100° for 5 min. Inorganic phosphate is then determined as described above.

Hexose phosphates

Hexose phosphates occur in minute amounts in cerebral tissues and the earlier methods, based on chemical methods after separation by the barium fractionation procedure described in a previous section of this chapter, are not very sensitive and are now not considered to be sufficiently specific. These methods have been superseded by enzymic methods which usually involve the addition of coupling enzymes to bring about a final enzymic stage which involves oxidation or reduction of nicotinamide nucleotides. In view of their low concentrations in cerebral extracts, spectrophotofluorometric enzymic methods, applied directly to neutralized acid extracts, are recommended: these are given in detail by Lowry & Passonneau (1972).

For determination of radioactivity present in hexose phosphates, enzymic methods can be applied under certain circumstances (Strang & Bachelard, 1971b). Alternatively chromotographic methods may be used. No one system of chromatography gives a complete separation of all the sugar phosphates present in cerebral extracts and none has been designed specifically to this end. The choice of matrix and solvents for two-dimensional separations must depend on the objectives of the study.

Two-dimensional paper chromatography was used by Dodd et al. (1971) to separate the hexose phosphates of cerebral extracts: the first solvent was butan-2-ol–formic acid–water (70:13·2:16·8, by vol.)

and the second was phenol–water–0·88 ammonia (80:20:3, w/v/v). Good separation of hexose phosphates is also obtained on ion exchange celluloses (MN- or ECTEOLA-cellulose) with various solvent systems (Lewis & Smith, 1969; see also Waring & Ziporin, 1964; Carminatti & Patterson, 1966; Verachtert et al., 1966).

Cyclic AMP

The earliest method of estimation of cyclic AMP (adenosine 3′,5′-cyclic monophosphate) was based on the original observation of its biological activity in activating liver glycogen phosphorylase (Robison et al., 1968). Subsequent methods have been based on cyclic AMP as the substrate for a specific phosphodiesterase, and on activation of a variety of protein kinases by cyclic AMP. The original observation of activation of phosphorylase is in fact another example of activation of a protein kinase, since it is the phosphorylase kinase system which is involved. Recently many sensitive and highly specific methods have been developed for estimation of cyclic AMP in tissues and body fluids, including cerebrospinal fluid. One of these which seems particularly suitable for body fluid analysis is the enzymic radio-isotopic displacement technique (Brooker et al., 1968). The samples of plasma or of cerebrospinal fluid are partially purified by treatment with $ZnSO_4$, $Ba(OH)_2$ and column chromatography using Dowex 50 W resin (Krishna et al., 1968) before estimation.

The use of specific binding proteins has also yielded a sensitive method, 'saturation analysis', for estimation in tissue extracts and in body fluids (Gilman, 1970). Cyclic AMP in neutralised perchloric acid extracts can be determined without the need for extensive purification (Brown et al., 1971). The method has been applied to the estimation of cyclic AMP in small samples of nervous tissues (Weller et al., 1972). These samples may contain factors which act to enhance or inhibit binding, but can be overcome by using unlabelled internal standards in the samples. Cyclic AMP-binding proteins are now available in assay kits from fine-chemical manufacturers. The luciferase assay system for ATP (above) has been applied by Johnson et al. (1970) to the estimation of cyclic AMP, using phosphodiesterase with myokinase and pyruvate kinase as coupling enzymes to produce ATP. This technique was found to be more sensitive than the enzymic isotopic displacement method.

The preceding description notes the major types of assay method that have been developed but so many techniques and variations have been reported that they now form a separate review subject (Greengard & Robison, 1972).

Creatine phosphate

A variety of chemical methods has been devised for separation and estimation of creatine phosphate. The most commonly applied have been based on the release of inorganic phosphate from creatine phosphate which occurs in acid-molybdate solutions at room temperature. The creatine phosphate has either been separated from other phosphates prior to the molybdate treatment by barium precipitation (Fig. 3.1) or is corrected by difference for inorganic phosphate, determined by extraction of the phosphomolybdate complex into organic solvents as described above. The value of the various modifications to this general approach has been appraised by Heald (1960). The enzymic methods described below are now more commonly applied.

The creatine content of the creatine phosphate has been determined after conversion of the creatine to creatinine, using the Jaffe reaction (McIlwain et al., 1951) and also by Ennor & Rosenberg (1952), who used the Barritt α-naphthaldiacetyl reaction. Values comparable with those based on phosphate determinations were obtained.

Enzymic methods
The method of Slater (1953) was adapted to enzymic determination of creatine phosphate in extracts by Kratzing & Narayanaswami (1953): a 'difference' method, it is based on the difference between the adenosine phosphates and the total energy-rich phosphates in the sample, determined after enzymic formation of dihydroxyacetone phosphate in the presence of glycerophosphate dehydrogenase and the concomitant oxidation of NADH.

Another difference method permits the simultaneous measurement of creatine phosphate and ATP in the same sample (Lowry & Passonneau, 1972). It can also be applied to measurement of creatine. The tissue extract is placed in a reagent mixture containing 25 mM-tris-HCl, pH 8·1, 1 mM-$MgCl_2$, 0·5mM-dithiothreitol, 0·5mM-$NADP^+$ and 1mM-glucose. Hexokinase (0·28 unit/ml) and glucose 6-phosphate dehydrogenase (0·07 unit/ml) are added and the mixture allowed to react until the change in extinction at 340 nm is complete. The ATP of the extract can be calculated from this. Then, ADP (0·5 mM) and creatine kinase (0·9 unit/ml, measured in the direction of creatine phosphate formation) are added. The further change in extinction is used to calculate the creatine phosphate present. Like all methods involving nicotinamide nucleotides, sensitivity is increased using fluorometry. A cycling method with greatly increased sensitivity has also been described (Lowry & Passonneau, 1972).

Both enzymic methods give results comparable to those from the chemical methods: the enzymic methods have the advantage of greater simplicity and sensitivity.

Nucleotides

Chemical methods. In the barium-insoluble factions (Fig. 3.1) the nucleotide di- and triphosphates are usually determined as inorganic phosphate after hydrolysis in N-HCl for 10 min at 100°C. Methods for adenosine diphosphate (e.g. Le Page, 1948), relying on the molar ratio of easily hydrolyzed phosphate to ribose, have been shown to be inaccurate when applied to neural tissues (see Kratzing & Narayanaswami, 1953); enzymic or chromatographic methods are to be preferred. Adenylic acid levels also, when determined by a chemical method (Le Page, 1948) are higher than those found by enzymic or chromatographic methods (Heald, 1960).

Enzymic methods. ATP is determined using the hexokinase, glucose 6-phosphate dehydrogenase method (Slater, 1953; Lowry & Passonneau, 1972) as described under creatine phosphate (above).

ADP can also be determined enzymically in the same tissue extracts using pyruvate kinase. The pyruvate so formed is converted to lactate, using lactate dehydrogenase, with oxidation of NADH. The spectrophotometric method is performed in triethanolamine or imidazole buffer at pH 7·0 to 7·6 with 2mM-$MgCl_2$, 75mM-KCl, 0·3mM phosphoenolpyruvate, 0·2mM-NADH, pyruvate kinase (0·3 unit/ml) and lactate dehydrogenase (0·4 unit/ml). The method is easily adapted to the more sensitive fluorometric technique (Lowry & Passonneau, 1972). AMP can be measured in the same reaction mixture if myokinase (0·4 unit/ml) and ATP (0·1mM) are present. The ADP formed (2 molecules per molecule of AMP present) is then measured by the above procedure.

The most sensitive method for estimating ATP is given by the fireflyluciferase system. Details of a method for measurement of the light emitted directly by liquid scintillation spectrometry are given by Stanley & Williams (1969): as little as 10^{-12} mole of ATP can be measured in non-neutralized perchloric acid extracts. With a commercial supply of luciferase now available (Boehringer u. Soehne, Mannheim) the method provides for simple, rapid and highly sensitive measurement of ATP and can be applied directly to cerebral extracts (I. Pull, personal communication).

Nicotinamide adenine nucleotides

As shown in Table 3.3, with the exception of NAD^+ (approx. 0·3 μmol/g tissue) the cerebral concentrations of these nucleotides are minute (0·03 μmol/g or less), so accurate estimation depends on 'cycling' methods. These depend on repeated enzymic oxidation and reduction of the coenzyme to build up the concentration of another reactant of one of the enzymes used, which is then estimated by standard techniques. The principle of enzymic cycling was applied by Glock & McLean (1955) to estimation of concentrations of NAD^+, NADH, $NADP^+$ and NADPH in tissue samples including those from the mammalian brain. Their method was based on spectrophotometric estimation of the cytochrome-*c* enzymically produced. Subsequent methods have utilized enzymic reactions more suited to the increased sensitivity of fluorometric measurement noted by Ciotti & Kaplan (1957). Thus Bassham *et al.* (1959) used alcohol dehydrogenase and glucose 6-phosphate dehydrogenase in combination with spectrophotofluorometry to measure NAD^+ and NADH; other workers have used a combination of glutamate dehydrogenase and either lactate dehydrogenase or glyceraldehyde 3-phosphate dehydrogenase (Lowry *et al.*, 1961; Matchinsky, 1971). More recently a combination of alcohol dehydrogenase and malic dehydrogenase has been used with improved cycling rates over a wider range of coenzyme concentrations (Kato *et al.*, 1973).

The following example is that described by Lowry *et al.* (1961). If NAD^+ is to be measured, the NADH present is destroyed by heating with alkali. Either NAD^+ or NADH is then estimated by cycling according to the following scheme (*A*), where lactate and α-oxoglutarate are included in the cycling reagent, and pyruvate and glutamate accumulate. One of the accumulated products is then estimated using the relevant dehydrogenase and fluorometry.

(*A*) Lactate $+$ NAD^+ $\underset{}{\overset{\text{lactate dehydrogenase}}{\rightleftharpoons}}$ pyruvate $+$ NADH $+$ H^+;

NADH $+$ NH_4^+ $+$ α-oxoglutarate $\underset{}{\overset{\text{glutamate dehydrogenase}}{\rightleftharpoons}}$ NAD^+ $+$ glutamate.

The number of cycles per unit time is controlled according to the amounts and proportions of the constituents of the cycling reagent. Not only does care have to be exercised in preventing breakdown of the labile coenzymes in the tissue extracts, but the cycling enzymes have to be purified from bound coenzymes. This is achieved by treatment with charcoal (Norite) as described by Kato *et al.* (1973).

A similar cycling system for estimating $NADP^+$ and NADPH may be carried out using glucose 6-phosphate dehydrogenase with glutamate dehydrogenase (Lowry & Passonneau, 1972) according to the following scheme (*B*).

(*B*) $NADP^+$ $+$ glucose 6-phosphate $\underset{}{\overset{\substack{\text{glucose 6-phosphate} \\ \text{dehydrogenase}}}{\rightleftharpoons}}$ NADPH $+$ H^+ $+$ 6-phosphogluconolactone;

NADPH $+$ NH_4^+ $+$ α-oxoglutarate $\underset{}{\overset{\substack{\text{glutamate} \\ \text{dehydrogenase}}}{\rightleftharpoons}}$ $NADP^+$ $+$ glutamate.

Other phosphate esters

The phosphate esters of ethanolamine, serine and choline occur in neural tissues and are described in Chapter 5.

ACID-INSOLUBLE PHOSPHATES

By far the greater part of the phosphorus of neural tissue remains insoluble on treatment of the tissue with acid extractants. Much of this insoluble phosphorus is accounted for by the phospholipids and nucleic acids, but even after these groups of compounds have been removed an appreciable amount of phosphorus remains bound in the insoluble residue. This residual phosphorus, now the subject of much research, contains phosphoproteins, lipid-protein complexes containing inositol phosphates and several as yet unidentified phosphorus-containing compounds.

Much of the work on the acid-insoluble phosphates in brain has arisen from attempts to measure nucleic acids in that tissue by determination of their phosphate moieties. Fractionation procedures are therefore conveniently considered in relation to methods available for nucleic acid determinations.

Nucleic acids

When a widely applied method for determining tissue nucleic acid phosphorus (Schmidt & Thannhauser, 1945) was critically examined with neural tissues it was found to give erroneous results for RNA (Logan et al., 1952). The reason for this was that the extraction procedure (extraction of the acid-soluble phosphates with TCA, removal of lipids from the precipitate with alcoholic solvents and hydrolysis of the residue with KOH) gave, in addition to some loss of nucleoprotein, phosphorus from compounds other than nucleoproteins and phosphoproteins, so the method could not be used to determine RNA phosphorus by difference. However, modifications of the Schmidt-Thannhauser procedure have recently been reported to be advantageous in determining total DNA in neural tissues (Penn & Suwalski, 1969; see below).

Current methods of extraction of the tissue depend on the objectives of the work: whether it is to separate nucleoproteins or nucleic acids in their native states or to analyse the nucleotide base ratios of the nucleic acids.

RIBONUCLEIC ACID

The various species of RNA which have been found in cerebral, as in other tissues (messenger RNA, ribosomal RNA and soluble or transfer RNA) are distinguished by a combination of criteria: size (as judged from sedimentation on sucrose density gradient centrifugation), elution profile on column chromatography, rate of labelling from radioactive precursors and base-ratio analysis. For such studies prior isolation and separation of the species in their native states is often required. It should be noted that the TCA extraction procedure of Schmidt & Thannhauser (1945) results in denatured nucleic acids (see below). For preparations of native nucleic acids, modifications of Kirby's (1955) phenol extraction technique are used. The most commonly applied is hot extraction using mixtures of phenol, sodium dodecyl sulphate and 8-hydroxquinoline (Jacob et al., 1966; Kimberlin, 1967). The nucleic acids are precipitated from the resultant aqueous solution with ethanolic salt solutions and separated by column chromatography. Jacob et al. (1966) separated cerebral RNA species by chromatography on columns of methylated albumen-kieselguhr and also by centrifugation through a gradient of sucrose (4 to 20%) in the presence of sodium chloride, sodium acetate and polyvinyl sulphate (to inhibit degradation by ribonuclease).

Oesterle et al. (1970) described modifications of the phenol extraction in which cold and hot extractions with phenol-m-cresol-8-hydroxyquinoline mixtures were used. The cerebral nucleic acids were precipitated using an ethanolic salt solution and then separated by gel filtration on coupled Sephadex G200 and G25 columns. This procedure yielded a fourth, high molecular weight, species of RNA, different from ribosomal, messenger or transfer RNA.

The most convenient method of estimating the amounts of RNA present are by ultraviolet spectrophometry at 260 nm or after hydrolysis (see below) by estimation for inorganic phosphate.

DEOXYRIBONUCLEIC ACID

Native DNA can be separated after applying the phenol extraction methods described above (see for example, Oesterle et al., 1970). In an appraisal of the methods for extracting denatured DNA for analysis, Penn & Suwalski (1969) found a modification of the original Schmidt-Thannhauser method to give the highest yields of cerebral DNA. The residue after extraction of the brain sample with 5% (w/v) TCA was incubated at 37°C overnight with 0·1 M-KOH containing dioxan (25% v/v). The neutralized material was treated again with 5% TCA and centrifuged. The residue, after neutralization with ammonia, centrifugation and freeze drying, provided the denatured DNA.

Analysis of nucleic acids

For quantitative estimation of the sugar and phosphate moieties of the nucleic acids, or for base-ratio analysis, RNA species are hydrolyzed in alkali: Oesterle et al. (1970) used 0·05 N-KOH at 100°C for 40 min and separated the nucleotide bases by chromatography on Dowex-1 formate columns; Jacob et al. (1966) hydrolyzed their RNA fractions in 0·3N-KOH at 37°C overnight. After neutralization the nucleotides were adsorbed onto charcoal and eluted with ethanolic ammonia. The constituent nucleotides were separated using paper electrophoresis. Inorganic phosphate is estimated by the colorimetric methods described previously.

DNA, which is resistant to the alkaline hydrolysis used for RNA, is hydrolyzed by heating with 12 N-perchloric acid at 100°C for 1 h

(Zamenhof et al., 1964; Penn & Suwalski, 1969; Mori et al., 1970). Estimation of the constituent bases, as for RNA, can be made by measuring the E_{260} or by phosphate analysis. Also the specific diphenylamine reaction for deoxyribose has been widely applied (Zamenhof et al., 1964; Penn & Suwalski, 1969) as described previously under Pentoses.

Nucleoproteins. Polysomes from cerebral tissues have been separated from crude microsomal preparations by centrifugation through sucrose density gradients, usually 0·3 to 1 M-sucrose containing a small amount of detergent (1% deoxycholate) to disrupt the membrane components of the preparation. The various species, separated on the gradient, have then been analysed for constituent RNA as described above (Jacob et al., 1967; Schneider & Roberts, 1968; Vesco & Giuditta, 1968).

Protein-phosphorus in residual fraction

The residual fraction remaining after extraction of the phospholipids (and, in the case of the Schneider procedure, the nucleic acids) contains components that release inorganic phosphate on treatment with mild alkali. These have been shown to be proteins containing phosphate bound by covalent linkage to serine-OH groups (Rodnight, 1971). To determine the total protein-phosphorus which is bound in the tissue in this way, the residual fraction (obtained as described above) is incubated with N-NaOH or KOH (1 ml/100 mg of tissue represented) for 18 h at 37°. The alkaline digest (1 ml) is cooled in ice and neutralized with 1·33 M-HClO$_4$ (1·6 ml) followed by the dropwise addition, with shaking, of 0·6 ml of silicotungstate reagent. (This is prepared by adding 5·7 g of sodium silicate nonahydrate and 79·4 g of Na$_2$WO$_4$2H$_2$O to 500 ml of water and 15 ml of H$_2$SO$_4$, refluxing for 5 h, filtering and diluting to 1 litre.) Omission of this may lead to erratic results. After standing in ice for 10 min, the mixture is centrifuged and inorganic phosphate determined on samples of the supernatant according to the methods described above. Values of protein-phosphorus in the guinea pig or rat cerebral cortex are of the order of 1 μmol P/g of fresh tissue.

REFERENCES

Ashwell, G. (1966) *Meth. Enzym.* **8**, 86.
Bachelard, H. S. (1965) *Analyt. Biochem.* **12**, 8; *Nature, Lond.* **205**, 903.
Bachelard, H. S., Daniel, P. M. Love, E. R. & Pratt, O. E. (1973) *Proc. R. Soc. B.*, **183**, 71.
Bachelard, H. S., Lewis, L. D., Ponten, U. & Siesjö, B. K. (1974) *J. Neurochem.* **22**, 395.
Bachelard, H. S. & Strang, R. H. C. (1973) In *Research Methods in Neurochemistry* ed. Marks, N. & Rodnight, R. Vol. 2, Chap. 12. New York: Plenum.
Barker, S. B. & Summerson, W. H. (1941) *J. biol. Chem.* **138**, 535.
Barnett, D., Cohen, R. D., Tassopoulos, C. N., Turtle, J. R., Dimitriadou, A. & Fraser, T. R. (1968) *Analyt. Biochem.* **26**, 68.
Bassham, J. A., Birt, L. M., Hems, R. & Loenig, U. E. (1959) *Biochem. J.* **73**, 491.

Berenblum, J. & Chain, E. (1938) *Biochem. J.* **32**, 286, 295.
Brooker, G., Thomas, L. J. & Appleman, M. M. (1968). *Biochemistry, Easton,* **7**, 4177.
Brown, B. L., Albano, J. D. M., Ekins, R. P., Sgherzi, A. M. & Tampion, W. (1971) *Biochem. J.*, **121**, 561.
Brunngraber, E. G., Brown, B. D. & Aguilar, V. (1969) *J. Neurochem.* **16**, 1059.
Busch, H. (1953) *Cancer Res.* **13**, 789.
Carminatti, H. & Passeron, S. (1966) *Meth. Enzym.* **8**, 108.
Chesler, A. & Himwich, H. E. (1943) *Archs Biochem.* **2**, 175.
Ciotti, M. M. & Kaplan, N. O. (1957) *Meth. Enzym.* **3**, 890.
Cohn, W. E. (1950) *J. Am. chem. Soc.* **72**, 1471.
Crabtree, G. W. & Henderson, J. F. (1971) *Cancer Res.* **31**, 985.
Davidson, E. A. (1966) *Meth. Enzym.* **8**, 53.
Di Benedetta, C., Brunngraber, E. G., Whitney, G., Brown, B. D. & Aro, A. (1969) *Archs Biochem. Biophys.* **131**, 403.
Dische, Z. & Borenfreund, E. (1957) *Biochim. biophys. Acta* **23**, 639.
Dodd, P. R., Bradford, H. F. & Chain, E. B. (1971) *Biochem. J.* **125**, 1027.
Edström, J. E. (1956) *Biochem. biophys. Acta* **22**, 378.
Edström, A. (1964) *J. Neurochem.* **11**, 309.
Ennor, A. H. & Rosenberg, H. (1952) *Biochem. J.* **51**, 606.
Ernster, L., Zetterström, R. & Lindberg, O. (1952) *Acta chem. scand.* **6**, 804.
Fawaz, E. N., Roth, L. & Fawaz, G. (1966) *Biochem. Z.* **344**, 122.
Fiske, C. H. & Subbarow, Y. (1925) *J. biol. Chem.* **66**, 375.
Folbergrova, J., MacMillan, V. & Siesjö, B. K. (1972) *J. Neurochem.* **19**, 2497, 2507.
Frohman, C. E., Orten, J. M. & Smith, A. H. (1951) *J. biol. Chem.* **193**, 277.
Gey, K. F. (1956) *Biochem. J.* **64**, 145.
Gilman, A. G. (1970) *Proc. natn. Acad. Sci. U.S.A.* **67**, 305.
Glock, G. E. & McLean, P. (1955) *Biochem. J.* **61**, 381.
Goldberg, N. D., Passonneau, J. V. & Lowry, O. H. (1966) *J. biol. Chem.* **241**, 3997.
Granholm, L., Kaasik, A. E., Nilsson, L. & Siesjö, B. K. (1968) *Acta physiol. Scand.* **74**, 398.
Greengard, P. & Robison, G. A. (1972) *Advances in Cyclic Nucleotide Research,* Vol. 2, N.Y.: Raven Press.
Guidotti, A., Cheney, D. L., Trabucchi, M., Doteuchi, M., Wang, C. & Hawkins, R. A. (1974) *Neuropharmacology,* **13**, 1115.
Hanes, C. S. & Isherwood, F. A. (1949) *Nature, Lond.* **164**, 1107.
Heald, P. J. (1960) *Phosphorus Metabolism of the Brain.* Oxford: Pergamon.
Hohorst, H.-J. (1963) In *Methods of Enzymatic Analysis,* ed. Bergmyer, H.-U., p. 266. London: Academic Press.
Horvath, C. & Lipsky, S. R. (1969) *Analyt. Chem.* **41**, 1227.
Jacob, M., Stevenin, J., Jund, R., Judes, C. & Mandel, P. (1966) *J. Neurochem.* **13**, 619.
Jacob, M., Samec, J., Stevenin, J. & Mandel, P. (1967) *J. Neurochem.* **14**, 169.
Johnson, R. A., Hardman, J. B., Broadus, A. E. & Sutherland, E. W. (1970) *Anal. Biochem.*, **35**, 91.
Kato, T., Berger, S. J., Carter, J. A. & Lowry, O. H. (1973) *Analyt. Biochem.* **53**, 86.
Kerr, S. E. (1935) *J. biol. Chem.* **110**, 625
Kerr, S. E. (1936) *J. biol. Chem.* **116**, 1.
Kerr, S. E. & Ghantus, M. (1936) *J. biol. Chem.* **116**, 9.
Kimberlin, R. H. (1967) *J. Neurochem.* **14**, 123.
Kinnersley, H. W. & Peters, R. A. (1930) *Biochem. J.* **24**, 711.
Kirby, K. S. (1955) *Biochem J.* **64**, 405.
Kratzing, C. C. & Narayanaswami, A. (1953) *Biochem. J.* **54**, 317.

Krishna, G., Weiss, B. & Brodie, B. R. (1968) *J. Pharm. Exp. Ther.* **163**, 379.
Le Baron, F. N. (1955) *Biochem. J.* **61**, 80.
Le Page, G. A. (1948) *Meth. med. Res.* **1**, 337.
Le Page, G. A. (1957) In *Manometric Techniques*, ed. Umbreit, Burris & Stauffer, p. 268. Minneapolis: Burgess.
Lewis, B. A. & Smith, F. (1969) In *Thin Layer Chromatography*, ed. Stahl, E., 2nd edn, p. 821. London: Allen & Unwin.
Lewis, L. D., Ljunggren, B., Ratcheson, R. A. & Siesjö, B. K. (1974) *J. Neurochem.* **23**, 673.
Lin, S. Cohen, H. P. & Cohen, M. M. (1958) *Neurology*, **8**, suppl. 1, 72.
Lindsay, J. R. & Bachelard, H. S. (1966) *Biochem. Pharmac.* **15**, 1045.
Logan, J. E., Mannell, W. A. & Rossiter, R. J. (1952) *Biochem. J.* **51**, 470, 480.
Lowry, O. H. & Passonneau, J. V. (1972) *A Flexible System of Enzymatic Analysis* London: Academic Press.
Lowry, O. H., Passonneau, J. V., Hasselberger, F. X. & Schulz, D. W. (1964) *J. biol. Chem.* **239**, 18.
Lowry, O. H., Passonneau, J. V., Schulz, D. W. & Rock, M. K. (1961) *J. biol. Chem.* **236**, 2746.
Ludoweig, J. & Benmaman, J. D. (1967) *Analyt. Biochem.* **19**, 80.
Mangold, H. K. (1969) In *Thin Layer Chromatography*, ed. Stahl, E. 2nd edn. p. 786. London: Allen & Unwin.
Margolis, R. U. (1969) *Handbook of Neurochem.* **1**, 245.
Margolis, R. U. & Margolis, R. K. (1972) In *Research Methods in Neurochemistry*, ed. Marks, N. & Rodnight, R.). Vol. 1, Ch. 11. New York: Plenum.
Martin, J. B. & Doty, D. M. (1949) *Analyt. Chem.* **21**, 965.
Matchinsky, F. M. (1971) *Meth. Enzym.* **18**, 3.
Mayman, C. I., Gatfield, P. D. & Breckenridge, B. McL. (1964) *J. Neurochem.* **11**, 483.
McIlwain, H. & Bachelard, H. S. (1971) *Biochemistry and the Central Nervous System*, 4th edn. London: Churchill.
McIlwain, H. & Tresize, M. A. (1956) *Biochem. J.* **63**, 250.
McIlwain, H., Buchel, L. & Cheshire, J. D. (1951) *Biochem. J.* **48**, 12.
Medina, M. A., Jones, D. J., Stavinoha, W. B. & Ross, D. H. (1975) *J. Neurochem.* **24**, 223.
Mori, K., Yamagami, S., Akahani, Y. & Kawakita, Y. (1970) *J. Neurochem.* **17**, 1691.
Mukerjee, H. & Ram, J. S. (1964) *Analyt. Biochem.* **8**, 393.
Oesterle, W., Kanig, K., Buchel, W. & Nickel, A.-K. (1970) *J. Neurochem.* **17**, 1403.
Penn, N. W. & Suwalski, R. (1969) *Biochem. J.* **115**, 563.
Piccoli, F., Camarda, R. & Bonavita, V. (1969) *J. Neurochem.* **16**, 159.
Pull, I. & McIlwain, H. (1972) *Biochem. J.* **126**, 965.
Racker, E. (1963) In *Methods of Enzymatic Analysis*, ed. Bergmyer, H.-U., p. 175. London: Academic Press.
Randerath, K. (1962) *Angew. Chem.* **74**, 780.
Randerath, K. & Randerath, E. (1967) *Meth. Enzym.* **12**, 323.
Robison, G. A., Butcher, R. W. & Sutherland, E. W. (1968) *A. Rev. Biochem.* **37**, 149.
Rodnight, R. (1971) *Handbook of Neurochem.* **5A**, 141.
Schmidt, M. J., Schmidt, D. E. & Robison, G. A. (1971) *Science, N.Y.* **173**, 1142.
Schmidt, G. & Thannhauser, S. J. (1945) *J. biol. Chem.* **161**, 83.
Schneider, D. & Roberts, S. (1968) *J. Neurochem.* **15**, 1469.
Shmukler, H. W. (1972) *J. Chromatogr. Sci.* **10**, 38.
Siesjö, B. K., Folbergrova, J. & MacMillan, V. (1972) *J. Neurochem.* **19**, 2483.
Singh, M. & Bachhawat, B. K. (1965) *J. Neurochem.* **12**, 519.

Singh, M., Chandrasekaran, E. V., Cherian, R. & Bachhawat, B. K. (1969) *J. Neurochem.* **16**, 1157.
Slater, E. C. (1953) *Biochem. J.* **53**, 157.
Somogyi, M. (1952) *J. biol. Chem.* **195**, 19.
Stanley, P. E. & Williams, S. G. (1969) *Analyt. Biochem.* **29**, 381.
Stone, W. E. (1948) *Meth. med. Res.* **1**, 353.
Strang, R. H. C. & Bachelard, H. S. (1971a) *J. Neurochem.* **18**, 1799.
Strang, R. H. C. & Bachelard, H. S. (1971b) *Analyt. Biochem.* **41**, 533.
Strang, R. H. C. & Bachelard, H. S. (1971c) *J. Neurochem.* **18**, 1067.
Strang, R. H. C. & Bachelard, H. S. (1973) *J. Neurochem.* **20**, 987.
Tower, D. B. (1958) *J. Neurochem.* **3**, 185.
Veech, R. L., Harris, R. L., Veloso, D. & Veech, E. H. (1973) *J. Neurochem.* **20**, 183.
Verachtert, H., Bass, S. T., Wilder, J. K. & Hansen, R. G. (1966) *Meth. Enzym.* **8**, 111.
Vesco, C. & Giuditta, A. (1968) *J. Neurochem.* **15**, 81.
Waring, P. P. & Ziporin, Z. Z. (1964) *J. Chromatogr.* **15**, 168.
Weller, M., Rodnight, R. & Carrera, D. (1972) *Biochem. J.* **129**, 113.
Wilson, W. S. & Thomson, R. Y. (1969) *Biochem. Pharmac.* **18**, 1297.
Wong, P. C. L. & Henderson, J. F. (1972) *Biochem. J.* **129**, 1085.
Zamenhof, S., Bursztyn, H., Rich, K. & Zamenhof, P. (1964) *J. Neurochem.* **11**, 505.

4. Lipids: Extraction, Thin-layer Chromatography and Gas-Liquid Chromatography

M. L. CUZNER and J. M. TURNBULL

Total extraction of lipids	62
Removal of non-lipids	62
Extraction of gangliosides	63
Extraction of polyphosphoinositides	63
Quantitative analysis of lipids	64
Thin-layer chromatography for analysis of brain lipids	64
Separation of phosphoinositides by t.l.c.	65
Gangliosides by t.l.c.	66
T.l.c. subfractionation of lipid classes	67
Detection reagents for lipids	68
Quantitative recovery of lipids	69
Rechromatography on alumina	70
Analysis of lipid component	70
Analytical techniques	70
Phosphorus determination	71
Cholesterol and cholesterol esters	71
Plasmalogen determination	73
Galactose determination	73
Gas–liquid chromatography for analysis of brain lipids	74
Apparatus	76
Preparation of samples for g.l.c. analysis of fatty acid esters	80
Formation of trimethylsilyl derivatives	82
Injection of sample; results	83
References	85

The lipids of biological origin range in molecular weight from about 300 to 3000, and have in common the property of dissolving in a range of organic solvents. Many neural tissues are rich in lipids through the high lipid content of myelinated nerve fibres; this applies especially to the white matter of the vertebrate brain (Table 4.1). The identification and accurate analysis of lipids has accelerated during the past ten years as a result of sophisticated technical developments, notably thin layer chromatography (t.l.c.) and gas–liquid chromatography (g.l.c.). A wide range of lipid techniques are comprehensively described in the volumes of Lowenstein (1969) and Kates (1972). The reader is referred also to the following books for general technical information about lipids: Marinetti (1967); Johnson & Davenport (1971) and Ansell & Hawthorne (1964).

LIPIDS: EXTRACTION, ETC.   61

Table 4.1 Lipid composition of tissues from the mammalian brain

Lipid	Human[a]			Rat[b]					Rat[c]			Ox[d]
	Grey matter	White matter	Myelin	whole brain	Whole brain	Myelin	Mitochondria	Microsomes	Neurons	Astrocytes	Oligodendroglia	Axons
	(μmoles/mg lipid)				(μmol/g fresh weight)				(% by weight of total lipid)			
Cholesterol	0·470	0·600	0·650	0·590	48·3	17·00	1·30	7·30	10·7	14·7	14·1	20·1
Ethanolamine phospholipid	0·290	0·188	0·190	0·300	24·5	7·17	1·55	3·34	18·4	19·4	14·0	14·6
Lecithin	0·284	0·162	0·135	0·263	21·5	4·18	1·45	3·44	40·1	36·3	29·4	18·3
Sphingomyelin	0·064	0·100	0·071	0·044	3·6	1·36	0·07	0·72	3·0	3·8	7·1	9·3
Phosphatidyl-inositol	—	0·017	0·009	0·037	1·5	0·71	0·19	0·32	5·0	3·9	4·1	3·4
Phosphatidyl-serine	0·099	0·082	0·085	0·085	6·9	2·50	0·24	1·01	3·8	5·0	4·7	5·6
Cardiolipin-phosphatidic acid	—	0·012	0·002	0·034	1·4	0·43	0·29	0·32	4·3	3·9	—	—
Cerebroside	0·071	0·240	0·255	0·168	11·1	5·60	0·07	1·01	} 2·0	} 1·7	7·3	12·9
Sulphatide	0·014	0·052	0·049	0·038	2·5	1·10	—	0·13			1·5	7·2
Ceramide	0·020	0·020	0·015	—	—	—	—	—	—	—	—	—
Total plasmalogen	—	0·200	0·180	0·163	10·7	7·85	—	1·39	—	—	—	10·0

[a] Data from O'Brien & Sampson (1965) who also report the water content of the grey matter to be 82·3% and of the white matter to be 75·2% of the fresh weight. The total lipid, as percentage of the dry weights of the tissues, were: grey matter, 39·6%; white matter, 64·6%; and myelin, 78·0%.
[b] Cuzner, Davison & Gregson, 1965; Cuzner & Davison, 1968.
[c] Poduslo & Norton (1972) who report the lipid content of neurons and astrocytes respectively to be 24% and 39% of the dry weights of the preparations.
[d] De Vries & Norton (1974), who report the lipid content of oligodendroglia and axons to be 29·5% and 13·4% respectively.

TOTAL EXTRACTION OF LIPIDS

Lipids are commonly extracted with chloroform–methanol mixtures followed by homogenization and filtration to remove insoluble residues (Folch, Lees & Sloane-Stanley, 1957). The lipid extract is washed with water or salt solution and the polar gangliosides separate into the aqueous phase, all other lipids remaining in the lower chloroform-rich phase. One group of lipids, the polyphosphoinositides, remain in the residue after extraction with neutral chloroform–methanol; these can be extracted with acidified chloroform–methanol mixtures (Dawson & Eichberg, 1965).

Autilio & Norton (1963) examined the solutes obtained by the action of 31 organic solvents on lyophilized bovine cerebral white matter. Four solvents (tetrahydrofuran, ethylcellosolve, n-amyl alcohol and pyridine) compared favourably with chloroform–methanol (2:1, v/v) in extracting nearly all the lipid at 4°. The solubility parameter theory has been used to demonstrate that the properties of chloroform and chloroform–methanol as lipid solvents are not unique (Schmid, Hunter & Calvert, 1973; Schmid, Calvert & Steiner, 1973). The instability of chloroform and chloroform–methanol mixtures commonly used in lipid studies has led these workers to propose the use of other binary solvent systems, such as toluene–ethanol, which are equally effective for extracting lipids from biological material.

Procedure

Accurately weighed neural tissue is extracted with 19 volumes of chloroform–methanol (2:1, v/v) per g of tissue. Extraction and homogenization with chloroform–methanol (1:1, v/v) is equally effective (Radin, 1969). The resulting fine suspension is filtered to obtain a clear lipid extract. If the protein residue, following extraction with chloroform–methanol is to be worked up later, it is better to centrifuge the precipitate. As the high density of chloroform–methanol (2:1, v/v) makes the sediment poorly compacting, methanol is added to give the proportions of chloroform–methanol of 1:2, v/v. This procedure obviates adsorption of lipid onto the filter paper.

Dry tissue should be rehydrated before adding chloroform–methanol, as the extraction of highly polar lipids will be incomplete in the absence of water (Radin, 1969). Subcellular fractions in sucrose or buffer solutions should be extracted with 19 volumes of chloroform–methanol per ml of suspension.

Removal of non-lipids

The filtrate obtained by the Folch extraction is shaken well with 0·2 vol. water or 0·1M-KCl. Separation into two phases occurs spontaneously in 4 to 6 h in the cold, or may be assisted by centrifugation. The upper phase should occupy about 40% of the total volume and contains the major part of the non-lipid contaminants such as glucose, salts, urea, amino acids and sucrose; it also contains the gangliosides. The lower phase contains the remainder of the lipids, except the polyphosphoinositides, and also contains protein that is by definition proteolipid protein (Folch et al., 1957). The protein can be removed by partitioning the lower phase with alkaline 'citrate upper phase', chloroform–methanol–0·1M-tripotassium citrate (3:48:47, by vol.) or by evaporating to dryness to denature the protein (Webster & Folch, 1961). The lipids can then be extracted with chloroform–methanol or a less polar solvent as the protein–lipid complexes are dissociated. Many factors influence the amount of protein soluble in the lower phase. When inorganic ions are removed during subcellular fractionation more protein enters the chloroform phase; the presence of sucrose has the same effect (Lees, 1966 and 1968). There are alternative methods for removing non-lipids from the lipid extract. The chloroform–methanol extract can be dialyzed in ordinary cellulose tubing against water. A second alternative method for removal of non-lipids employs liquid-liquid partition through Sephadex G-25 (Wells & Dittmer, 1963; Siakotos & Rouser, 1965).

Extraction of gangliosides

Suzuki (1965) found that gangliosides were not completely extracted with chloroform–methanol (2:1, v/v) alone and recommended further extraction with an equal volume of chlorofom–methanol (1:2, v/v), containing 5% water. The pooled extracts are readjusted with chloroform to give a chloroform–methanol 2:1 ratio and partitioned twice with 0·2 vol. of 0·2% KCl followed by one partition with water (Vanier et al., 1971). The partition with water is essential for quantitative recovery from the lower phase of the least polar gangliosides (Suzuki, 1965). The combined upper phase is concentrated in vacuo to a manageable volume, followed by dialysis in the cold. The content of the dialysis sac is then lyophilized to obtain a white fluffy powder.

Extraction of polyphosphoinositides

There is a rapid turnover and disappearance of polyphosphoinositides of the brain within minutes of the death of an animal, and so it is advisable to freeze the brain immediately and extract the frozen tissue (Dawson & Eichberg, 1965). The frozen tissue is first extracted with 10 vol. of chloroform–methanol (1:1, v/v) and the residue washed with chloroform–methanol (2:1, v/v). The residue is then extracted three times at 37° with four times its packed volume of chloroform–methanol (2:1, v/v) containing 0·25% (v/v) of conc. HCl. The extract is filtered and washed with 0·2 vol. of N-HCl and the lower phase and interface contain the polyphosphoinositides.

QUANTITATIVE ANALYSIS OF LIPIDS

The number and concentration range of lipid classes in the brain were first established using the classical methods of solvent precipitation and partition (see McIlwain & Rodnight, 1962). Later, column chromatographic techniques were used to separate most lipid classes in quantitative yield. The major components of brain were shown to be cholesterol, cerebroside, sulphatide, phosphatidyl choline, phosphatidyl ethanolamine and its corresponding plasmalogen, phosphatidyl serine and sphingomyelin, with smaller amounts of phosphatidyl inositol, di- and triphosphoinositides, phosphatidic acid, diphosphatidyl glycerol, gangliosides, ceramide, triglyceride and sterol esters (Rouser, Kritchevsky, Yamamoto & Baxter, 1972). The structural formulae of these lipids can be found in the first chapter of Johnson & Davenport (1971) and in McIlwain & Bachelard (1971), and information on their distribution in the brain is given in Table 4.1.

A quantitative assessment, involving a minimal number of chemical assay methods, of a wide spectrum of individual lipids can be obtained after thin-layer chromatography of a total lipid extract. Column chromatographic techniques, which are better suited to bulk preparation and purification of individual lipids, will be discussed in the following chapter.

THIN-LAYER CHROMATOGRAPHY FOR ANALYSIS OF BRAIN LIPIDS

The thin-layer chromatogram is essentially an unrolled column. Most thin-layer methods for the separation of lipid classes employ the principle of adsorption chromatography. As differences in the nature and position of polar substituents are emphasized in adsorption chromatography, the individual phospholipids and galactolipids are effectively separated but not the individual neutral lipids. Reversed-phase partition methods are used for the resolution of fatty acids, di- and triglycerides and cholesterol esters. The subclasses and molecular species of the polar lipids can be separated in many cases by the modification of the adsorbent by special additives.

Application of thin-layer chromatography (t.l.c.) for different types of lipid analysis has been the subject of many reviews and chapters in books; see Randerath (1966), Stahl (1969) and Truter (1963). The following section is a description of the thin-layer chromatographic techniques used by the authors for the separation of brain lipids, and includes comment on alternative methods which may be more useful in different experimental conditions.

Preparation of chromatoplates. Glass plates (20 × 20 cm) are coated with a layer, 0·5 mm thick, of Silica Gel G (Merck) using a Desaga or Camlab spreading apparatus. The adsorbent is prepared by shaking a specific amount of silica gel with approximately twice its weight of water or buffer solution in a stoppered flask or bottle. The slurry is allowed to thicken slightly (2 or 3 min) before spreading the plates. Although there is a wide range of adsorbents available for t.l.c., e.g. alumina, Florosil and DEAE cellulose, the two most commonly used adsorbents are Silica Gel G, which has $CaSO_4$ added as binder, and Silica Gel H which contains no binder. The plates are dried and stored at room temperature. Immediately before use, they are pre-run in chloroform, activated for 30 min at 110° and marked into lanes, using a template.

Lipid samples are dissolved in chloroform or chloroform–methanol (2:1, v/v) and diluted accurately to a concentration in the range 20–30 mg/ml. The optimum concentration of individual lipid standards is 2–5 mg/ml. For qualitative work the lipids are applied as spots (5–10 μl), 2–3 cm from the bottom edge of the plate, For quantitative work it is necessary to chromatograph approximately 5 mg of a mixed lipid extract, which can be applied as a row of spots close to each other. The individual lipids will then appear as bands after separation. The lipid samples can be applied using either a micropipette or a Hamilton microsyringe. The solvent is allowed to evaporate completely after application and the chromatograms are developed in closed tanks at room temperature to within 5 cm from the top of the plates. It is advisable to add an antioxidant, 0·005% 4-methyl-2,6-di-tert-butyl phenol, to either the lipid extracts or the developing solvent; it does not interfere with the separation of the lipids.

Developing solvents. Three developing solvents have been used for routine separation of lipids: (1) hexane–ether (85:15, v/v) for the separation of neutral lipid classes; (2) chloroform–methanol-15N-NH_4OH (17:7:1, by vol.) for the separation of cerebroside and individual phospholipids; and (3) chloroform–methanol–water (17:7:1, by vol.) for the separation of sulphatide, cerebroside and the major phospholipids (Table 4.2). Separations in the polar solvent systems are impaired if the air is humid. Under these conditions it is best to develop the plates in a fume cupboard.

Separation of phosphoinositides by t.l.c.

As the calcium salts of phosphoinositides in general and of the naturally-occurring triphosphoinositide in particular are insoluble in water, the lack of mobility on Silica Gel G plates is due to the $CaSO_4$ binder. If potassium oxalate is used with Silica Gel H for the preparation of plates, any calcium present is sequestered and polyphosphoinositides then migrate on thin layer plates (Gonzalez-Sastre & Folch, 1968). Glass plates are coated with layers prepared from a slurry of 30 g Silica Gel H in 80 ml of 1% potassium oxalate. Plates are dried and activated before use by placing them at 110° for 30 min. After

Table 4.2 R_F Values of lipids

Lipid	Relative band speeds with developing solvents:		
	Hexane–ether (85 : 15, v/v)	Chloroform–methanol–15N-NH$_4$OH (17 : 7 : 1, by vol.)	Chloroform–methanol–water (17 : 7 : 1, by vol.)
Cholesterol ester	0·86	0·99	0·99
Triolein	0·45	0·99	0·99
Cholesterol	0·10	0·95	0·95
Diolein	0·07	—	—
Monolein	*	—	—
Cerebroside	*	0·64 / 0·55	0·67
Cardiolipin	*	0·55	0·75
Ethanolamine phospholipid	*	0·45	0·51
Sulphatide	*	0·45	0·34
Lecithin	*	0·34	0·27
Sphingomyelin	*	0·18	0·18
Ethanolamine lysophospholipid	*	0·18	0·18
Phosphatidyl inositol	*	0·11	0·18
Lysolecithin	*	0·11	0·13
Phosphatidyl serine	*	0·06	0·13

*Remain at origin.

lipid samples are applied to the plate, the chromatograms are developed with *n*-propanol-4N-NH$_4$OH (2:1, v/v) or with chloroform–methanol–4N-NH$_4$OH (9:7:2, by vol.). Spots can be visualized with iodine or bromothymol blue prior to quantitative analysis.

Gangliosides by t.l.c.

The polar gangliosides remain at the origin of a thin-layer chromatogram when any of the solvent systems so far mentioned are used. One technique giving excellent resolution is the descending system of Korey & Gonatas (1963). The *n*-propanol–water (7:3, v/v) solvent is conducted from a trough to the top of the plate by means of a paper wick. Another thin-layer system employs silica gel plates of double length (20 × 40 cm) with chloroform–methanol–2·5N-NH$_4$OH (60 : 40:9, by vol.) as solvent (Ledeen, 1966). Two successive ascending runs of approximately seven h apiece, with an hour's drying period in between, result in resolution of the four major and some of the minor gangliosides. The thin layer chromatographic behaviour of the different ganglioside species in four commonly used t.l.c. systems has been catalogued by Penick, Meisler & McCluer (1966).

T.l.c. subfractionation of lipid classes

The lipid classes separated by t.l.c. as described in the preceding pages consist of compounds whose molecules differ primarily in the fatty acids they contain. Thin-layer techniques have been adapted in many laboratories for subfractionation of the lipid classes.

Separation according to degree of unsaturation. Unsaturated lipids form complexes with silver ions. This property is utilized by chromatography of lipids on silica gel plates impregnated with silver nitrate (Morris, 1966; Malins, 1966). Lipids which have been separated firstly into classes, are further separated according to the degree of unsaturation of the component fatty acids. Geometrical isomers (*cis–trans* isomers) also have different R_F values on silver–nitrate impregnated plates. The technique has been used successfully for many classes of compounds: methyl esters of fatty acids, cholesterol esters of fatty acids, triglycerides, sterols and lecithins (Skipski & Barclay, 1969). Other types of complex formation have been used in combination with t.l.c. Of particular interest to the neurochemist is the separation on sodium borate-impregnated plates of glycolipids, such as brain cerebrosides, on the basis of differences in the carbohydrate moiety (Renkonnen & Varo, 1967).

Separation according to chain length of fatty acid components. Lipids can be separated according to the chain length of their fatty acids by reversed-phase partition t.l.c. The technique is most effective following a prior separation of the lipids according to their degree of unsaturation. In reversed-phase chromatography the stationary phase is hydrophobic, e.g. undecane, paraffin or silicone oil, and the mobile phase is hydrophilic (Vereschchagin, 1964). Kieselguhr is a good supporting material because it has low adsorptive activity but silica gel can also be used. Solvent systems have been tabulated by Mangold (1961). The following lipid classes have been differentiated using reversed-phase t.l.c.: cholesterol esters, free fatty acids and their methyl esters, triglycerides and lecithins (Skipski & Barclay, 1969).

Subclasses of polar lipids. According to Renkonnen (1971), 'native polar lipids tend to associate and form multimolecular aggregates in organic solvents; this counteracts their chromatographic separation.' Individual phospholipids can be fractionated into alkenyl–acyl, alkyl–acyl and diacyl compounds, on Silica Gel G plates, if the polar group is first masked (Arvidson, 1967). This is achieved by splitting the phosphate bond with phospholipase D or the complete polar group with phospholipase C, followed by methylation or acetylation. The resulting dimethyl phosphatides or diglyceride acetates are chromatographed on

Silica Gel G plates, using first hexane–ether as developing solvent, followed by toluene. The three subclasses can be further characterized by running on silver-impregnated silica gel and reversed phase t.l.c.

Detection reagents for lipids

Notes follow on some of the available reagents.

Bromothymol blue to detect lipids non-specifically. Plates are sprayed with a solution containing 400 μg bromothymol blue/ml in 0·01N-NaOH. The lipids appear as deep yellow bands which turn blue if exposed to ammonia vapour.

Iodine vapour to detect lipids non-specifically. Dry plates are exposed for about 1 min in a glass or plastic tank, containing iodine vapour produced from iodine crystals.

Ninhydrin to detect aminophosphatides. Dry plates are sprayed with a solution of 0·2 g ninhydrin in 95 ml of *n*-butanol plus 5 ml of 10% (v/v) acetic acid and heated for 1 min in an oven at 110°. Pink-violet spots appear on a white background.

Molybdic acid to detect phosphatides. Dry plates are sprayed with a solution of 3 g ammonium molybdate in 25 ml of H_2O plus 30 ml of 1N-HCl and 15 ml of 60% (v/v) perchloric acid. After heating for 5–10 min in an oven at 110°, phosphatides show up as blue spots on a white background. Cholesterol and galactolipids appear as blue-black spots (Hanes & Isherwood, 1949; Skidmore & Entenman 1962).

Dragendorf reagent to detect choline. Dry plates are sprayed with a mixture of 4 ml of solution 1, 1 ml of solution 2, and 20 ml deionized water. Solution 1 contains 1·7 g $Bi(NO_3)_3.5H_2O$ diluted to 100 ml with 20% (v/v) acetic acid. Solution 2 contains 40 g KI in 100 ml deionized water. As the plates are dried at room temperature choline-containing compounds produce orange spots. This spray is rather unspecific as ethanolamine phospholipid and cerebrosides give faint orange spots also (Wagner, Hörhammer & Wolff, 1961).

Ferric chloride–sulfosalicylic acid to detect phosphate groups. Dry plates are sprayed with a solution of 7 g sulfosalicylic acid, 0·1 g $FeCl_3.6H_2O$, and 25 ml water diluted to 100 ml with 95% ethanol. White fluorescent spots appear on a purple background as plates are dried at room temperature (Wade & Morgan, 1955).

Orcinol to detect galactolipids. Dry plates are sprayed with a solution of 100 mg orcinol plus 1 ml of 1% $FeCl_3$ and 40 ml of 2N-H_2SO_4. Each plate is covered with a plain glass plate and heated in an oven at 110° for 35 min. After uncovering the plates and heating for a further 3 min, cerebroside and sulphatide appear as purple spots on a white background (Honegger, personal communication).

Resorcinol to detect gangliosides. Dry plates are sprayed with a solution of 20 mg. resorcinol in 10 ml 4M-HCl, 1mM with $CuSO_4$.

Each plate is covered with a plain glass sheet and heated in an oven at 150° for 10 min.

Quantitative recovery of lipids

This requires localization of the lipids, followed by their elution.
Detection of lipids. The developed, dried thin-layer plates are exposed to iodine vapour for about a minute, and the visible lipid zones marked with any ordinary pin. The iodine staining is then allowed to fade for 2–4 h or overnight. Alternatively the plates are sprayed with bromthymol blue and allowed to dry.
Elution of silicic acid columns. Each rectangular zone enclosing a spot is scraped off with a thin spatula and the silicic acid transferred to a small glass column (1 × 10 cm) with a reservoir, which holds up to 30 ml of solvent. After each silicic acid area is removed from the plate, a small ball of fat-free cotton wool, soaked with chloroform and held in forceps, is used to wipe the exposed area of glass plate and then transferred to the top of the small glass column. If the iodine-stained plate is allowed to fade for only 2–4 h, all the columns, with the exception of the one containing cholesterol, are first eluted with 5 ml chloroform to remove adsorbed iodine. This is not necessary if plates are allowed to fade overnight. The neutral lipid silicic acid zones of both iodine and bromthymol blue-stained plates, are eluted with 15 ml chloroform. The polar lipid zones are eluted with 30 ml chloroform–methanol–water (7:7:1, by vol.), which also elutes bromothymol blue. The dye does not interfere with phosphorus estimations, but must be washed from galactolipid eluates prior to assay (Cuzner & Davison, 1973). Iodine-stained galactolipids cannot be assayed by the anthrone method. Eluates are taken to dryness under a stream of air at a temperature of 60° or below.

Comparable control areas should be removed from a blank lane and eluted for cholesterol, galactolipid and phospholipid estimations. Alternative methods for elution of lipids from thin-layer plates and subsequent analysis are described by Skipski & Barclay (1969) and Parker & Peterson (1965).

Elution of lipid from unstained plates. For preparative thin-layer chromatography and for the estimation of individual plasmalogen lipids, plates are not stained before elution. Lipid samples are spotted in the middle of the plate and lipid markers spotted on two small outside lanes. After development, the middle of the plate is covered with a plate of glass or wide strip of cellophane, and the thin-layer chromatogram stained with iodine in petroleum ether, b.p. 100–120 (1% w/v) or with bromthymol blue. Visible lipid zones are marked off and the glass plate or cellophane removed; using the outer lanes as guides, the lipids in the middle lanes of the thin layer chromatogram are marked off.

Rechromatography on alumina

The lysocompounds of lecithin and ethanolamine phospholipid have the same R_F values as phosphatidyl inositol and sphingomyelin respectively, in the developing solvent, chloroform–methanol–15N-NH_4OH (17:7:1, by vol.). In each case the two lipids can be further separated on small alumina columns. Small glass columns (1 × 10 cm) with reservoirs of 100 ml capacity are packed with 7–10 g of alumina (BDH, alumina for chromatography) and the single silicic acid areas containing two lipids are transferred to the top of the alumina. Lysolecithin and phosphatidyl inositol are separated by elution of the former with 80 ml of chloroform–methanol (1:4, v/v) and the latter with 100 ml of chloroform–ethanol–water (2:5:2, by vol.). Similarly sphingomyelin is eluted with 100 ml of chloroform–methanol (1:1, v/v), followed by elution of lysophosphatidyl ethanolamine with chloroform–ethanol–water (2:5:2, by vol.). The above elution volumes were worked out from a standard lipid mixture containing 0·38 μmol lysophosphatidyl ethanolamine, 0·83 μmol sphingomyelin, 0·50 μmol lysolecithin, and 0·27 μmol phosphatidylinositol (Cuzner & Davison, 1967).

Analysis of lipid components

The analytical techniques described below are applicable as follows.

In mixed lipid extracts. The original unseparated lipid solutions can be analysed for total phospholipid by phosphorus assay (Martland & Robison, 1926), total plasmalogen by iodine absorption method (Williams, Anderson & Jasik, 1962), total galactolipid by galactose assay using the anthrone method (Radin, 1958) and for total cholesterol by the Liebermann-Burchard reaction (Sperry & Webb, 1950; Davison et al., 1958).

Analysis of separated lipids from thin layer plates. Phosphatidyl serine, lysolecithin, phosphatidyl inositol, sphingomyelin, lysophosphatidyl ethanolamine, lecithin, ethanolamine phospholipid and cardiolipin can be determined by phosphorus assay. Serine plasmalogen, lecithin plasmalogen and ethanolamine plasmalogen are analysed by the iodine absorption method. Sulphatide and cerebroside can be determined by the anthrone method. Cholesterol and cholesterol ester can be assayed by Liebermann-Burchard reaction or by the method of Hanel & Dam (1955).

ANALYTICAL TECHNIQUES

A relatively complete analysis of brain lipids can be accomplished after thin-layer separation with the aid of four basic analytical techniques, to be described in this section. Techniques used for the identification and analysis of individual lipids will be dealt with in the following chapter.

Phosphorus determination

The method of estimating lipid-bound phosphorus involves the digestion of a given lipid fraction with a 60% (w/v) perchloric acid–10N-sulphuric acid mixture at a temperature between 250° and 350°, whereby the organic matter is completely oxidised and the phosphorus present is quantitatively converted into inorganic orthophosphate. The orthophosphate content of the digest is estimated by the method of Bell & Doisy (1920), as modified by Martland & Robison (1926). (see also Chap 3).

Acid hydrolysis. To dry lipid samples, containing not more than 5 µmol phospholipid in tubes graduated to both 10 and 15 ml., 1 ml of 60% (w/v) perchloric acid 10N-sulphuric acid (1:1, v/v) is added and the tubes are heated on a sand bath at a temperature between 250° and 350° until clear (usually about 30–50 min. After cooling to room temperature, the samples are diluted to approximately 5 ml and assayed directly.

Reaction. The reagents used are: (1) ammonium molybdate, 10% w/v aqueous solution; (2) reducing agent: sodium sulphite, 20g and quinol, 0·5 g made to 100 ml with water.

The standard, containing 0·1 mg of phosphorus/ml, is prepared by dissolving 43·94 mg KH_2PO_4 in 10 ml water. Blanks and standards are treated in an identical manner to the samples. To the diluted samples, 1 ml of ammonium molybdate solution is added followed by 1 ml of reducing agent. The tubes are well mixed and allowed to stand at room temperature for 30 min. Volumes are made up to 10 or 15 ml and the absorbance read against water in a Unicam S.P. 600 spectrophotometer at 760 nm. For smaller quantities of phospholipid (0·1–0·5 µmol), the method can be adapted to a micro scale by reducing all quantities to one-tenth, with a final volume of 1·5 ml. The molar extinction coefficient of phosphorus under these conditions is $3·95 \times 10^3$.

Cholesterol and cholesterol esters

SPERRY AND WEBB'S METHOD

The Liebermann-Burchard (1889) reaction is given by all unsaturated sterols. Cholesterol reacts with an acetic anhydride–sulphuric acid mixture producing a characteristic green colour, the intensity of which is proportional to the concentration of sterol (Sperry & Webb, 1950; Davison et al., 1958). Free cholesterol is precipitated from ethanolic solution by digitonin (Bladon, 1958). This property is utilized to determine free cholesterol in the presence of esterified cholesterol.

Total cholesterol. A suitable quantity of total lipid extract, containing approximately 1 µmol of cholesterol is pipetted into a centrifuge tube and heated for 60 min., on a water bath at 50° after addition of 0·1 ml of 10% (w/v) alcoholic KOH. The sample is cooled and neutralized with 30% acetic acid using phenolphthalein as indicator. The cholesterol is extracted with 5 ml of chloroform-methanol (2:1, v/v) after the addition to each tube of 1 ml of water. Two phases are formed and the lower phase is retained for cholesterol assay.

Free cholesterol. The appropriate quantity of total lipid extract is dried in a centrifuge tube and 1 ml of ethanol-ether (3:2, v/v) is added, followed by 1 ml of 0·5% digitonin in 50% ethanol. The sample is left overnight, centrifuged and the precipitate washed twice with ether. The precipitate is finally dissolved in 1 ml glacial acetic acid.

Assay after thin-layer chromatography. Cholesterol and cholesterol ester are separated on thin-layer plates using hexane-ether (85:15, v/v). Cholesterol is eluted with chloroform into suitable test tubes and after evaporation of the solvent, assayed directly. The reagents used are the Liebermann-Burchard colour reagent, with 20 parts by vol. of ice-cold acetic anhydride plus 1 part conc. H_2SO_4.

For the assay, dry cholesterol samples are dissolved in 1 ml portions of glacial acetic acid. Cholesterol standard is prepared by dissolving recrystallised cholesterol in glacial acetic acid. The final volume of each sample before the addition of the colour reagent is 1 ml. To the acetic acid solutions of cholesterol, 4 ml portions of ice-cold colour reagent are added. The samples are mixed and allowed to stand in the dark at room temperature for 25 min. The absorbances are read in the Unicam S.P. 600 spectrophotometer at a wavelength of 620 nm. This method gives accurate readings over a range of 0·3 to 3 μmol of cholesterol. The molar extinction coefficient of cholesterol under these conditions is $1·5 \times 10^3$.

HANEL AND DAM'S METHOD

Cholesterol and its esters in glacial acetic acid when treated with acetyl chloride and zinc chloride develop a bright red colour (Tchugaev, 1909). Amounts of cholesterol as low as 0·004 mg/ml of reaction mixture can be determined with an accuracy of about $\pm$ 3%. Since it is much more sensitive than the Liebermann-Burchard reaction, it is used for the determination of small amounts of cholesterol in biological material. The present method is a modification developed by Hanel & Dam (1955). The reagents used are chloroform, acetylchloride and a $ZnCl_2$ solution prepared as follows. Anhydrous $ZnCl_2$ sticks, 40g, are quickly crushed, weighed in a dry 250 ml bottle and mixed with 153 ml. of glacial acetic acid. The mixture is stoppered and kept at 80° for 2½ hr. with occasional shaking. After being cooled to room temperature, the mixture is filtered through a glass filter and kept in a dark glass-stoppered bottle. This reagent can be kept until it takes up sufficient moisture to become turbid, when shaken with $CHCl_3$.

Assay. Dried cholesterol samples are dissolved in 2 ml of chloroform. To the samples, 1 ml of $ZnCl_2$-acetic acid reagent plus 1 ml of acetyl chloride are added. The samples are well mixed and kept at 65° for 15 min. After being cooled in an ice bath, the tubes are made up to 5 ml with chloroform and read within 30 min at 528 nm. The absorbances are linear over the range 0·05 to 0·40 μmol of cholesterol or sterol ester and the molar extinction coefficient is 1·00 × 10^4.

Plasmalogen determination

The plasmalogen content of lipids can be estimated by chemical assay of the enol ether group according to the method of Williams, Anderson & Jasik (1962). In 1948 Siggia & Edsberg reported that vinyl ethers could be determined specifically by measuring the uptake of iodine in 50% methanol. The present method couples the iodination reaction with a spectrophotometric measurement of the iodine remaining after reaction. As little as 0·01 to 0·02 μmol of enol ether can be determined accurately and rapidly using this method.

The reagents used are: (1) methanol; (2) 3M-KI, Analar: this is stable for a week if kept cold in a dark bottle; (3) 94 mM-sodium citrate pH5·5; (4) 0·5 mM-iodine in 3M-KI: this is stable for a month if kept in a dark bottle; (5) n-butyl acetate.

Procedure. Triplicate lipid samples containing from 0·02–0·125 μmol of plasmalogen are pipetted into centrifuge tubes and the solvent evaporated under nitrogen at room temperature. One of the tubes serves as a sample blank, and the other two for measuring iodine uptake by the plasmalogen in the lipid. After evaporation of the solvent 0·9 ml of methanol is added; the contents are mixed, allowed to stand for 5 min and mixed again. This ensures solution of the lipid sample. Then 3·2 ml of sodium citrate is added and the tube contents mixed well. To the sample blank tube is added 0·9 ml of potassium iodide, and to the other two tubes are added 0·4 ml of potassium iodide and 0·5 ml of iodine solution. The tubes are shaken vigourously for 10 s and allowed to stand for 40 min at room temperature. To each tube is added 5 ml of n-butyl acetate; the tubes are shaken vigorously for 10 sec and then centrifuged for 10 min. The n-butyl acetate layer of each sample is read at 363 nm against a reagent blank. The molar extinction coefficient of iodine under these conditions is 2·12 × 10^4. Total plasmalogen can also be determined after conversion of the vinylic ether side-chain to an aldehyde by acid hydrolysis. The aldehyde can be assayed by the *p*-nitrophenyl hydrazone method or with Schiff's reagent *p*-rosaniline (Dittmer & Wells, 1969).

Galactose determination

The method for analysis of cerebroside and sulphatide is based on the conversion of the hexose to the furfural derivative in strong sulphuric acid followed by the formation of a coloured complex with anthrone (Radin, Brown & Lavin, 1956; Dittmer & Wells, 1969). The reagents used are: (1) 1mM-hexose as standard, made by diluting a 100mM stock solution kept frozen; (2) 3N-H_2SO_4; (3) diethyl ether; (4) anthrone reagent: dissolve 0·5 g anthrone, recrystallized from benzene–petroleum ether in 25 ml concentrated sulphuric acid and allow it to stand 4 h at room temperature. This stock solution can be kept approximately 2 weeks in the freezer. Before use, make a 1:15 dilution with 75% sulphuric acid in water and cool in the freezer to $-10°$.

Procedure. Pipette samples with from 0·3 to 1·0 μmol of hexose, and a set of four standards in this range, into screw-capped tubes. Take to dryness in a vacuum oven at 80°. Add 0·5 ml of 3N-H_2SO_4, tightly cap, and heat in a boiling water bath for 2 h. Cool and extract three times with 0·5 ml of diethyl ether. Heat the extracted hydrolyzate for a few minutes in a boiling-water bath to remove the last traces of ether and then cool in the freezer for 30 min. While the contents of the tube are still frozen add 5 ml of cold anthrone reagent, mix, cap, and heat in a boiling-water bath for 10 min. Cool to room temperature and read the absorbance at 625 nm. The extraction with ether is not necessary if the lipid samples are purified galactolipids. In the authors' laboratory, purified individual galactolipid samples were solubilized in orthophosphoric acid for 15 min at 90°, prior to reaction with anthrone reagent.

Relatively large amounts of cerebroside and sulphatide are needed for the anthrone method. Sphingosine can be assayed by the methyl orange method of Lauter & Trams (1962) after hydrolysis of purified galactolipid; less than 0·2 μmol of sphingosine can be accurately measured. Precise values for the cerebroside, sulphatide and ganglioside content of brain have been obtained with the hydrolytic TNBS assay of Yamamoto & Rouser (1970). Trinitrobenzene sulphonic acid (TNBS) reacts with the amino group of the long-chain base, sphingosine. Gas–liquid chromatographic methods for the assay of small samples of lipid hexose are reported in detail by Sweeley & Vance (1967).

Comment on thin-layer chromatography

The development of t.l.c. has shortened the time required to analyze a lipid extract as well as reducing the amount of lipid necessary to obtain a complete analysis. The technique is easily mastered and as identification of lipids is simplified the number of chemical assay methods can be reduced to a minimum. Lipid analyses of white and grey matter, of subcellular fractions from whole brain and of different cell types were given in Table 4.1. Results can be expressed in terms of total lipid content, or based on dry or wet weight. The interpretation of results is affected by the method for calculation, e.g., if during development results are expressed 'per whole brain', the rate of deposition of lipid can more readily be assessed.

GAS–LIQUID CHROMATOGRAPHY FOR ANALYSIS OF BRAIN LIPIDS

Gas–liquid chromatography (g.l.c.) is a procedure for the separation, identification and quantitation of various constituents of complex mixtures; for its general description see Burchfield & Storrs (1962), Littlewood (1971), Marinetti (1967) and Pattison (1973). In all chromatographic procedures the constituents to be separated are distributed between two different phases—a stationary phase and a mobile phase. In g.l.c. the *stationary phase* is a viscous liquid coated onto an inert support which is packed into a length of tubing called a column. The *moving phase* is a gas. The nature of the liquid and the types of gas vary, depending on what is to be separated.

Several major classes of brain lipids can be characterized by performing g.l.c. with the lipid itself or with one or more of its constituents. Thus cholesterol and cholesterol esters are examined as such; from galactolipids the fatty acids, long chain bases and sugars can be examined; from phospholipids, the fatty acids, bases and glycerol; and from gangliosides, the sialic acids, sugars and fatty acids. Most constituents of brain lipids which are characterized by g.l.c. are not easily vaporized, and therefore prior to application to the column suitable, more volatile, derivatives are made.

Fatty acids of low boiling point can be estimated directly without recourse to making such derivatives (James & Martin, 1952; Hrivnák & Soják, 1972). Long-chain fatty acids are estimated as their methyl esters and their separation and analysis has been well documented (James, 1959). The fatty acids of brain tissue have been thoroughly investigated in many laboratories with the use of g.l.c. Patterns of fatty acids from beef-brain galactolipids (O'Brien & Rouser, 1964), rat-brain galactolipids (Kishimoto & Radin, 1959) and the phospholipids from normal human brain have been established by several laboratories. A definitive paper is that of Svennerholm (1968) on human phospholipid fatty acids; Horning *et al.* (1964) described the calibration of a g.l.c. system for the estimation of higher fatty acids. The development of g.l.c. has proved valuable in investigating the biosynthesis of odd-numbered fatty acids (Hajra & Radin, 1962), and small changes in fatty acid composition with age (Svennerholm & Ställberg-Stenhagen, 1968). Several workers have reported the analysis and characterization of the aminolipid alcohol bases by g.l.c. methods (Sweeley & Moscatelli, 1959; Rosenberg & Stern, 1966).

In g.l.c. of the monosaccharides associated with cerebrosides, glycolipids, and glycoproteins, the *O*-methylglycosides can be estimated as their trifluoroacetate derivatives (Zanetta, Breckenridge & Vincendon, 1972). However, the carbohydrate moiety of such lipid complexes is more often estimated as the trimethylsilyl derivative of the methyl ester (Sweeley *et al.*, 1963; Carter & Gaver, 1967). The sialic acid component of gangliosides can also be analysed by g.l.c.; sialic acids are converted into their methylketoside methyl esters which are then chromatographed as the trimethylsilyl derivative (Yu & Ledeen, 1970). Tissue-culture lipids can now be characterized by g.l.c. methods even where the amount of material available for analysis is very small (Di Jong, Mora & Brady, 1971).

APPARATUS

The column is the item of importance and is kept at a controlled temperature. At one end of the column is an isolated injection port for introducing the mixture to be analysed and for vapourising it. At the other end is a detector system to detect the separated constituents. The detector system is connected to a recorder and preferably to an integrator so that the materials separated are automatically measured. For isothermal runs the column is maintained at the desired temperature throughout the separation. When the molecules to be separated consist of hydrocarbon chains of quite disparate length, i.e. fatty acids C_{12} and fatty acid C_{28}, an isothermal separation is not feasible as the long-chain fatty acids require much higher temperatures to bring them off the column. At this high temperature the short-chain fatty acids are lost in the solvent front. A temperature-programmed separation is then appropriate, and this involves a low initial oven temperature which is increased in a stepwise fashion over the period of the run.

Columns. Various types of packed columns are used for separating lipid constituents from the brain. Table 4.3 shows a small selection of polar and non-polar column packings used in neurochemical research.

Packed columns are relatively insensitive to the subtle differences among positional isomers; such separations are usually made on capillary columns. These columns are long and have high resolution potential. Their length is an important factor in their ability to separate the constituents from very small amounts of sample; they are often employed in steroid separations. In capillary column separations on temperature programmed runs a single column only is needed as there is a very small background signal. A pressure controller for the carrier gas flow rate is still required. The rate of temperature increase is usually 2–3°C per min, much smaller than in packed columns due to the high thermal mass of the capillary columns. The supporting materials used in columns often have some active polar sites; a reduction in polarity is achieved by silane treatment of the inert material.

Differential detectors are most commonly used in g.l.c. The magnitude of the signal from these detectors is proportional to the amount of material in the detector, or the mass of the material which passes through the detector per unit time. The two principal types are the thermal conductor cell and the flame ionization detector. For fatty acid esters a flame ionization detector which is sensitive to carbon is appropriate. An account of detectors and their usefulness is given by Littlewood (1971).

Carrier gas. The choice of carrier gas depends on the detector type and to a small extent on the nature of the sample. Nitrogen, helium and argon are the gases most frequently used. The flow of carrier gas must be

LIPIDS: EXTRACTION, ETC.

Table 4.3 Some column packings, stationary phases and supports commonly used in g.l.c. of complex lipids from the brain

(a) Liquid phases	(b) Inert support	Amount used: 100a/b (w/w)	Mesh size	Temperature optima	Types of compounds analysed
Non-polar materials					
Apiezon L	Chromosorb W: acid-washed and 'silanized'	4	80–100	250	Hydroxy fatty acid esters
Silicone gum (OV1)	Gas Chrom. G	2 to 3	80–100	300	Steroids
Methyl silicone (SE-30)	Gas Chrom. Q	1 to 3	100–120	250	Fatty acid methyl esters (not of the 18:0, 18:1, 18:2, 18:3 acids)
	Chromosorb W	2·5	100–120	210	Gangliosides as trimethylsilyl derivatives
	Gas Chrom. P	3	100–120	165	Trimethylsilyl derivatives of carbohydrates
Polar materials					
XE-60	Chromosorb G	2·5	80–100	230	Steroids
Diethylene glycol succinate (DEGS)	Chromosorb W	20	80–100	190	Fatty acids; methyl esters
Ethylene glycol succinate silicone (EGSS-X)	Gaschrom. Por Q	10	80–100	225	Unsaturated fatty acids methyl esters
Carbowax 20M polyglycol	Chromosorb G: acid-washed and silanized	10	60–80 80–100 100–120	200	Aldehydes; alcohols; hydrocarbons

fully variable over a range determined principally by column diameter. Filter drier assemblies, filled with molecular sieve material and silica gel, are usually placed in the gas lines to remove residual organic impurities and water.

Notes on setting up the instrument. These apply particularly to the Perkin Elmer F11 instrument (Perkin Elmer, Beaconsfield, Buckinghamshire), using a flame-ionization detector system. It is advisable to check the gas lines; red tubing designates hydrogen. Nuts should be screwed tight, then a further half-turn with a spanner to prevent constriction of the tube by the ferule. Once the column is fitted the carrier gas should be set at 30 ml/min, and leaks checked for, using a dilute Teepol solution and a Pasteur pipette. The recorder should then be zeroed. If the recorder gain is correctly adjusted the pen should always return to a set position without excessive chatter. The amplifier should next be balanced by disconnecting the signal input from the front of the analyser unit. If the zero control is fully clockwise and the sensitivity knob switched to 1×10^4, the balance control system is then adjusted so that there is no movement of the recorder pen.

To light the flame the H_2 supply is set to 60 ml/min, the air to 300 ml/min, the ignition button is then pressed, ignition is indicated by a pop, and the recorder pen moves across the recorder bed. Effective ignition can be confirmed by holding a cold spanner over the detector chimney. The flame is difficult to see: it is very pale blue. When a flame-ionization detector is used, optimum flame conditions should be established before samples are analysed and gas pressures adjusted accordingly.

There is a useful rule of thumb for the conversion of pounds per square inch to ml per min when the analyser unit is fairly new: air flow rate of 20 ml/min is equivalent to 1 p.s.i.; N_2 flow rate of 1·5 ml per min is equivalent to 1 p.s.i.; H_2 flow rate of 2 ml per min is equivalent to 1 p.s.i. These are approximate values and more accurate flow rates can be determined by attaching a simple bubble flow meter to the bulk-head connectors on the analyser unit. Obviously restrictor characteristics will change with time, so the rule of thumb values apply principally to new analyser units. The oven should be heated to the required temperature for at least half an hour before the sample is injected. The injection port temperature should be set at about 150° above oven temperature.

On older models of gas chromatographs, the temperature of the oven was not independent of the injection port temperature. A cooling period must be allowed between duplicate programme runs to ensure that the injection port temperature has stabilised. Optimum carrier flow rates should be established for a simple compound of average chain length. The retention time of any compound is inversely related to temperature of the column, being shorter at greater column temperature. In seeking improved resolution both column temperature and gas flow rate can be varied. Once the column is set up, suitable standards must be used to calibrate the column.

Standards for calibration of the system for fatty-acid ester analysis. A series of methyl ester mixtures which are commonly used for calibration are called NIH (National Institutes of Health) mixtures, three of which are designated KA, KD, and HlO4 (Applied Science Laboratories Inc., U.K. Distributors: Field Instruments, Tetrapak House, Orchard Road, Richmond, Surrey). Several characteristics of the analytical system can be established using suitable mixtures; these are: the load limits of the detector system, the relative load–response relationship for saturated fatty acid methyl esters, linearity of response in relation to molecular weight, and column efficiency and the detector response to unsaturated esters.

The mixture KA consists of C_{14}, C_{16} and C_{18} compounds in different proportions and is suitable for establishing the first parameter. The ampoules (50 mg) are opened, made up to a suitable volume with spectroscopic grade hexane to a concentration of approx. 10 mg/ml, then 1 or 2µl samples are injected into a prepared gas chromatograph. Column efficiency and detector response are determined using KD (see Fig. 4.1).

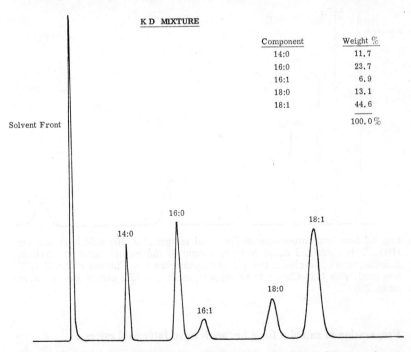

Fig. 4.1 Standard chromatogram for a 2µl sample of a fatty acid ester mixture KD. This consisted of the five components listed. Column EGSS-X 10% was used, with as support Gas–Chrom Q, mesh 100–120; carrier gas N_2, oven temp. 160°C.

The mixture HlO4 is a suitable standard for investigating the quantitative response to fatty acids of different chain lengths (see Fig. 4.2). If an integrator is connected to the detector system, it is activated at zero time when the sample is injected, and the digital data can be correlated with peaks coming off at specific time intervals. The areas under the peaks are determined and the weight percentage of the different compounds calculated and compared with specification. Retention time data for simple standard mixtures may be plotted as log retention time along an X axis and carbon chain length along a Y axis. Such graphs are useful in assessing the time taken for the mixtures to be analysed. They may serve to indicate possible carbon chain lengths of unidentified compounds giving peaks in chromatograms of the mixtures of fatty acid esters which are under analysis.

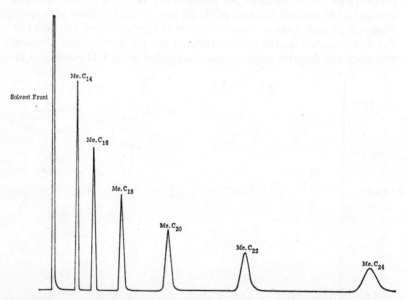

Fig. 4.2 Standard chromatogram for a $2\mu l$ sample of a fatty acid ester mixture HIO_4. This contained equal parts by weight of the methyl esters of myristic, palmitic, stearic, arachidic, behenic and lignoceric acids. Column EGSS-X 10% was used, with Gas–Chrom Q as support, mesh 100–120; carrier gas N_2, oven temp. 200°C.

Preparation of samples for g.l.c. analysis of fatty acid esters

The formation of methyl esters of fatty acids and acetoxy derivatives of hydroxy fatty acids is the first step in the analysis. It is followed by g.l.c. of the esters on suitably standardized columns.

Methanolysis. One of the probelms in methylating is the loss of volatile esters. Most methylations are acid-catalysed esterifications, employing, for example, dry HCl in excess alcohol. The time and temperature required for effective esterification depends on the lipid class. To prevent autooxidation of unsaturated fatty acids, many chemists esterify under N_2 or add an antioxidant. Methanolysis of unsaturated fatty acids and triglycerides is sometimes carried out with a boron trihalide–methanol reagent. This is achieved by heating the lipid system with 14% boron trifluoride in methanol at 100° C for 90 min, followed by hexane extraction of the methyl esters (Morrison & Smith 1964; Moscatelli, 1972). Esters free from contaminants are produced by this method (Klopfenstein, 1971). However, when boron trifluoride–methanol is the system employed in methylation, cyclopropane esters may be lost (Brain *et. al.*, 1972). Some losses of highly unsaturated esters after 90 min at 120° have also been reported. Methanolic HCl is a suitable reagent for producing methyl esters of fatty acids, sugars and sialic acids. A reagent kit is available (Applied Science Laboratories) which is totally anhydrous and effectively forms the methyl esters of fatty acids; this contains acetylchloride, provided in ampoules of 5 ml. To produce a 2·5% w/v solution of methanolic HCl, 100 ml of pure methanol is added slowly; the hydrogen chloride produced dissolves in the methanol to yield methanolic HCl. An alternative is to bubble hydrogen chloride gas into methanol to produce methanolic HCl. This reagent does not keep and should be used within two days.

Reaction method. The lipid sample is distributed into special methanolysis tubes. These are sealed Pyrex assemblies (Coleman Instruments, Maywood, Illinois, U.S.A.) with metal clips for holding the cap and the tube together (see Fig. 4.3; Kishimoto & Radin, 1965).

The methanolysis reagent (2·5 ml) is added to approx. 5–10 mg of the lipid in each methanolysis tube. The tubes are sealed and left overnight at 75°. After this hydrolysis the fatty acid esters are extracted as follows. The tubes are cooled in ice, the seal opened, and 2·5 ml of hexane added. After shaking and allowing to separate, hexane in the top layer contains the FAE. Two further extractions with 2·5 ml hexane are carried out; the remaining methanolysis products are in the lower layer, which is discarded. This is followed by the acetoxylation of the hydroxy FAE; the hexane solution is evaporated under a gentle stream of N_2 in suitable nitrogen evaporation apparatus (Beckman Instruments, P.O. Box 6100, Anaheim, California, U.S.A.), the lipid transferred into conical-tipped tubes with several 0·5 ml portions of hexane, and evaporated to dryness under N_2. An isopropenyl acetate–*p*-toluene sulphonic reagent is added (0·1 ml, containing 4 mg *p*-toluene sulphonic acid per ml of isopropenyl acetate). The tubes are shaken in a vortex mixer, corked tightly, and heated for 30 min at 60°C.

In the *separation of normal and acetylated hydroxy (i.e. acet-oxy) FAE of cerebrosides* the acetylation mixtures are spotted directly on to

Fig. 4.3 Methanolysis reaction tube used in preparation of methyl esters from lipid samples.

t.l.c. plates (0·5 mm Silica Gel G). The plates are developed in hexane-ether (85:15, v/v) sprayed with 0·04% bromothymol blue in 0·01 N-NaOH, and allowed to dry. The FAE bands are scraped into glass-stoppered tubes, 10 ml of ether is added, the tubes are shaken in a vortex mixer, centrifuged, and equal volumes of both normal and acetoxy FAE solutions are distributed into tubes containing known amounts of internal standard.

Formation of trimethylsilyl (TMS) derivatives

The new reagent recommended for formation of these TMS derivatives is made as follows. Hexamethyl disilazane (2·6 ml) is added to dry pyridine (2 ml) and the solution is thoroughly mixed; trimethyl chlorosilane (1·6 ml) is added to this mixture and shaken. The opaque solution which is obtained is centrifuged and the clear colourless supernatant is aspirated off and stored in the dark. It will keep for several weeks (Carter & Gaver, 1967). A commercial product, Sil-Prep, for preparing TMS derivatives can be obtained from Applied Science Laboratories. It contains hexamethyldisilazane–trimethyl chlorosilane–pyridine, 3:1:9 by vol.

For the formation of TMS derivatives of β N-acetylneuraminic acid, dried ganglioside (10 μg aliquots) is methylated as described. The cooled solutions are extracted three times with hexane to remove fatty acid esters. The methanolic solutions are then evaporated to dryness under N_2. The TMS reagent (50 μl, prepared as above according to Carter & Gaver, 1967) is added. The sample is vortexed and left for 15 min. Approximately 4 μl are injected for g.l.c. (Yu & Ledeen, 1970). The TMS derivatives of long-chain bases may be synthesized as follows: 50 μl of TMS reagent are added to dry, solvent-free long-chain base (150 μg) and the sample is then injected into a suitable column such as SE 30 (see Table 4.3).

Injection of sample; results

For g.l.c., all solutions are evaporated to dryness under N_2 and subsequently resuspended in small portions (10–15 μl) of hexane or carbon disulphide of Spectrosol grade. The sample, 1–5 μl, is then injected into the column. The volume used is determined by the conditions of the separation, but the smaller the volume the better. A fatty acid ester mixture yields a chromatogram with a series of peaks, the areas of which are computed as described below. If an internal standard is added to the system, it must be a fatty acid which does not normally occur in the system under investigation. Methyl nonadecanoate is often used in approximately 7–15 μg amounts, and it is added to the conical-tip tubes prior to the g.l.c. analysis of the normal fatty acid mixture. An identical amount is added to the acetoxy fatty ester mixture. The ratios of normal fatty acid to hydroxy fatty acid can then be calculated. If only the relative percentages of the constituent normal fatty acids are required, internal standards are not necessary; analyses should be carried out in triplicate.

Analysis of results

Data on fatty acid ester mixtures are usually presented as tables which report the range of carbon chain-lengths obtained, and the relative percentages of the fatty acid constituents.

These latter data are obtained by computing the area under each peak, summating the individual areas and calculating the individual areas as a percentage of the whole. In the quantitative analysis of long-chain bases, gangliosides and fatty acids the procedure adopted is the preparation of a standard curve for a pure sample, taking known concentrations of the sample. A constant amount of the material used as internal standard is distributed into g.l.c. tubes, the sample is added and g.l.c. run with the mixtures. The ratio: (sample area)/(internal standard area), plotted against sample concentration, should be linear. Interpolation can then be made to determine unknown concentrations of the material under investigation. Two parameters which are useful in the identification and quantitation of substances are the relative detector response, and the relative retention time.

The relative detector response (r.d.r.) indicates adsorptive losses on the column; these are important considerations in obtaining quantitative results. It is estimated by:

$$\text{r.d.r.} = \frac{\text{Area given by sample}}{\text{Area given by internal standard}} \times \frac{\text{Weight of internal standard}}{\text{Weight of sample}}$$

The r.d.r. values should not change significantly over the range of sample amounts likely to be encountered in analysis (Yu & Ledeen, 1970): see Table 4.4. The r.d.r. can be measured periodically as a means of checking the performance of the detector used. The r.d.r. values for the same sample, employing different columns, are useful aids in identifying unknown samples.

Table 4.4 Relative detector response (rdr) for two kinds of sialic acid

Weight ratio of sialic acid: internal standard	rdr	
	β-NANA	β-NGNA
10.0	0.932	0.950
8.0	0.937	0.950
6.0	0.925	0.949
4.0	0.938	0.949
2.0	0.912	0.913
1.0	0.924	0.919
0.8	0.918	0.920
0.6	0.900	0.897
(Average ±2 SD:	0.923 ± 0.028	0.931 ± 0.044)
0.4	0.851 ± 0.016	0.824 ± 0.066
0.2	0.814 ± 0.030	0.765 ± 0.014
0.1	0.697 ± 0.010	0.648 ± 0.022

The experiment used the same weight ratios of sialic acid to internal standard for N-acetyl and N-glycolyl neuraminic acids. The compounds were chromatographed as trimethylsilyl derivatives on OV-1. Three samples of each mixture were injected three times, and peak areas were measured by an electronic integrator. The two averages shown (average ±2 SD) were calculated for all samples having weight ratios between 10.0 and 0.6 (Yu & Ledeen, 1970).

Results showed that the r.d.r. values for the two were not signicantly different. Further, it may be seen that the values are relatively constant between weight ratios of 10 and 0.6. The quantity of sialic acids estimated was 0.05–5 μg; internal standard, 0.5 μg.

β-NANA: methyl-β-ketoside methyl ester of N-acetylneuraminic acid.
β-NGNA: methyl-β-ketoside methyl ester of N-glycolyneuraminic acid.

Retention time measurements and qualitative analysis. Retention values in g.l.c. are sensitive to such variables as the concentration and composition of liquid phase, the size of the support and the pretreatment of the support material. The conditions of the separation run, such as temperature and gas pressure also influence the retention time on the column. To provide an easy means of identifying an unknown compound, Kovats (1958) proposed an index system based on the n-alkanes as reference substances: for they are chemically inert, soluble in common stationary phases and are non-polar. This index, I, is defined as:

$$I = 100\,[n(\log R_x - \log R_z)/(\log R_{z+n} - \log R_z) + Z]$$

where R_x = retention time of unknown substance x; R_z = retention time of normal alkane having z carbon atoms; R_{z+n} = retention time of the normal alkane having $(z+n)$ carbon atoms, n being the difference between the number of carbon atoms in the n-alkanes concerned.

A further procedure for the qualitative analysis of a mixture is to chromatograph a simple mixture containing an unknown component. The retention times and maximum peak heights for each peak are measured. Then a small portion of the material suspected to be the unknown component is added, and a new chromatogram obtained for the enriched mixture. If a further peak is obtained then the selected material was not present in the original mixture. If, on the other hand, one of the peaks has increased in height relative to the other peaks, but its retention time value is the same as in the first chromatogram, then the identity of the unknown is most probably that suspected.

Data on retention times relative to the internal standard, can be obtained by employing two or more columns in the analysis of a mixture. Knowing the polar characteristics of the different columns, useful information about the constituents of the mixture can be obtained by investigating which constituents emerge first in relation to internal standard from the different columns. Relative retention times can then be usefully employed in the identification of an unknown constituent. A compilation of gas-chromatographic retention data is available (McReynolds 1966).

REFERENCES

Ansell, G. B. & Hawthorne, J. N. (1964) *Phospholipids. Chemistry, Metabolism and Function.* Amsterdam: Elsevier.
Arvidson, G. A. E. (1967) *J. Lipid Res.* **8**, 155.
Autilio, L. A. & Norton, W. T. (1963) *J. Neurochem.* **10**, 733.
Bell, R. D. & Doisy, E. A. (1920) *J. biol. Chem.* **44**, 55.
Bladon, P. (1958) In *Cholesterol: Chemistry, Biochemistry and Pathology,* ed. Cook, R. P., p. 15. New York: Academic Press.
Brain, B. L., Gracy, R. W. & Scholes, V. E. (1972) *J. Chromatogr.* **66**, 141.
Burchfield, H. P. & Storrs, E. E. (1962) *Biochemical Applications of Gas Chromatography.* New York: Academic Press.
Carter, M. E. & Gaver, R. C. (1967) *J. Lipid Res.* **8**, 391.
Colowick, S. P. & Kaplan, N. O. (1969) (Eds.) *Methods in Enzymology.* New York: Academic Press.
Cuzner, M. L., Davison, A. N. & Gregson, N. A. (1965) *J. Neurochem.* **12**, 469.
Cuzner, M. L. & Davison, A. N. (1967) *J. Chromatogr.* **27**, 388.
Cuzner, M. L. & Davison, A. N. (1968) *Biochem. J.* **106**, 29.

Cuzner, M. L. & Davison, A. N. (1973) *J. neurol. Sci.* **19**, 29.
Davison, A. N., Dobbing, J., Morgan, R. S. & Payling-Wright, G. (1958) *J. Neurochem.* **3**, 89.
Dawson, R. M. C. & Eichberg, J. (1965) *Biochem. J.* **96**, 634.
De Vries, G. H. & Norton, W. T. (1974) *J. Neurochem.* **22**, 259.
DiJong, I., Mora, P. T. & Brady, R. O. (1971) *Biochemistry* **10**, 4039.
Dittmer, J. C. & Wells, M. A. (1969) *Meth. Enzym.* **14**, 482.
Folch, J. P., Lees, M. & Sloane-Stanley, G. H. (1957) *J. biol. Chem.* **226**, 497.
Gonzalez-Sastre, F. & Folch-Pi, J. (1968) *J. Lipid Res.* **9**, 532.
Hajira, A. K. & Radin, N. S. (1962) *J. Lipid Res.* **3**, 327.
Hanel, H. K. & Dam, H. (1955) *Acta chim. scand.* **9**, 677.
Hanes, C. S. & Isherwood, F. A. (1949) *Nature, Lond.* **164**, 1107.
Horning, E. C., Ahrens, E. H. Jun., Lipsky, S. R., Mattson, F. H., Mead, J. F., Turner, D. A. & Goldwater, W. H. (1964) *J. Lipid Res.* **5**, 20.
Hrivnak, J. & Sojak, L. (1972) *J. Chromatogr.* **68**, 55.
James, A. T. (1959) *J. Chromatogr.* **2**, 552.
James, A. T. & Martin, A. J. P. (1952) *Biochem. J.* **50**, 679.
Johnson, A. R. & Davenport, J. B. (1971) *Biochemistry and Methodology of Lipids.* New York: Wiley-Interscience.
Kates, M. (1972) In *Laboratory Techniques in Biochemistry and Molecular Biology*, ed. Work, T. S. & Work, E. Vol. 3. Amsterdam: North Holland.
Kishimoto, Y. & Radin, N. S. (1959) *J. Lipid Res.* **1**, 72.
Kishimoto, Y. & Radin, N. S. (1965) *J. Lipid Res.* **6**, 435.
Klopfenstein, W. B. (1971) *J. Lipid Res.* **12**, 773.
Korey, S. R. & Gonatas, J. (1963) *Life Sciences* **1**, 296.
Kovats, E. (1958) *Helv. chim. Acta* **41**, 1915.
Lauter, C. J. & Trams, E. G. (1962) *J. Lipid Res.* **3**, 136.
Ledeen, R. (1966) *J. Am. Oil Chem. Soc.* **38**, 708.
Lees, M. B. (1966) *J. Neurochem.* **13**, 1407.
Lees, M. B. (1968) *J. Neurochem.* **15**, 153.
Littlewood, A. B. (1971) *Gas Chromatography Principles, Techniques & Applications.* 2nd edn. New York: Academic Press.
Lowenstein, J. M. (1969) **14** p.1-702 in Colowick & Kaplan (1969)
Malins, D. C. (1966) *Prog. Chem. Fats* **8**, 301.
Mangold, H. K. (1961) *J. Am. Oil Chem. Soc.* **38**, 708.
Marinetti, G. V. (1967) (Ed.) *Lipid Chromatographic Analysis.* Vol. 1. New York: Marcel Dekker.
Martland, M. & Robison, R. (1926) *Biochem. J.* **20**, 847.
McIlwain, H. & Rodnight, R. (1962) *Practical Neurochemistry*, p. 62. London: J. & A. Churchill.
McIlwain, H. & Bachelard, H. S. (1971) *Biochemistry and the Central Nervous System.* Edinburgh and London: Churchill-Livingstone.
McReynolds, W. O. (1966) *Gas Chromatographic Retention Data*, Evanston, Illinois: Preston Technical Abstracts Co.
Morris, L. J. (1966) *J. Lipid Res.* **7**, 717.
Morrison, W. R. & Smith, L. M. (1964) *J. Lipid Res.* **5**, 600.
Moscatelli, E. A. (1972) *Lipids* **7**, 268.
O'Brien, J. S. & Rouser, G. (1964) *J. Lipid Res.* **5**, 339.
O'Brien, J. S. & Sampson, E. L. (1965) *J. Lipid Res.* **6**, 537.
Parker, F. & Peterson, N. F. (1965) *J. Lipid Res.* **6**, 455.
Pattison, J. B. (1973) *Gas Chromatography. A Programmed Introduction to GLC.* London: Heyden.
Penick, R. J., Meisler, M. H. & McCluer, R. H. (1966) *Biochim. Biophys. Acta* **116**, 279.
Poduslo, S. E., & Norton, W. T., (1972) *J. Neurochem.* **19**, 727.
Radin, N. S. (1958) *Meth. biochem. Analysis* **6**, 163.

Radin, N. S. (1969) *Meth. Enzym.* **14,** 245.
Radin, N. S., Brown, J. R. & Lavin, F. B. (1956) *J. biol. Chem.* **219,** 977.
Randerath, K. (1966) *Thin-Layer Chromatography.* New York: Academic Press.
Renkonnen, O. (1971) *Prog. Thin-Layer Chromatogr. & Related Meth.* **2,** 143.
Renkonnen, O. & Varo, P. (1967) In *Lipid Chromatographic Analysis,* ed. Marinetti, G. V., Vol. 1, p. 41. New York: Marcel Dekker.
Rosenberg, A. & Stern, N. (1966) *J. Lipid Res.* **7,** 122.
Rouser, G., Kritchevsky, G., Yamamoto, A. & Baxter, C. F. (1972) *Adv. Lipid Res.* **10,** 261.
Schmid, P., Calvert, J. & Steiner, R. (1973) *Physiol. Chem. & Physics* **5,** 157.
Schmid, P., Hunter, E. & Calvert, J. (1973) *Physiol. Chem. & Physics* **5,** 151.
Siakotos, A. N. & Rouser, G. (1965) *J. Am. Oil Chem. Soc.* **42,** 913.
Siggia, S. & Edsberg, R. L. (1948) *Analyt. Chem.* **20,** 762.
Skidmore, W. D. & Entenman, C. (1962) *J. Lipid Res.* **3,** 471.
Skipski, V. P. & Barclay, M. (1969) *Meth. Enzym.* **14,** 530.
Sperry, W. M. & Webb, M. (1950) *J. biol. Chem.* **187,** 97.
Stahl, E. (1969) (Ed.) *Thin-Layer Chromatography—A Laboratory Handbook.* Berlin: Springer Verlag.
Suzuki, K. (1965) *J. Neurochem.* **12,** 629.
Svennerholm, L. (1968) *J. Lipid Res.* **9,** 570.
Svennerholm, L. & Ställberg-Stenhagen, S. (1968) *J. Lipid Res.* **9,** 215.
Sweeley, C. C. & Moscatelli, E. A. (1959) *J. Lipid Res.* **1,** 40.
Sweeley, C. C., Bentley, R., Makita, M. & Wells, W. W. (1963) *J. Am. chem. Soc.* **85,** 2497.
Sweeley, C. C. & Vance, D. E. (1967) In *Lipid Chromatographic Analysis,* ed. Marinetti, G. V., Vol. 1, p. 465. New York: Marcel Dekker.
Truter, E. V. (1963) *Thin-Film Chromatography.* London: Cleaver-Hume.
Vanier, M. T., Holm, N., Ohman, R. & Svennerholm, L. (1971) *J. Neurochem.* **18,** 581.
Vereshchagin, A. G. (1964) *J. Chromatogr.* **14,** 184.
Wade, H. E. & Morgan, D. M. (1955) *Biochem. J.* **60,** 264.
Wagner, H., Hörhammer, L. & Wolff, P. (1961) *Biochem. Z.* **334,** 175.
Webster, G. R. & Folch, J. P. (1961) *Biochim. biophys. Acta* **49,** 399.
Wells, M. A. & Dittmer, J. C. (1963) *Biochemistry* **2,** 1259.
Williams, J. N., Anderson, C. E. & Jasik, A. D. (1962) *J. Lipid Res.* **3,** 378.
Yamamoto, A. & Rouser, G. (1970) *Lipids* **5,** 442.
Yu, R. K. & Ledeen, R. W. (1970) *J. Lipid Res.* **11,** 506.
Zanetta, S. P., Breckrenridge, W. C. & Vincendon, G. (1972) *J. Chromatogr.* **69,** 291.

5. Individual Lipids: Preparative and Analytical Methods

M. L. CUZNER

Column chromatography of lipids	88
Lipids and lipid constituents	92
Cholesterol	92
Neutral lipids, exlcuding cholesterol	93
Acyl ester determinations	93
Glycerol determination	93
Cerebroside and sulphatide	94
Minor galactolipids	95
Ethanolamine phospholipids	95
Serine phospholipids	96
Inositol phosphatides	97
Inositol determination	98
Lecithins	99
Choline determination	99
Sphingomyelin	100
Minor phospholipids	100
Phosphatidic acid and cardiolipin	101
Gangliosides	101
References	102

The composition of an extract of the total lipids of the brain can be determined with a high degree of accuracy and specificity by using a combination of thin-layer chromatography and gas–liquid chromatography. Although these excellent tools enable small lipid samples to be very completely subfractionated and analyzed there are more appropriate techniques for the preparation and purification of lipids in quantity. The following sections will list selected column chromatographic methods for separating and purifying individual lipids, including chemical methods for determining components of lipid molecules.

When the aim is the preparation in bulk of purified lipids from brain, the problem is simplified by a preliminary separation of different lipid groups using fractional precipitation. The following scheme (Fig. 5.1) of Ansell & Hawthorne (1964) is applicable.

COLUMN CHROMATOGRAPHY OF LIPIDS

The three most commonly used adsorbents for the separation of brain lipids by column chromatography are alumina, silicic acid and Florosil. The ion-exchange resins, DEAE- and TEAE-cellulose, are also used to obtain effective fractionation of a total lipid extract. Neutral

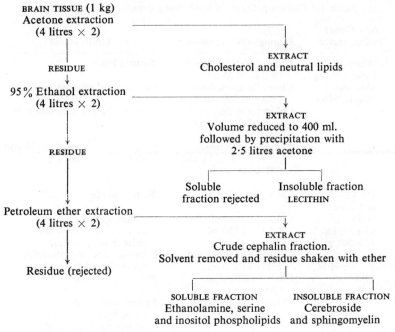

Fig. 5.1 Preliminary solvent fractionation of lipids.

lipid, the majority of which in nervous tissue is cholesterol, is eluted with an initial volume of chloroform from all four adsorbents. The phospholipids and glycolipids are then eluted individually by gradient or stepwise elution.

Aluminium oxide (alumina). Alumina used for column chromatography is the basic type which has a pH of near 7·5 to 8 (BDH, alumina for chromatography). Thus the more acidic phospholipids, phosphatidyl serine, ethanolamine and inositol are adsorbed, allowing the non-reactive choline phospholipids (present as zwitterions at all pH values) to be eluted first, followed by the galactolipids (Davison & Wajda, 1959) (Table 5.1). The flow rate can be quite high with alumina and still allow a suitable resolution. The loading factor for this adsorbent is in the range of 10 mg lipid per g alumina.

Silicic acid. The separation of individual polar lipids on silicic acid with increasing amounts of methanol in chloroform is perhaps the most frequently used procedure in lipid column chromatography (Lea, Rhodes & Stoll, 1955; Wren, 1960). The acidic phospholipids are eluted first followed by the choline phospholipids. The variability of preparations makes it difficult to specify one procedure for elution of silicic acid columns, but a general outline, which can be adapted by the individual investigator, is given in Table 5.2. As the flow rate on silicic acid columns

Table 5.1 Elution patterns of lipids using column chromatography

Adsorbent; loading factor	Eluting solvent; volume	Lipids eluted
1. Alumina, 50 g; lipid, 500 mg Davison & Wajda, 1959	Chloroform–methanol 98:2 (v/v) 55 ml	Neutral lipids
	Chloroform–methanol 1:1 (v/v) 600 ml	Lecithin and sphingomyelin
	Chloroform–methanol–water 7:7:1 by vol. 700 ml	Cerebroside and sulphatide
	Chloroform–ethanol–water 2:5:2 by vol. 500–1000 ml	Ethanolamine phospholipid and phosphatidyl inositol
2. DEAE-cellulose, acetate form, 15 g; lipid, 100–600 g Rouser et al., 1969	Chloroform 750 ml	Neutral lipids
	Chloroform–methanol 9:1 (v/v) 750 ml	Lecithin, sphingomyelin, cerebroside, lysolecithin, galactosylglycerides.
	Chloroform–methanol 7:3 (v/v) 750 ml	Ethanolamine phospholipid, ceramide polyhexosides
	Glacial acetic acid 750 ml	Phosphatidyl serine
	Chloroform–methanol–ammonia–salt: 750 ml*	Phosphatidic acid phosphatidyl inositol, phosphatidyl glycerol, cardiolipin and sulphatide
3. Florosil, 25 g lipid, 500 mg Radin et al., 1956	Chloroform 625 ml	Neutral lipids
	Chloroform–methanol 19:1 (v/v) 700 ml	Ceramide
	Chloroform–methanol 7:3 (v/v) 700 ml	Cerebroside and sulphatide
	Chloroform–methanol 2:1 (v/v) saturated with water: 800 ml	Phospholipids

The volume of eluting solvent stated in the second column is for use with the quantities of material quoted in the first column.

*Chloroform–methanol (4:1, v/v) made 10–50 mM with respect to ammonium acetate or potassium acetate to which is added 20 ml of freshly prepared 28% (by wt) aqueous ammonia per litre.

can be very slow, a preparation with a uniform mesh of 100–200, called Unisil (Clarkson Chemical Co. Inc., Williamsport, Penn., U.S.A.) is most widely used. Prior to the separation of individual phospholipids on Unisil a brain lipid extract can be quantitatively fractionated into three groups on the same adsorbent. The solvent sequence chloroform, acetone and methanol elutes neutral lipids, glycolipids and phospholipids respectively (Rouser et al. 1969). The loading factor for Unisil is approximately 10 mg lipid/g adsorbent.

Table 5.2 Elution of lipid classes and individual lipids on silicic acid

Separation; loading factor	Volume of eluting solvent	Lipids eluted
1. Group separation: adsorbent, 30 g, lipid, 300 mg Rouser et al., 1969	Chloroform 175 ml	Neutral lipids
	Acetone 700 ml	All glycolipids with the exception of gangliosides
	Methanol 175 ml	Phospholipids
2. Neutral lipid separation: adsorbent, 20 g lipid 200 mg Hirsch & Ahrens, 1958	Petrol ether–ether 99:1 (v/v): 400 ml	Cholesterol esters
	Petrol ether–ether 94:6 (v/v): 500 ml	Triglycerides
	Petrol ether–ether 92:8 (v/v): 300 ml	Cholesterol
	Petrol ether–ether 75:25 (v/v) 300 ml	Diglycerides
	Ether: 300 ml	Monoglycerides
3. Phospholipid separation: adsorbent, 20 g lipid 200 mg Hanahan, 1960	Chloroform: 300 ml	Neutral lipids
	Chloroform–methanol 9:1 (v/v): 500 ml	Cardiolipin and phosphatidic acid
	Chloroform–methanol 6:1 (v/v): 600 ml	Ethanolamine phospholipid
	Ethylacetate–methanol 3:2 (v/v): 600 ml	Phosphatidyl serine and phosphatidyl inositol
	Ethylacetate–methanol 1:1 (v/v): 600 ml	Lecithin
	Chloroform–methanol 1:9 (v/v): 700 ml	Sphingomyelin

The separations, loading factors and solvent volumes quoted refer to Unisil (see text).

Florosil. Florosil is a co-precipitated mixture of silica gel and magnesia, a suspension of which in water has a pH of 8·5 (Radin, 1969). This adsorbent in columns gives high flow rates and is particularly useful for the isolation of ceramide and the galactolipids, cerebroside and sulphatide (Radin, Brown & Lavin, 1956). The loading factor is 20–25 mg lipid/g. Florosil. The phospholipids, particularly the choline-containing ones, are tenaciously adsorbed but they are partially eluted with water-saturated chloroform–methanol mixtures (Table 5.1).

DEAE–cellulose. The fractionation of lipids on ion-exchange columns is based on their ionic properties and the successively eluted fractions according to Rouser et al. (1969) are as follows: (1) non-acidic with one anionic and cationic group without an exchangeable proton (e.g. lecithin); (2) non-acidic with one anionic and cationic group with an exchangeable proton (e.g. phosphatidyl ethanolamine); (3) weakly acidic (e.g. free fatty acids); (4) more polar weakly acidic (e.g. phosphatidyl serine, gangliosides); and (5) strongly acid (from phosphate or sulphate groups). The coarser preparations of DEAE–cellulose are most suitable as flow-rates are reduced by the use of finer grades. The preparation recommended by Rouser et al. (1969) is Selectacel DEAE, regular grade (Brown & Co. Berlin, New Hampshire U.S.A.) which has a loading factor of 20 mg/g adsorbent. TEAE–cellulose has a much higher capacity than DEAE for lipids with carboxyl groups as the only ionic groups and is used for the fractionation of fatty acids and gangliosides (Rouser et al., 1969). The general schemes shown in Tables 5.1 and 5.2 allow for the fractionation of a total lipid extract into relatively pure components. The purification of individual lipids and the chemical reactions of their component groups will be considered in the next section.

LIPIDS AND LIPID CONSTITUENTS

Cholesterol

There are traces of cholesterol ester, diglyceride and triglyceride in adult brain but the major neutral lipid is cholesterol, which accounts for 4–5% of brain wet weight. The spinal cord and brain of cattle constitute the most important raw material for the industrial production of cholesterol. In neural tissues cholesterol is associated with specific structural components notably the myelin sheath (Davison & Peters, 1970). Cholesterol esters are present in brain during development (Adams & Davison, 1959) and also appear in cerebral tissues in the areas of demyelination formed during the course of diseases, such as multiple sclerosis and subacute sclerosing panencephalitis (Ramsay, 1973).

Cholesterol can be crystallized from 95% ethanol (v/v) or glacial acetic acid after alkaline hydrolysis of a total lipid extract or after preliminary separation of neutral lipids on a silicic acid or alumina column.

Procedure. A total brain lipid extract is prepared as described in the previous chapter. A neutral lipid fraction can be prepared by column chromatography (p. 90) prior to crystallization of the cholesterol in acetic acid or ethanol. Alternatively phospholipids can be removed by alkaline hydrolysis and the sphingolipids remain in the liquor during the crystallization process. The total brain-lipid extract is evaporated to dryness, weighed and dissolved in 25 ml chloroform–methanol (2:1, v/v) per g lipid. To the dissolved lipid is added 12 ml chloroform and 6 ml 0·5N-methanolic NaOH/g lipid. After stirring and shaking for 60 minutes at room temperature, the extract is neutralized with acetic acid to pH6 (approximately 6 ml 0·5N-acetic acid). Water is then added to make two phases and the lower phase is retained for the preparation of cholesterol and sphingolipids. The chloroform-rich lower phase is taken to dryness and dissolved in hot glacial acetic acid to the point of saturation. The solubility of cholesterol in acetic acid at 25° is 0·62 g/100 ml. Upon cooling cholesterol crystallizes out and the sphingolipids remain in the liquor. The crystallization should be carried out quickly to avoid acetylation of cholesterol (Fieser, 1953).

Neutral lipids excluding cholesterol

Neutral lipids can be effectively separated on a silicic acid column with increasing concentrations of ether in petroleum ether (Table 5.2). As they are present in such small amounts in normal brain, cholesterol esters and di- and triglycerides from nervous tissue are most effectively purified by preparative thin-layer chromatography (Shenstone, 1971). Plates are spread with a 1 cm thick layer of adsorbent and the lipid is applied as a long continuous strip, leaving space on either edge to spot standards. After development in hexane–ether (85:15, v/v) or other suitable solvent systems (Skipski & Barclay, 1969), the centre of the plate is covered and the standards visualized. The lipid areas of the bulk sample can be marked out with reasonable accuracy and the lipids eluted from the silica gel. Alternatively layers containing phosphorescent material will indicate the positions of compounds with conjugated double-bond systems (Shenstone, 1971).

Cholesterol ester is assayed by the method of Hanel & Dam (1955), given in detail on page 72. Purity can be checked by alkaline hydrolysis of the ester and precipitation with digitonin of the resultant cholesterol. In the author's laboratory di- and triglycerides were assayed for acyl ester content by the method of Shapiro (1953).

Acyl ester determination

The method for the estimation of fatty acyl esters is based on their ability to react with hydroxylamine quantitatively in an alcoholic alkaline solution. The ester is converted stoichiometrically to a hydroxamic acid, which forms a colour complex with ferric chloride (Hestrin, 1949; Shapiro, 1953). This method uses as reagents: (1) 2M-hydroxylamine hydrochloride in 50% (v/v) ethanol; (2) 3·5 N-NaOH; (3) 3·3 N-HCl; (4) 0·37 M-$FeCl_3$ in 0·1 N-HCl.

Procedure. Dry samples containing not more than 6 μmol of acyl ester are dissolved in 1·5 ml of ethanol–diethyl ether (3:2, v/v). To the samples add 0·5 ml of the hydroxylamine hydrochloride and 0·5 ml of NaOH. The tubes are well mixed and allowed to stand for 20 min at room temperature. Then add 0·5 ml of the HCl to each tube mix well and finally add 0·5 ml of the $FeCl_3$ and again mix well. The absorbance of the samples is read at 520 nm in a Unicam S.P. 600 spectrophotometer. A calibration curve is obtained using triolein as the standard. The molecular extinction coefficient for the method is $8·7 \times 10^2$.

Glycerol determination

In order to check the purity of the di- and triglycerides, the ratio of acyl ester groups to glycerol can be estimated. The glycerol method of

Dittmer & Wells (1969) is based on the determination of formaldehyde produced by oxidation of glycerol with periodate. The formaldehyde is reacted with chromotropic acid in a strong sulphuric acid solution and the violet-coloured product is determined spectrophotometrically.

The reagents used are: (1) standard 0·25 mM-glycerol or glycerol phosphate; (2) 4 N-HCl; (3) 2 N-HCl; (4) 0·1 M-sodium periodate, stored refrigerated in an amber bottle; (5) 10% w/v sodium bisulphite; (6) 0·5g chromotropic acid, disodium salt (disodium 4,5-dihydroxy-2,7-naphthalene disulphonate) in 50 ml of water; add 200 ml of 12·5 N-H_2SO_4, preparing freshly just before use; (7) 10% w/v thiourea.

Procedure

Hydrolysis. Transfer lipid samples with up to 0·5 μmol of glycerol to tubes with constrictions, and remove the solvent in a vacuum oven. Add 2 ml of 2N-HCl to each tube. Prepare standards containing 1 ml of glycerol and 1 ml of 4N-HCl, and blanks containing 2 ml of 2N-HCl. Seal the tubes by heating and collapsing the glass at the constriction and heat the tubes in an oven at 100–105° for 96 hours.

Periodate oxidation. Open the cooled sealed tubes and add 0·5 ml of 0·1M-$NaIO_4$ to each tube. Mix and after 5 min add 0·5 ml of 10% w/v sodium bisulphite to stop the reaction.

Formaldehyde determination. Transfer 1 ml samples of the oxidized hydrolysate to clean test-tubes and add 5 ml of chromotropic acid reagent. Mix and heat in a boiling water bath for 30 min. Cool to room temperature and add 1 ml of 10% thiourea. Mix and read the absorption at 570 nm.

Cerebroside and sulphatide

Cerebroside is both a sphingolipid and a galactolipid (*N*-acyl, *O*-galactosyl sphingosine) and is greatly enriched in neural tissues; 2·4% of the whole brain wet weight is cerebroside. The two predominant fatty acids of cerebroside and sulphatide are C-24, mono-unsaturated and hydroxy-saturated acids (Svennerholm & Ställberg-Stenhagen 1968). The ratio of hydroxy- to normal fatty acids increases slightly during myelination and in adult human brain the ratio for cerebroside and sulphatide is 2 and 0·7 respectively. The hexose constituent is always galactose in normal brain, but glucosyl ceramide accumulates in the central nervous system under certain pathological conditions (McIlwain & Bachelard, 1971). Sulphatides have a sulphate group present on the 3 position of the galactose and in normal brain the ratio of cerebroside to sulphatide is approximately 3·7 to 1.

Preparation of cerebroside and sulphatide. Cerebrosides and sulphatides are completely separated from neutral and phospholipids on Florosil columns (Table 5.1). The combined galactolipids can be completely separated on a Unisil column. The loading factor is 30 mg galactolipid/g. Unisil and the first eluting solvent is chloroform-methanol (94:6, v/v).

Fractions of 5 ml/g Unisil are collected and approximately the first nine fractions will contain only cerebroside. Sulphatide begins to appear in following fractions and when the last traces of cerebroside are eluted the solvent is changed to chloroform–methanol (85:15, v/v) and sulphatide is then eluted. Fractions should be monitored by t.l.c. to check for purity of the lipids and to check for contamination by minor galactolipids. The hexose content of the purified lipids can be assayed by the anthrone method (p. 74). Sulphatide which is anionic, forms a complex stoichiometrically with the cationic dye Azure A. This method for determining sulphatide has a lower limit of 0·002 μmol (Kean, 1968).

Minor galactolipids

Small amounts of ceramide monoglycosyl and diglycosyl diglycerides, cerebroside esters and ceramide polyhexosides are present in the brain (Rouser et al., 1972). These minor galactolipids can be separated and purified by a combination of column and thin-layer chromatography. A preliminary separation on Unisil separates ceramide, which is eluted with neutral lipids from the rest of the galactolipids, which are eluted with acetone. Ceramide can be subsequently separated from neutral lipids by t.l.c. in the developing solvent, chloroform–methanol (9:1, v/v). The remaining galactolipids are separated on a DEAE cellulose column as follows: (1) monglycosyl diglyceride is eluted with chloroform–methanol (98:2, v/v); (2) diglycosyl diglyceride and cerebroside are eluted with chloroform–methanol (9:1, v/v); (3) ceramide polyhexosides are eluted with chloroform–methanol (7:3, v/v); and (4) sulphatide is eluted with chloroform–methanol–ammonia–salt. Diglycosyl diglyceride can be separated from cerebroside on t.l.c. in the developing solvent, chloroform–methanol–water (24:7:1, by vol.) and the ceramide polyhexosides are separated by t.l.c. on Silica Gel H in chloroform–methanol–water (65:25:5, by vol.).

Ethanolamine phospholipids

The two major categories of phospholipids in neural tissues are lecithins and ethanolamine phospholipids. In white matter all the ethanolamine phospholipid is in the vinyl ether plasmalogen form (Webster, 1960), whereas in grey matter it accounts for less than 50%. In ethanolamine phospholipid from human grey matter, stearic acid accounts for 26% of the total fatty acid content and polyunsaturated acids account for 53% of the total (O'Brien & Sampson, 1965). Oleic acid is the major C-18 fatty acid in white matter ethanolamine phospholipid and only 28% of the total fatty acids are polyunsaturated (Alling et al. 1971).

Preparation. Ethanolamine phospholipids can be purified on a silicic acid column or on a DEAE-acetate cellulose column (Tables 5.1 and 5.2). If the former system is used any interfering glycolipids can be removed by eluting the column first with acetone. Acidic phospholipids are eluted first with chloroform–methanol (9:1, v/v) followed by elution of ethanolamine phospholipids. At this stage it is necessary to cut smaller fractions as phosphatidyl inositol and serine overlap with the ethanolamine phospholipid elution. Better recovery is obtained if a mixed ethanolamine and serine phospholipid fraction from a Unisil column is rechromatographed on a silicic acid–silicate column (Rouser, O'Brien & Heller, 1961). The column is prepared by passing chloroform–methanol (4:1, v/v) containing 1% concentrated aqueous ammonia through the Unisil bed. When the lipid mixture is applied, ethanolamine phospholipid is eluted with chloroform–methanol (4:1, v/v), and serine phospholipid with methanol. As alkaline pH helps to protect plasmalogens from breaking down, this method is particularly advantageous for preparing ethanolamine plasmalogen from white matter.

Serine phospholipids

Phosphatidyl serine is concentrated in the brain in comparison to extra-neural tissues and is all in the diacyl form of the phospholipid. The C-18 fatty acids stearic and oleic together account for 46% and 75% respectively of the human grey and white matter total fatty acid content; the polyunsaturated acid C-22:6 is present in grey matter phosphatidyl serine as 37% of the total fatty acid content (O'Brien & Sampson, 1965). This acidic lipid is strongly bound, along with phosphoinositides and sulphatide, to brain proteins. White matter proteolipid protein, prepared by emulsion centrifugation, contains 30% phospholipid of which half is phosphatidyl serine (Folch, 1963; Uda & Nakazawa, 1973). Braun & Radin (1969) have shown the *in vitro* combination of water-soluble, delipidated proteolipid protein with phosphatidyl serine and other anionic lipids to form insoluble complexes. The encephalitogenic basic protein also binds phosphatidyl serine and the other acidic phospholipids *in vitro* (Palmer & Dawson, 1969; London & Vossenberg, 1973).

Preparation. Phosphatidyl serine can be prepared on a silicic acid (Unisil column) or on a silicic acid–silicate column but complete separation from phosphatidyl inositol complicates the purification. The two phospholipids are separated on a DEAE–acetate cellulose column (Table 5.1); phosphatidyl serine is eluted with glacial acetic acid and phosphatidyl inositol remains on the column.

Phosphatidyl serine and ethanolamine are assayed for their lipid amine group by the ninhydrin method.

Lipid amine determination

The ninhydrin procedure of Moore & Stein (1948) was modified by Lea & Rhodes (1954) for application to the estimation of phosphatidyl ethanolamine and serine, without preliminary hydrolysis of the phospholipid. The method uses as reagents: (1) redistilled methyl cellosolve;

(2) 0·2 M-citrate buffer pH5; (3) ninhydrin reagent: 0·2 g of ninhydrin dissolved in 5 ml of methyl cellosolve and 8 mg of $SnCl_2$ in 5 ml of the citrate buffer, these two solutions being made up freshly and mixed just before use.

For the determination, dried lipid samples containing between 1 and 5 μg of nitrogen are dissolved in 1 ml methyl cellosolve. To each sample is added 1 ml of ninhydrin reagent. The sample tubes are fitted with reflux condensers and placed in a boiling water bath for 20 min. When cool each sample is made up to 7 ml with methyl cellosolve and the absorbance read at 570 nm. Serine and ethanolamine solutions can be used to obtain a calibration curve and the molecular extinction coefficient under these conditions is 8×10^3.

Inositol phosphatides

Three groups of phosphoinositides are present in brain. Monophosphoinositide is a phosphate diester of L-myoinositol and a diglyceride, and has a wide tissue distribution. The di- and triphosphoinositides have additional phosphate groups at the 3 and 4 positions of the inositol and apart from the brain, the only tissue that contains an appreciable amount of polyphosphoinositide is the kidney (Dawson & Eichberg, 1965a). Only the monophosphoinositide is readily extracted from brain tissue by chloroform–methanol mixtures; a mineral acid, such as hydrochloric acid, must be added to the solvent to extract the polyphosphoinositides, which are present in the brain as calcium and magnesium salts.

The molar ratio of triphosphoinositide to diphosphoinositide ranges from 3 to 6 in different species and triphosphoinositide is found in fractions rich in myelin (Dawson & Eichberg, 1965b). Metabolic changes in these compounds are very rapid in the brain and there is extensive disappearance of polyphosphoinositides within minutes of death. Thus brain tissue should be frozen *in situ* and immediately extracted with the chloroform–methanol solvent. The fatty acid pattern of all three phosphoinositides is similar—the predominant unsaturated fatty acid is arachidonic and 35% of the total fatty acid content is accounted for by stearic and 20% by oleic acid. The phosphoinositides like other anionic lipids form strong complexes with brain proteins *in vitro* (Palmer & Dawson 1969).

Preparation. The phosphoinositides can be purified on DEAE–cellulose if they are first converted to sodium salts (Hendrickson & Ballou, 1964). In the case of the polyphosphoinositides, the neutralized, lyophilized lipid is suspended in 2% disodium EDTA, adjusted to pH 7 with N-NaOH, dialyzed against water and washed through Chelex-100 resin (sodium form). The suspension is freeze-dried and 1 g of lipid in chloroform–methanol–water (20:9:1, by vol.) is applied to a DEAE–cellulose column, $3·6 \times 40$ cm. A linear gradient elution pattern with the above-mentioned solvent mixture in the mixing chamber and 0·6 M-ammonium acetate in the solvent resevoir separates the lipids to successive fractions as follows: the first fraction A, contains the calcium and magnesium salts of triphosphoinositide; fraction B contains phosphatidylinositol and phosphatidylserine; fraction C contains phosphatidylserine alone, fraction D, diphosphoinositide and fraction E, triphosphoinositide.

Analytical methods for free inositol have been based on microbiological assay with yeast strains that have an absolute requirement for inositol and on chemical reaction with periodate. Gas chromatographic methods for the analysis of both inositol and glycerol are described in detail by Roberts (1967).

Inositol determination

The method is based on periodate oxidation of inositol (Agranoff, Bradley & Brady, 1958) after hydrolysis of the phospholipid (Dittmer & Wells, 1969). For the hydrolysis are required 2N-HCl and standard 0·5mM-inositol. Dry samples and standards containing 0·05 to 0·5 μmol of inositol in test tubes are prepared as described for the glycerol assay. The samples are hydrolyzed in 2 ml of 2N-HCl in sealed tubes as for the glycerol assay. The tubes are cooled and opened, then the water is removed in a stream of air in a boiling water bath. For both assays it is important to remove completely the HCl; to ensure this, the samples are redissolved in 0·5 ml of water and dried either in a stream of air or in a desiccator over NaOH for 12 h.

Procedure. The periodate method requires as reagents: (1) 0·3N-barium hydroxide; (2) zinc sulphate, a 5% w/v solution; (3) Amberlite IRA-400 in the hydroxyl form; (4) M-Na or K acetate, pH 7·4; (5) 0·1M-sodium metaperiodate.

The dried sample is dissolved in 1 ml of water and 1·0 ml of 0·3N-Ba(OH)$_2$ is added. Heat the tubes at 100° for 15 min. Cool them and add 1 ml of the zinc sulphate. Centrifuge the mixture for 10 min. Remove 2 ml aliquots of the supernatant and add 1 g IRA-400 resin. Allow to stand 5 min. and transfer 1 ml aliquots to screw-capped tubes that are wrapped in aluminum foil. If carbohydrates are absent from the sample, dissolve the dried hydrolyzate in 3 ml of water and remove 1 ml aliquots and proceed as described below. Add 1 ml of 1M-acetate buffer and 0·3 ml of 10mM-periodate. Mix the contents well and allow to stand at room temperature for 30 min. During this time any glycerol present is completely oxidized, but the inositol is unaffected. Read the absorbance of each tube at 260 nm, then return the samples to their tubes and heat at 65° for 2 h. Cool the tubes to room temperature and read again at 260 nm. The decrease in absorbance is a measure of inositol content. It is necessary to run a standard curve with each determination since the iodate produced during oxidation has a small absorption at 260 nm.

Lecithins

In quantitative terms the predominant phospholipid in extra-neural tissues is phosphatidyl choline or lecithin. It is an integral constituent of lipoproteins, red blood-cell membranes, and of the electron-transport chain in mitochondrial membranes. Lecithin is also a major lipid constituent in brain but the levels are not as high as those of ethanolamine phospholipids, particularly in the myelin sheath (Cuzner et al., 1965). In dedifferentiated developing myelin, lecithin levels exceed those of ethanolamine phospholipid, but as the myelin compacts and matures, this ratio is reversed (Cuzner & Davison, 1968). The predominant fatty acids of both human grey and white matter lecithin are palmitic and oleic acids, and the proportion of long-chain polyunsaturated fatty acids is low (O'Brien & Sampson, 1965).

Preparation. Lecithin, lysolecithin and sphingomyelin are eluted together from an alumina column. Neutral lipid is removed by elution with chloroform–methanol (98:2, v/v), followed by elution of choline phospholipids with chloroform-methanol (1:1, v/v). The combined lipid fraction is reduced in volume, precipitated with acetone, dissolved in chloroform-methanol and loaded onto a silicic acid (Unisil) column (Tables 5.1 and 5.2). Lecithin is eluted first with chloroform–methanol (7:3, v/v), followed by sphingomyelin with chloroform–methanol (1:9, v/v).

Choline determination

In the method described here, the choline formed on hydrolysis is precipitated from solution as the reineckate, which is then measured spectrophotometrically (Dittmer & Wells, 1969). The reagents used are: (1) barium hydroxide, saturated aqueous solution; (2) ethanol, 95% by vol.; (3) standard choline, 0·5mM-choline iodide; (4) thymolphthalein, 1% solution in ethanol; (5) acetic acid, glacial; (6) ammonium reineckate, 5% solution in methanol; (7) n-propanol saturated at 0° with choline reineckate; (8) acetone, reagent grade.

Procedure. Duplicate samples and standards containing 0·1–0·5 μmol are taken to dryness in screw-capped tubes; ethanol, 0·1 ml, is added, warming if necessary to dissolve the sample. Saturated Ba(OH)$_2$, 1ml, is added, the tubes closed and heated in a boiling-water bath for 12–16 h. The samples are cooled, and with 1 drop of thymolphthalein added are neutralized by adding glacial acetic acid until 1 drop just eliminates the blue colour. The contents of the tubes are filtered through small pieces of Whatman No. 1 filter paper into glass-stoppered centrifuge tubes. The tube and the filter paper are washed with three 0·5 ml portions of water. To each of the combined filtrates 1 ml of 5% ammonium reineckate in methanol is added. The contents of the tubes are mixed well and stored in an ice bath for at least 4 h. The samples are spun at 0° in a refrigerated centrifuge. The supernatant solutions are removed and to the pellets are added 0·5 ml of ice-cold n-propanol saturated with choline reineckate, followed by good mixing and centrifuging. The washing is repeated twice. Finally, the pellet is dissolved in 5 ml of acetone, the tubes capped, centrifuged to remove insoluble material, and allowed to come to room temperature while protected from light. The absorbance at 327 nm is then measured.

Sphingomyelin

Sphingomyelin has a ubiquitous distribution in mammalian tissues and represents approximately 10% of whole brain total phospholipid. Like all sphingolipids it is resistant to alkaline hydrolysis and this property can be put to advantage in the preparation of the lipid. Of the sphingolipids only sphingomyelins differ according to their derivation from white or grey matter: C18 fatty acids preponderate in grey and mono-unsaturated C24–26 acids in white matter. The hydroxy acids present in cerebrosides and sulphatides are absent (O'Brien & Rouser 1964).

Preparation. If a total lipid extract is subjected to mild alkaline hydrolysis, the resulting material contains cholesterol, cerebroside and sulphatide, sphingomyelin, and the alkyl ether and cyclic acetal derivatives of glycerophosphoryl ethanolamine. Sphingomyelin can then be purified on a single alumina column (Table 5.1). Cholesterol is eluted with chloroform–methanol (98:2, v/v) followed by sphingomyelin, which is eluted with chloroform–methanol (1:1, v/v). Cerebroside and sulphatide can also be eluted and further fractionated on a Unisil column. The derivatives of ethanolamine phospholipid are eluted from the alumina column with chloroform–ethanol–water (2:5:2, by vol.) Alternatively, sphingomyelin can be separated from lecithin and lysolecithin on a silicic acid column, following a preliminary group separation on alumina.

Minor phospholipids

There are only trace amounts of the three lysophospholipid derivatives of ethanolamine, choline and serine, in normal neural tissue, and these may arise by breakdown of the respective diacyl or alkenyl–acyl compounds. The preparation of lyso compounds on a large scale is best carried out by enzymatic hydrolysis of the purified diacyl compound with phospholipase A1 or A2 followed by chromatography on a Unisil column (Ansell & Hawthorne, 1964). Preparation and purification of these compounds from brain can be accomplished using preparative t.l.c. (see page 91).

Basic plates, 1mm thick, are prepared by mixing 40 g Silica Gel H, without binder, with 100 ml of $0 \cdot 1\text{M-Na}_2\text{CO}_3$. Up to 35 mg of total lipid extract can be spotted on a 20 × 20 cm plate. For preparative work, glass plates, 40 × 20 cm, are often used, thus doubling the lipid load. When the plates are developed in the solvent, chloroform–methanol–acetic acid–water (25:15:4:2, by vol.), lysophosphatidyl serine just moves off the origin, lysolecithin is clearly separated, and lysophosphatidyl ethanolamine or lysoethanolamine plasmalogen run with sphingomyelin. The latter two can then be separated on a small alumina column (see page 89).

Phosphatidic acid and cardiolipin

The acidic phospholipids, phosphatidic acid and cardiolipin are both present in small amounts in the brain. In guinea-pig brain cardiolipin represents 1·4% of total lipid phosphorus and is enriched in the mitochondrial fraction. Most of the phosphatidic acid, which totals 0·9% of lipid phosphorus, is found in the myelin fractions (Eichberg, Whittaker & Dawson, 1964). As with lysophospholipids, phosphatidic acid can be prepared by treating purified lecithin with phospholipase D, followed by chromatography on silicic acid (Ansell & Hawthorne, 1964). Likewise, the small amounts of phosphatidic acid and cardiolipin present in the brain are clearly separated on preparative thin-layer plates (Skipski & Barclay, 1969).

The lipids are separated on basic plates in two developing solvents: firstly, in pyridine–petroleum ether (3:1, v/v), which is permitted to run to the top of the chromatograms. After drying in a vacuum oven at room temperature for 30 min, the plates are developed in chloroform–methanol–pyridine–$2\text{M-NH}_4\text{OH}$ (35:12:6·5:1, by vol.) to within 2 cm of the first solvent front. Phosphatidic acid and cardiolipin are clearly separarated from each other and from the galactolipids. A prior group separation of a total brain-lipid extract on a Unisil column would concentrate the acidic phospholipid fraction, and allow for a clear separation by thin-layer chromatography.

Gangliosides

Gangliosides are a series of glycosphingolipids containing sialic acid and were isolated from beef brain in 1942 by Klenk. Chromatography on silicic acid and cellulose has led to the subdivision of the group into mono-, di- and trisialogangliosides (Svennerholm, 1963a, 1964). For detailed structures of individual gangliosides the reader is referred to Gurr & James (1971). The hexose directly linked to ceramide is glucose, indicating a divergence from the cerebroside biosynthetic pathway. The fatty acid pattern also differs from those of cerebrosides and sulphatides in that 80–90% of the fatty acids are stearic (Svennerholm & Ställberg-Stenhagen, 1968).

Gangliosides comprise 6% of the lipids in grey matter 0·6% in white matter and are primarily located in axons and dendrites but not in the neuronal perikarya. The axolemma-free ox axon preparation (Table 5.3) has a low ganglioside content 0·03%, in contrast to a value of 0·53% for unmyelinated cerebellar axons of the cat, which were isolated with their ganglioside-rich membrane intact (Dekirmerjian et al., 1969). Synaptosomes and microsomes are enriched in gangliosides (Whittaker, 1969). The pattern of individual gangliosides is uniform throughout the subcellular fractions, except in myelin where the monosialoganglioside, G_{MI}, is predominant (Suzuki, Poduslo & Norton, 1967).

Table 5·3 Ganglioside content of brain

Human[b]		Rat[c]		Ox[d]	
Part	Content[a]	Part	Content[a]	Part	Content[a]
Grey matter	0·565	Whole brain	0·370	—	—
White matter	0·074	Neurons	0·069	Oligo dendroglia	0·074
Myelin	0·019	Myelin	0·060	Axons	0·033

[a]Values are expressed as mg. total N-acetyl neuraminic acid (NANA) per mg. dry weight. For further details see Eichberg et al. (1969).
[b]Suzuki (1965).
[c]Poduslo & Norton (1972)
[d]De Vries & Norton (1974)

Preparation. For preparative purposes gangliosides are extracted from acetone-dried grey matter with chloroform–methanol (1:2, v/v). The acetone–extracted tissue is first treated with chloroform–methanol (2:1, v/v) to remove all other lipids, followed by extraction with chloroform–methanol (1:2, v/v) to obtain a crude ganglioside preparation (Svennerholm, 1963a). The mono- and disialogangliosides can then be separated on silicic acid columns, eluting with chloroform–methanol (3:2, v/v) and (1:2, v/v) respectively. The loading factor is 1 g total gangliosides/100 g silicic acid with 1 litre eluting volumes. Svennerholm (1963b) also separated individual monosialogangliosides on a silicic acid column with the same solvent systems, collecting 15–25 ml fractions. The disialogangliosides were fractionated on a carefully monitored paper-roll column, eluted with n-propanol–H_2O (7:3, v/v and 6:4, v/v). Similar ganglioside fractions were prepared on silicic acid columns, using a gradient elution technique (Penick, Meisler & McCluer, 1966).

REFERENCES

Adams, C. W. M. & Davison, A. N. (1959) *J. Neurochem.* **4**, 282.
Agranoff, B., Bradley, R. N. & Brady, R. O. (1958) *J. biol. Chem.* **233**, 1077.
Alling, C., Vanier, M. & Svennerholm, L. (1971) *Brain Res.* **35**, 325.
Ansell, G. B. & Hawthorne, J. N. (1964) *Phospholipids. Chemistry, Metabolism and Function,* p. 91. Amsterdam: Elsevier.

Braun, P. E. & Radin, N. S. (1969) *Biochemistry* **11,** 4310.
Cuzner, M. L. & Davison, A. N. (1968) *Biochem. J.* **106,** 29.
Cuzner, M. L., Davison, A. N. & Gregson, N. A. (1965) *J. Neurochem.* **12,** 469.
Davison, A. N. & Peters, A. (1970) *Myelination,* p. 101. Springfield, Illinois: Charles C. Thomas.
Davison, A. N. & Wajda, M. (1959) *J. Neurochem.* **4,** 353.
Dawson, R. M. C. & Eichberg, J. (1965a) *Biochem. J.* **96,** 634.
Dawson, R. M. C. & Eichberg, J. (1965b) *Biochem. J.* **96,** 644.
Dekirmenjian, H., Brunngraber, E., Lemkey-Johnston, N. & Larramendi, J. M. H. (1969) *Exp. Brain Res.* **8,** 97.
Derry, D. M. & Wolfe, L. S. (1967) *Science, N.Y.* **158,** 1450.
De Vries, G. H. & Norton, W. T. (1974) *J. Neurochem.* **22,** 259.
Dittmer, J. C. & Wells, M. A. (1969) *Meth. Enzym.* **14,** 482.
Eichberg, J. Hauser, G. & Karnovsky, M. L. (1969) in *The Structure and Function of Nervous Tissues* (Bourne, G. H., ed.) vol. 3, p.185. New York: Academic Press.
Eichberg, J., Whittaker, V. P. & Dawson, R. M. C. (1964) *Biochem. J.* **92,** 91.
Fieser, L. F. (1953) *J. Am. chem. Soc.* **75,** 4395.
Folch, J. P. (1963) In *Brain Lipids and Lipoproteins and the Leucodystrophies,* ed. Folch & Bauer, p. 18. Amsterdam: Elsevier.
Gurr, M. I. & James, A. T. (1971) *Lipid Biochemistry.* London: Chapman & Hall.
Hanahan, D. J. (1960) *Lipid Chemistry,* p. 35. New York: Wiley.
Hanel, H. K. & Dam, H. (1955) *Acta chim. scand.* **9,** 677.
Hendrickson, H. S. & Ballou, C. E. (1964) *J. biol. Chem.* **239,** 1369.
Hestrin, S. (1949) *J. biol. Chem.* **180,** 249.
Hirsch, J. & Ahrens, E. H. (1958) *J. biol. Chem.* **233,** 311.
Kean, E. L. (1968) *J. Lipid Res.* **9,** 319.
Lea, C. H. & Rhodes, D. N. (1954) *Biochem. J.* **56,** 613.
Lea, C. H., Rhodes, D. N. & Stoll, R. D. (1955) *Biochem. J.* **60,** 353.
London, Y. & Vossenberg, F. G. A. (1973) *Biochim. biophys. Acta* **307,** 478.
McIlwain, H. & Bachelard, H. S. (1971) *Biochemistry and the Central Nervous System,* p. 346. Edinburgh & London: Churchill Livingstone.
Moore, S. & Stein, W. H. (1948) *J. biol. Chem.* **176,** 367.
O'Brien, J. S. & Rouser, G. (1964) *J. Lipid Res.* **5,** 339.
O'Brien, J. S. & Sampson, E. L. (1965) *J. Lipid Res.* **6,** 537.
Palmer, F. B. & Dawson, R. M. C. (1969) *Biochem. J.* **111,** 637.
Penick, R. J., Meisler, M. H. & McCluer, R. H. (1966) *Biochim. biophys. Acta* **116,** 279.
Poduslo, S. E. & Norton, W. T. (1972) *J. Neurochem.* **19,** 727.
Radin, N. S. (1969) *Meth. Enzym.* **14,** 245.
Radin, N. S., Brown, J. R. & Lavin, F. B. (1956) *J. biol. Chem.* **219,** 977.
Ramsay, R. B. (1973) *Biochem. Soc. Trans.* **1,** 341.
Roberts, R. N. (1967) In *Lipid Chromatographic Analysis,* ed. Marinetti, Vol. 1, p. 447. New York: Marcel Dekker.
Rouser, G., O'Brien, J. S. & Heller, D. J. (1961) *J. Am. Oil Chem. Soc.* **38,** 14.
Rouser, G., Kritchevsky, G., Yamamoto, A., Simon, G., Galli, C. & Bauman, A. J. (1969) *Meth. Enzym.* **14,** 272.
Rouser, G., Kritchevsky, G., Yamamoto, A. & Baxter, C. F. (1972) *Adv. Lipid Res.* **10,** 261.
Shapiro, B. (1953) *Biochem. J.* **53,** 663.
Shenstone, F. S. (1971) In *Biochemistry and Methodology of Lipids,* ed. Johnson & Davenport, p. 171. New York: Wiley-Interscience.
Skipski, V. P. & Barclay, M. (1969) *Meth. Enzym* **14,** 530.
Suzuki, K. (1965) *J. Neurochem.* **12,** 969.

Suzuki, K., Poduslo, S. E. & Norton, W. T. (1967) *Biochim. biophys. Acta* **144**, 375.
Svennerholm, L. (1963a) *Acta chim. scand.* **17**, 239.
Svennerholm, L. (1963b) *J. Neurochem.* **10**, 613.
Svennerholm, L. (1964) *J. Lipid Res.* **5**, 145.
Svennerholm, L. & Ställberg-Stenhagen, S. (1968) *J. Lipid Res.* **9**, 215.
Uda, Y. & Nakazawa, Y. (1973) *J. Biochem.* **73**, 755.
Webster, G. R. (1960) *Biochim. biophys. Acta* **44**, 109.
Whittaker, V. P. (1969) In *Handbook of Neurochemistry*, Vol. 2, p. 327, ed. Lajtha. New York: Plenum Press.
Wren, J. J. (1960) *J. Chromatogr.* **4**, 173.

6. Preparing Neural Tissues for Metabolic Study in Isolation

H. McILWAIN

Slicing and handling tissues in aqueous fluids	106
Methods available	107
Recommended method	107
Tissue handled with minimum aqueous fluid	116
Methods available	117
Recommended method	119
Preparation of tissue by chopping	121
Circumstances when chopping is advantageous	122
Apparatus	124
Procedure	125
Dispersing chopped tissue in fluid	126
Analysis	127
Cell suspensions from neural tissues	128
Earlier methods	128
Neuronal perikarya and glial clumps, by dissection or collection	129
Tissue disintegration for bulk separation of material originating from different cell-types	130
References	131

In metabolic studies with isolated neural tissues, the experimental material is less complex than an intact organ, but still has considerable structural complexity; it does not have its normal blood supply, and materials normally exchanged with the blood are exchanged with a fluid which surrounds or flows over the tissue. Essential materials must thus diffuse from outside the tissue sample, and this inevitably limits the size of the sample which can be satisfactorily examined as an isolated tissue. In practice, oxygen is the most critical metabolite whose diffusion limits tissue thickness; as calculated by Warburg (see Umbreit et al., 1964), even during incubation in 90–100% O_2, a tissue of respiratory rate 150 μmol O_2/g.h should be not more than about 0·4 mm in thickness for adequate oxygenation at all parts of its interior. This requirement applies whether or not respiratory measurements are being made; it is indeed more important when the measurements are not made. The limitation in thickness implies that in preparing many parts of the mammalian nervous system for metabolic studies, two distinct categories of processes are involved: (1) removal from the body and preliminary dissection; and (2) the further dissection, slicing or chopping to yield the necessary thin sections.

(1) Removal from the body and dissection to yield the chosen tissue are described in Chapters 1 and 12. Particular aspects which should be

emphasized in connection with metabolic work, are as follows. The dissection should cause as little damage as possible to the tissue. Crushing or 'smearing' it, and handling or dragging instruments across its surface are to be avoided. So also should contact with any material other than a chosen isotonic solution. As little time as possible should elapse between stopping blood supply to the tissue, and placing the final cut specimens at 37° under chosen experimental conditions. In this intervening time the tissue is best kept or manipulated at 0–15°. Freezing would damage its structure. When room temperature is above 15°, a cold room, cold box, or other means should be used to keep the tissue within 0–15° during preparation.

(2) Certain tissues require no more than their removal from the body, as the retina, which is already a sheet 0·15 to 0·3 mm in thickness. Others however require cutting to give sections about 0·3 or 0·4 mm in thickness; note that, as is described in Chapter 12, functionally intact electrophysiological systems of much complexity can be retained within a section 0·3 mm thick when it is taken from a selected part of the brain or a rat or guinea pig. The greater part of the present account concerns the processes of slicing and chopping which are involved in preparing such sections.

SLICING AND HANDLING TISSUES IN AQUEOUS FLUIDS

This is the technique most widely used in preparing tissues for metabolic studies. Reasons for its general adoption are: (A) the tissue *in vivo* is already bathed by aqueous fluids; (B) the tissue after its preparation is commonly placed in aqueous fluids for metabolic studies; and (C) tissues are most easily handled while they are in or moistened with aqueous fluids. This is true from a purely manipulative point of view and to an extent which can be appreciated only by trying to handle tissues with and without aqueous fluids. Such fluids (i) lubricate the tissue, aiding the motion of a cutting blade. (ii) They help to separate two pieces of tissue, so that removal of a slice from a block of tissue is facilitated; they also help to separate the tissue and the implements with which it has been held or cut. (iii) Tissues float freely in aqueous fluids; tissue density is usually only a little greater than that of the fluid, and the tissue is thus supported.

On the other hand, these practical advantages and conveniences make it all the more necessary to emphasize that the use of aqueous fluids constitutes one way only of preparing a tissue for metabolic study and that any adequate investigation of a tissue involves comparison of the results of preparing the tissue in this way, with results obtained by other procedures. In particular, it is often advantageous to cut in the absence of added aqueous fluids (see below); lubrication then depends on pre-existing cerebrospinal fluid which remains available provided that manipulation is sufficiently rapid.

Methods available

Most methods employ the thin cutting blades made for safety razors, or available as a stage in the manufacture of the razor blades. Razors themselves, or the razor-like blades of microtomes used in histological work, are only occasionally employed.

(i) The cutting blade may with practice and judgement be used entirely freehand, but then yields slices of limited size, which are thinner at their edges than at their centre (Fig. 6.1, a). This is true even when, as was often the case, the blade was caused to follow the surface of the piece of tissue in the fashion in which a knife is used in peeling an apple. Until about 1935 or 1940, tissue slices were usually prepared in this way.

(ii) A larger flat surface may be formed on a block of tissue, such as that obtained from the cerebral hemispheres, by pressing lightly on the block with a piece of glass. The blade may then be moved so as to cut the tissue below the glass, while observing its progress from above (Fig. 6.1, b).

(iii) It is however much preferable to control the thickness of the slice by ridges on the edge of the glass slip, so that this becomes a template or guide (Fig. 6.1, c; Fig. 6.2). This is the basis of the first recommended method which is described in detail below.

(iv) The Stadie-Riggs cutter. This keeps the blade at a fixed distance from the tissue, as in method (iii); but it also fixes the blade in a position parallel to the tissue which is being cut and so increases the rubbing and shearing to which it is subjected (see Fig. 6.2). The apparatus is also relatively cumbersome and the slice cannot easily be floated from it into fluid. For these reasons it is not recommended for relatively fragile neural tissues, as cerebral cortex. An adjustable microtome which cuts by means of a fine nylon thread has been described by Franck (1970).

(v) Fixed-blade cutters. These also are more elaborate and cumbersome than the blade and guide which is recommended. With the fixed blade cutters, the blade is mounted on a support and the tissue is moved by hand across and along the blade. Slice thickness is conditioned by the position of the cutting edge of the blade, which is just above a table from which the tissue slides. In the instrument of Majno & Bunker (1958) the progress of cutting cannot be seen and the instrument must be inverted to obtain the slice. A model which obviates these difficulties has been made (McIlwain, unpublished), but the blade and guide of method (iii) remain the method of choice.

Recommended method

APPARATUS

Removing tissue samples from an animal, their slicing, weighing, and the commencing of a metabolic experiment with the samples usually all occupy only 3–30 min. Slices are cut at the same bench as is used for the

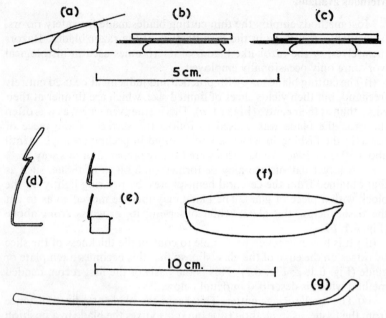

Fig. 6.1 Above, slicing cerebral cortex: (*a*) freehand; (*b*) below a piece of glass; and (*c*) with blade and guide. Below, apparatus used in handling the sliced tissue; (*d*) rider for torsion balance shaped to take a slice of 20–150 mg fresh weight; (*e*) the rider in use for draining adhering medium from a slice; (*f*) shallow dish for floating and trimming slice; (*g*) spatula.

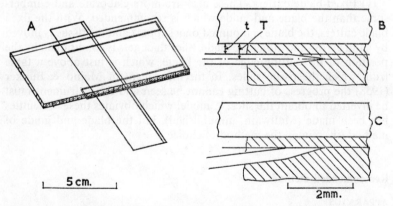

Fig. 6.2 *A:* The blade and glass guide recommended for tissue-slicing. *B:* Relationship between tissue and blade when cut by a blade held in the same plane as a guide; the tissue is compressed from thickness T to t. *C:* As *B*, with an angle between blade and guide.

metabolic experiment, with the apparatus for the metabolic experiment already prepared; and at an adjacent bench the tissue is removed from the animal. Materials and apparatus needed specifically for slicing and for trimming, weighing and transferring the slices, are as follows.

Blade and guide. The blades are strips of steel 0·26 mm thick, 12 cm long and 1·8 cm broad, with one long edge sharpened for cutting. They are supplied by Gillette Surgical, Great West Road, Isleworth, Middlesex, protected by grease or a silicone preparation and wrapped individually in treated paper. For use, one is unwrapped, washed with chloroform on a cloth or filter paper, and then with ethanol and warm tap water and dried with filter paper. It should then be wettable with water, and will usually serve in some 30 experiments over a period of a month before it is discarded. Immediately after cutting slices, the blade should be wiped dry with filter paper and replaced in its paper wrapping, or 'painted' with medicinal liquid paraffin which requires removal before the blade is reused.

The guide is of glass, about 7·5 × 3·5 × 0·2 cm and prepared as described in Note 2 below.

Cutting table (Fig. 6.3). This provides a small steady support, a few inches above the working bench, on which specimens can be placed while they are being dissected and sliced. The base is of brass, iron or steel, weighing 1·5–2 kg and is painted, varnished or stainless. To its top is fixed a sheet of hard plastic or glass, about 6 mm thick and 7·5 cm square. The plastic may be attached by screws which do not reach the upper surface, or it or the glass may be cemented to the base. After use it is promptly cleaned, washed with water and dried with filter paper.

Balance, dishes, spatulae. A torsion balance of capacity 250 mg and accuracy of 0·5 mg is usually the most suitable. The dial of the balance should be about 30 to 50 cm above the level of the bench and if the stand of the balance itself does not bring this about, the balance should be put on a small block or box; it must be level.

The tissue is hooked over a small wire rider while it is picked from fluid, drained, weighed, and transferred to an experimental vessel. The rider illustrated in Fig. 6.1, *d* is designed for this purpose. Made from aluminium or light alloy wire, it weighs about 50 mg and hooks on the arm of the torsion balance with the tissue hanging free.

Small shallow porcelain dishes of the shape illustrated in Fig. 6.1 are most suitable for receiving the tissue after cutting. It is preferable that they be glazed black inside; this facilitates the manipulation of most neural tissues, for the thin slices involved are whitish or translucent. Two dishes 7–8 cm in diameter and 1 cm deep, with about 15 ml of fluid in each, are used in an experiment in which two workers prepare 6 or 8 slices of 50–150 mg from a guinea pig neocortex.

While floating in fluid in the dish, slices are trimmed to size, or cut in order to reject an undesired portion, by pressing them against the base of the dish with a scalpel or with a spatula which has a fairly sharp edge.

The thin spatula of Fig. 6.1 is useful for this purpose, and also for manipulating the floating slice into special apparatus; a broad spatula with a chisel-like edge may also be used for cutting the slice. The cutting edge of the spatula or scalpel should be slightly curved, as indicated in Fig. 6.1, and should be used with a rocking motion in cutting; this is because the bottom of the dish is rarely entirely flat.

Salines for metabolic work are described in the following chapter. Circles of hardened filter paper (Whatman no. 50, 'extremely tough when wet'; 5·5 cm diameter) are used on the cutting table and scraps or 10 cm squares of ordinary filter paper (Whatman no. 1) are used for wiping apparatus.

PROCEDURE

1. *The apparatus* just described is assembled, and all preparations are made for the subsequent metabolic experiment in which the tissue is to be used. The fluid medium is prepared and just before slicing, the shallow dishes are filled with it to a depth of 3–5 mm. A circle of hardened filter paper is held at one point at its edge and quickly dipped almost completely in the medium, excess fluid flicked from the paper by a jerk of the wrist, and the paper put on the cutting table. At this point and throughout the cutting the paper should be completely flat and just wet enough to stick firmly to the table. If it is wet but will slide along the cutting table on being pushed, then the paper is too wet. Excess fluid can be removed from the hardened paper at any stage by pressing onto it a pad of filter paper scraps. A filter paper moistened in the same way is placed in a Petri dish which is to receive the organ to be sliced.

2. *Cutting*. The following description is based on obtaining a few slices from each cerebral hemisphere of a rat or guinea pig. For obtaining maximal yield of cerebral slices, for large slices, for subcortical white matter, and for other parts of the brain and nervous system, Chapter 12 should be consulted.

The animal is stunned, the brain removed (Chapter 1) and a hemisphere placed on its outer convexity on the cutting table. The subcortical structures which are uppermost are cut off in a horizontal plane with the narrow spatula (Fig. 6.1, *g*), placing a finger on the opposite side of the tissue while this is done so that the spatula cuts bluntly towards it. The hemisphere is rolled over so that its outer convexity is uppermost, and brought on the hardened filter paper to the position on the cutting table shown in Fig. 6.3. The glass guide is taken between the left thumb and forefinger, the thumb above and about at the centre of the 'handle' end and pointing along the length of the guide. The first joint of the forefinger is below the guide and at right angles to the thumb. The guide is dipped in the fluid medium, excess fluid flicked from it by a jerk of the wrist, and the guide rested horizontally with pressure only a little more than its own weight, on the surface of the hemisphere. The blade is picked up with the right hand, the thumb again above but with its axis

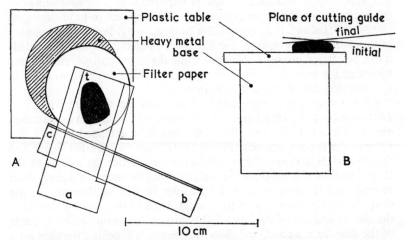

Fig. 6.3 Slicing tissue with blade and guide. *A:* In plan; the tissue is placed on a circle of moist filter paper on the cutting table. The guide is held between the left thumb (above) and forefinger at *a*, and the blade at *b* between the right thumb (above) and fingers. When the blade is used in sweeping out air bubbles its corner *c* is initially placed at *t*. *B:* Elevation of cutting table and tissue, as seen from the left hand side of *A* and showing the change in plane of the guide during cutting.

at about 45° to that of the blade, and the forefinger below at right angles to the thumb. About 5 cm of the distant end of the blade is dipped in the fluid, the excess flicked off, and the blade brought under the guide and almost parallel to it, as shown in Fig. 6.3. The cutting edge of the blade touches the ridge at each side of the guide; the plane of the blade is at a few degrees to that of the guide (Fig. 6.2, *C*).

The blade is now advanced into the tissue with an ordinary cutting action, moving it from side to side across the tissue while it is going in. The side to side motion is typically of 2–5 mm, at 1–2 cuts/s, with forward motion of 2 to 5 mm per cut. Negligible resistance should be felt to this motion; if it is, the blade or guide are not properly positioned or they or the tissue are too dry. The motion of the blade is stopped when it ceases to cut the tissue. The slice which has been cut will already be attached mainly to either the blade or the guide, because of the angle which separates them. This angle is increased and the blade and guide removed from the block of tissue. The tissue is washed from the blade or guide by holding whichever carries it in the dish of fluid medium, and giving the blade or guide a few sharp jerks, when the slice should float free. Excess fluid is then flicked from the blade and guide, and a second slice cut from the block of tissue; a third may follow, and a fourth is often obtainable from the cerebral hemisphere of the guinea pig before white matter is reached.

3. *Weighing*. When tissue weight is required it may be determined at the beginning or end of an experiment, and by direct or indirect means (see Note 3, below). If determination by direct weighing at the beginning of an experiment is desired, the following procedure is recommended; it can be carried out by a colleague while the investigator is cutting and trimming the successive slices, and then introduces little or no delay in commencing the subsequent experiment.

After a slice is cut, any undesired portions are removed while it is floating in the dish of fluid medium, using a scalpel or spatula as described above. While the trimmed slice is floating fully extended in fluid, the rider of the torsion balance (Fig. 6.1*d*) is taken in the right finger and thumb and its horizontal arm put in the fluid under the middle of the slice. The rider is lifted from the fluid carrying the slice as a double sheet looped over the arm of the rider; further folding or clumping of the slice should be avoided as it traps fluid. Held in this way on the rider, the slice is drained of fluid first by holding its lower tip against the side of the dish for a second, and then by repeatedly touching the slice on a clean glass surface. A clock glass 20 cm in diameter is suitable for draining all 6 to 8 slices of a typical experiment. The touching should be repeated about 8 to 15 times in some 10 s; first the tip of the hanging slice is brought to the glass, then the slice laid on one side of the glass while still being retained, looped, on the rider; then picked up, the tip brought to an adjacent position on the glass, and the slice laid on its other side (Fig. 6.1, *d* and *e*).

When the slice leaves on the glass negligible fluid and only a very slight imprint which dries almost immediately, draining is complete and the slice and rider are weighed together on the torsion balance. Still handled by the rider, the slice is put below the fluid of an experimental vessel and freed from the rider by a few side to side shakes, which should be continued until the slice opens. The next slice is picked up with the same rider, drained and weighed.

4. There is an additional manoeuvre during slicing which increases the area of the slice cut, and minimizes damage to the tissue by friction from the blade. This is illustrated in Fig. 6.3*B* and consists of altering the angle between the guide and the cutting table as the blade advances, while keeping the same angle between blade and guide. Initially, the 'handle' end of the guide is lowest, and it is raised as the cutting proceeds; the result is similar to following the surface of an apple in peeling it with a knife. In tissue slicing, the operation is valuable not only because the block of tissue is convex, but also because the blade and guide are most effectively used with an angle of a few degrees between them (Fig. 6.2).

5. When cutting a nearly flat (and not convex) piece of tissue, it will be found valuable to sweep out air bubbles between tissue and guide, before slicing. This is because any air bubbles will cause holes in the slices cut. Bubbles are removed by first applying the blade and guide to

the tissue in a way different from that described above. The guide is first brought down on the tip of the blade as shown at t in Fig. 6.3. Both are then lowered on the tissue so that the tip of the blade touches the part of the tissue distant from the person cutting; he then draws the blade towards himself while lowering the 'handle' end of the guide. When the blade has been drawn past the tissue its cutting edge is placed against both of the lateral ridges of the guide, and cutting begun as described above.

NOTE 1: TEACHING THE RECOMMENDED PROCEDURE

Many beginners will adopt an effective method almost immediately, but if it must be taught the following sequence is suggested. (i) Practise placing the blade and guide, wet with saline, in the relative positions described (see text and Figs. 6.2 and 6.3). The blade should first be taken in the right hand and held at one end in a horizontal plane. The guide is held at its 'handle' end and its other end rested, flat, on the distant end of the blade. The slight additional upward pressure needed to bend the blade a little, should be noted; it is equivalent to an additional weight of about 5 g. The guide itself weighs 8–9 g and these forces are all that are required between blade and guide. The non-cutting edge of the blade, which is toward the person cutting, is now lowered about 0·5 mm from the guide. This leaves the cutting edge in contact with the guide, as indicated in Fig. 6.2C. Contact of the blade with both of the lateral ridges of the guide should be confirmed by rocking the guide on the blade, and the guide left in the position in which it is in contact with both ridges. Comparable 'feeling' of the correct relative position of blade and guide is later done, almost unconsciously, before or during the cutting of each slice, and is then not so much an actual rocking as a momentary shifting of the slight downward pressure on the guide, from one side of it to another so that any lack of contact is detected and rectified.

(ii) The guide should then be rested horizontally on a small pad of cotton wool wet with saline. While the guide is resting there, the blade should be brought below the guide and by the stages already described, manoeuvred into the correct position for cutting (but not cut). In a trial with 0·1 g of absorbent cotton wool wet with 1·2 ml of fluid, excessive pressure was made evident by the fluid being squeezed out.

(iii) The operations should be carried out with a piece of cerebral tissue. A rabbit brain will provide six trial blocks. It is of some, but only limited, value to practise first with other tissues. Liver or kidney yield satisfactory slices when the guide is applied with pressures which would damage cerebral tissues.

NOTE 2: PREPARATION OF CUTTING GUIDE; TISSUE THICKNESS

The guide is made from glass microscope slides and coverslips of standard sizes and good quality. The slides, 76×38 mm ($3 \times 1·5$ in) and 1–1·2 mm thick, with their edges ground smooth, and also no. 0 to no. 3 coverslips, 64 × 44 mm (2·5 × 1·75 in) are put one by one into detergent cleaning fluid and after some hours are washed well and dried in an oven at 105°. The coverslips are placed on a flat clean surface (opalite glass) and cut to strips 5–6 mm × 64 mm using a steel rule and a diamond. Between each cutting the opalite and the rule must be wiped free of fine glass slivers with a soft tissue.

A Canada balsam–xylene solution, approx. 2:1 (by vol.), is prepared and with a glass rod four small spots of the solution are placed at each edge of a slide in the positions to be occupied by the strips of coverslip (Fig. 6.2A). A strip is taken with forceps and lowered on to each set of balsam spots. These should flow to a complete film between slide and strip, with a little excess fluid appearing at the edges. If necessary, further spots of balsam–xylene are placed on the affixed strips, and further strips placed above them. The spots of fluid are transferred with a rod already drained from balsam and are not added as drops. The strips of coverslip are aligned with each other and with the edge of the slide, and the slide carrying them put horizontally in an oven at 105° overnight. Excess balsam is then wiped off with a tissue moist with xylene. If the balsam has retracted and left a gap under a strip of coverslip, the gap is filled with xylene–balsam solution. The slides are left for a further day at 105°. Measurement has shown the thickness of their ridges to be stable for a year or more after this treatment.

The guides are conveniently made in batches of 6 to 12, and are then calibrated with an engineer's micrometer gauge. Measurements are made to 0·005 mm: of the thickness of the slide at its centre, and of the thickness of the slide plus coverslips at three points on each ridge. The difference between measurements at the centre and at the ridge gives the thickness of the ridge and is the main factor conditioning the thickness of slices cut with the guide. The six measurements of thickness of ridge should agree to within 0·01–0·02 mm in a satisfactory guide. No. 0 coverslips are about 0·10 mm in thickness, and no. 3 about 0·25 mm.

The thickness of slices cut with the guide and blade can be judged by Fig. 6.2, B and C, to be conditioned by the depth of the recess in the guide and by the thickness of the blade. However, it is less than the sum of the depth and half the thickness of the blade by an amount which depends on the angle between the blade and guide. This, and the pressure exerted by the guide on the tissue, are individually-conditioned matters which should be standardized as far as possible by an investigator, who should then measure the thickness of the slices which he cuts. The following measurements were made with slices of guinea pig cerebral cortex, cut with blades of thickness 0·26 mm in the fashion recommended and using two different guides. Immediately after cutting the slices were weighed on a torsion balance to the nearest milligram; they were then floated in a petri dish of isotonic saline above paper ruled in millimetre squares. The data showed that with one guide, slices were 0·32 mm and with the other, 0·34 mm in thickness. By keeping the blade parallel to the guide, slices were about 0·06 mm thicker. The bevelled part of the blade carrying the cutting edge is approximately 1·2 mm long, and the angle between blade and guide in cutting by the recommended method is therefore about 0·05 radians, or 3°.

For cutting tissues without added fluid, guides are used with an inserted coverslip to support the tissue (see below, Fig. 6.4). It is then necessary to use a guide with a deeper recess. Thus guides with a recess of 0·38–0·4 mm have been used, with a loosely inserted no. 0 coverslip, in preparing tissue slices 0·33–0·35 mm in thickness. A convenient size for the loose, inserted coverslip is 2×3 cm; a no. 0 coverslip of this size weighs 150–170 mg.

NOTE 3: TISSUE WEIGHT

Tissue weight may be determined at many stages during the handling of tissue for metabolic experiments. The best standards of reference are the fresh weight or dry weight of tissue immediately on its removal from an animal, before contact with any extraneous fluid. When tissue is weighed at other stages, the relationship of the weights so obtained to the weight under one of these reference conditions should be determined, as is indicated below.

(i) *Initial wet weight*. This term is used, in distinction to fresh weight, to indicate the weight of the tissue measured as described above, after it has been in contact with aqueous fluids but has been adequately drained from them. The aqueous fluids are introduced to facilitate the handling of the tissue; it is best that the slice be drained from them before transfer to the chosen experimental conditions and consequently it is common practice to weigh the tissue at this stage. The weights so obtained are remarkably reproducible; it is a good exercise to drain the tissue on a glass surface as described above, to return the tissue to the dish of saline, and then to redrain and reweigh it. The weight of an 80 mg slice should be reproducible to within one milligram.

It is especially valuable to weigh at this stage if the tissue is subsequently to be analysed for labile constituents, for this may demand rapid transfer to fixing agents, and preclude weighing at the later stage. The dry weight of cerebral cortical tissue as a fraction of its initial fresh weight after preparation at 14–16°, has been found to be 0·14–0·15. A block of such tissue rapidly removed from an animal has a dry weight about 0·19 of its fresh weight, and the change to 0·14–0·15 is due mainly to the tissue imbibing fluid from the medium in which it has been floated (Chapter 2). Not all this fluid is intracellular; when tissue elements, for example dendrites, take up fluid and become more nearly circular in cross-section, interstices between them necessarily increase in volume and contribute to total fluid uptake. Quantitative comparisons between fresh and wet weights are given by Varon & McIlwain (1961).

(ii) *The fresh weight* corresponding to the initial moist weight can be calculated from the ratios quoted, determined by an observer under his own conditions of manipulation. It is also quite feasible to prepare tissue slices without contact with fluid, as is indicated below, and they can then be weighed directly. Also, when tissue is to be chopped (see below) the block of tissue chosen for chopping is weighed directly before contact with fluid. It should be noted that slices or blocks of tissue which have been prepared quickly in the absence of fluid, if placed in contact with fluid even briefly will take up an amount of fluid similar to that taken up by tissue sliced in fluid. When dissection is unduly long, cerebrospinal fluid can be absorbed by surrounding tissue.

(iii) *The final wet weight* of the tissue at the end of an experiment, drained as in (i), is not recommended as a measure of the quantity of tissue. This is because further fluid absorption frequently accompanies an experiment, but to a degree which varies with the metabolic conditions to which the tissue has been exposed (Varon & McIlwain, 1961, analysed these different processes). Thus two slices in an experiment which tested the metabolic effects of different substrates might well have swollen to different degrees. Also, the tissue is usually more fragile at the end of an experiment than at the beginning, and draining is less satisfactory. The final wet weight is however important to the study of tissue fluids.

(iv) *The final dry weight* of tissue is also not recommended as a measure of tissue quantity, though it has been widely employed. Tissue slices usually suffer loss of substance to the surrounding fluid during metabolic experiments, and this again occurs to an extent which varies with metabolic experimental circumstances. In some experimental conditions slices may also absorb measurable quantities of media constituents.

(v) *Indirect methods.* In some circumstances it is desired to place freshly sliced tissue rapidly in fluid without weighing, and to fix the tissue rapidly for analysis at the end of an experiment. As an indirect measure of tissue weight under these conditions, protein content may be used. This is not entirely satisfactory as some protein is lost to the fluid during incubation; however, this can be determined or computed, and analysis for protein is compatible with analysis for many labile constituents. In some cases, slices at the end of an experiment were transferred to ice-cold trichloroacetic acid in homogenizer tubes as recommended for determining labile phosphates, centrifuged, and after taking the supernatant for the desired determinations, the precipitates were used for protein estimation by the Folin-Ciocalteu method. By using the same method to determine the protein content of other slices which had been weighed at the commencement of the experiments, the initial wet weight or the fresh weight of the first slices could be determined from their protein content.

TISSUE HANDLED WITH MINIMAL AQUEOUS FLUID

The thin sections of tissues used for metabolic work are usually most easily handled in aqueous media, and such fluids form their normal route of material exchange with their environment. Consequently, when other techniques are employed it is usually for a particular purpose or in specific instances, and examples of this are now quoted.

(i) To obtain the true fresh weight of a tissue sample, or to minimize exchange of tissue constituents and media constituents at the beginning of an experiment, tissue may be prepared and weighed without fluid. This can be advantageous even if, subsequently, the tissue is placed in aqueous media.

(ii) Tissue prepared in absence of added fluid, may, further, be incubated with little or no aqueous media. The added volume, for instance, may be less than twice that of the tissue itself rather than the 50–300 times which is more generally employed. Here it should be noted that the volume of blood and of extracellular fluid associated with a tissue at any one time, is *in vivo* no more than about 0·4 of its bulk. The smaller volumes may be used *in vitro* in order to minimize the loss of materials from the tissue which usually occurs when tissues are placed in the larger bulk of aqueous fluid. Such loss can be considerable, despite the existence of processes of assimilation to the tissue. Thus of the 7–8 μmol of creatine/g of cerebral cortex which is found as the free, non-phosphorylated, compound *in vivo*, the greater part is lost to salines and about 2·5 μmol/g remains when the external creatine is about 0·1 μmol/ml.

(iii) The smaller fluid volume also allows certain products of tissue metabolism to reach stable values relatively rapidly; and in some cases the rates of formation so obtained correspond more closely to those normal to the tissue *in vivo*, than do those observed in the presence of excess fluid. This is the case with lactic acid in the cerebral cortex (Rodnight & McIlwain, 1954). On the other hand, however, the smaller fluid volume is likely to bring problems of buffering and avoidance of the abnormal accumulation of metabolites. These matters receive further attention in the subsequent description of superfusion methods (Chapter 7).

(iv) The smaller fluid volumes can be markedly advantageous when electrical observation of the tissue is in progress. Such indeed constitutes the orthodox fashion of handling peripheral nerve from lower organisms; and with tissues of greater metabolic requirement, as mammalian ganglia (q.v.), arrangements have been made for dipping the specimen in nutrient fluids at intervals, but draining from fluids when electrical measurements are being made. It should be noted, however, that the need for minimal fluid in such experiments is related to the type of electrode used, and that with micropipette electrodes, insulated except at a minute tip, much fluid can be employed (see Chapter 9).

Methods available

Bow cutter and coverslip support. Tissues may be cut with a specially prepared narrow blade using the recessed glass guide already described. This is usually the method of choice and is given in detail below.

In addition, the guide is furnished with a coverslip which supports the tissue slice during its transfer and weighing (Fig. 6.4).

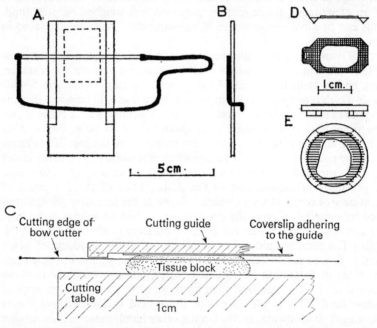

Fig. 6.4 *A-C:* Tissue preparation using the bow cutter and coverslip support. (*A*) plan and (*B*) end elevation, showing their position when used in slicing; the interrupted line shows the position of the attached coverslip. *C:* Vertical section passing through the cutting edge of the blade when it is in the position shown in *A*. *D, E:* Supports for slices used in metabolic experiments in which the tissues are not immersed in aqueous fluids (see text and Fig. 7.1).

Chopping. If the addition of a small volume of fluid about 10% of the tissue volume is permissible, the tissues may be chopped. This procedure is described in a subsequent section.

A major difficulty in cutting tissue in the absence of fluid lies in the extent to which the tissue adheres to apparatus with which it comes into contact. The extent to which aqueous fluids of orthodox techniques lubricate the tissue and give gentle support, is made very evident when attempts are made to dispense with them. A tissue slice inherently presents a large area during manipulation; it is therefore of advantage to reduce the area of instruments which must be moved across it. Distorting and shearing forces are so minimized. In ordinary cutting procedures the glass guide rests on the full area of the part of the block of tissue which is to become the slice; the guide however need not move in relation to the tissue. The blade must so move, and it is thus of advantage to reduce its area by making it narrower; but the blades which are most desirable in being narrowest, are found to be too flexible to yield dependable slices. Increased thickness (rather than breadth) of blade could give increased rigidity, but thicker blades are undesirable because they tend to crush either the slice cut or the uncut portion of tissue which remains.

For these reasons, blades which are of the usual thickness but which are narrow, have been mounted in a holder which gives them the necessary rigidity. The type of mounting which has been chosen is that of the carpenter's bow-saw or the fret-saw, in which a flexible blade is kept rigid under tension as is the string of a bow. This gives the bow-cutter (Fig. 6.4).

Recommended method

APPARATUS

Bow cutter. The narrow blades are obtained as follows, from the strips of razor blading described above. (i) The strip is scored with a diamond about 1·2–1·5 mm from its cutting edge, and gripped at this line between steel bars in a vice or in a bending machine so that the cutting edge is protected. The projecting portion of the blade is then bent, and a proportion of blades break at the scored line, leaving between the bars the narrow blade desired. (ii) Alternatively, the blade, held between bars in a vice as described, is cut by abrasion. The edge of a copper sheet is fed with emery powder and used to grind through the blade.

The frame of the bow cutter is made from silver-steel wire, bent to a shape which will become that of Fig. 6.4 when compressed. The wire and the blade are tinned with solder at their ends, and the frame fixed between nails driven to a block of wood, so that its ends are compressed. The blade and frame are brought together and soldered. When released from the nails, the completed cutter should hold the blade under tension equivalent to a weight of about 500 g; the blade should be quite flat and straight.

Immediately after the cutter has been used its blade should be washed or wiped free from tissue juice or debris, and dried. If it is not to be used again for some days, the blade should be lightly oiled or greased.

Guide and coverslip support during cutting. It is an important feature of the present method that it protects the tissue slice during preparation and initial manipulation, by providing a rigid support. The part of the tissue sample which is to yield the slice does not make contact with the guide, but with an already-weighed coverslip loosely attached to it (Fig. 6.4*A*). A set of these coverslips is prepared ahead of the main experiment, and weighed. They may be 2 × 3 cm rectangles of no. 0 coverslip glass, and will then weigh about 160 mg each. The cut tissue is weighed while still attached to the coverslip, and then usually washed free from the coverslip with whatever fluid is being used in the subsequent part of the experiment.

Supports for slices during metabolism. In most cases these are not needed, and slices are suspended in aqueous fluids so that oxygen reaches them from all sides. They then show little or no tendency to adhere to wet glass surfaces. In absence of aqueous fluids slices stick to surfaces, and if it is desired that they be adequately oxygenated they must either be thinner than usual, or must be kept extended on a support which allows access of oxygen to their under side as well as to their upper surface. In some earlier work, slices were placed on moist filter paper or sintered-glass surfaces; they may also be left, as cut, on the coverslip support described above. For adequate oxygenation, slices in such situations should be only one-half the thickness of those prepared for normal use. The permissible thickness for cerebral cortex would then be about 0·15–0·2 mm. Cerebral tissues sliced to this thickness are extremely difficult to transfer in absence of aqueous fluid.

Supports allowing oxygenation are made from either metal or nylon gauze, and are described in Chapter 7 and shown in Fig. 6.4, *D* and *E*. In addition, certain of the tissue-holding electrodes described in Chapter 8 are suitable, whether or not the tissues are to receive electrical pulses during the experiments.

Other apparatus. The following have already been described: cutting table, torsion balance, and hardened filter paper; the recessed glass cutting guide is prepared as above and in addition is permanently coated with a water-repellant silicone film.

PROCEDURE

Preparations are first made for a subsequent experiment in which the tissue is to be used. The supports for handling the tissue are weighed.

The block of tissue to be cut is placed on the cutting table in the position shown in Figs. 6.3 and 6.4. Moistened filter paper may be used between the tissue and the table to keep the tissue in position, as the paper does not transfer fluid to the part of the tissue which is being cut. It is preferable that the upper surface of the tissue be slightly convex, and not concave.

A weighed coverslip is attached to the recessed portion of a glass guide, in the position shown in Fig. 6.4, by a microlitre quantity of water. This is provided by a microlitre pipette, by the tip of a fine glass rod, or by breathing on the guide; a slight film of condensed water-vapour and light pressure with a finger-nail is sufficient to make the attachment. The glass guide with its coverslip is then placed above the tissue in the position shown in Fig. 6.3, contact being first made between the guide and the part of the tissue distant from the operator, the handle end of the guide being raised. This end is then gently lowered; a combination of slight longitudinal force and downward pressure about equal to the weight of the guide eliminates air pockets. The bow cutter is then introduced round the guide at its handle end; the blade is below the guide and the bow above.

The slice is cut with the normal combination of forward and side to side motion. By raising the handle end of the guide during cutting, the convexity of the tissue is followed; this also encourages the slice to separate from the tissue block from which it is being cut, and to adhere to the coverslip attached to the guide. While held on the guide, the tissue is inspected and any necessary trimming performed with a scalpel. The coverslip carrying the tissue is then separated from the guide by pushing the coverslip along until it projects from the guide, when it is taken up by a torsion-balance rider for weighing.

The weighed tissue may for some purposes be studied while still attached to the coverslip (see Chapter 7). For most metabolic studies, however, it is washed from the coverslip to already-incubating oxygenated fluid by a jet of the fluid from a Pasteur pipette. The incubating fluid may be already in the vessel in which metabolic observations will be made. Alternatively, the tissue may be washed from the supporting coverslip into a preincubation bath of warm, oxygenated fluid, as is shown in Fig. 7.2.

Tissues prepared in this fashion have been used in experiments described in Chapters 7 to 9, and by McIlwain & Snyder (1970) and Pull & McIlwain (1972).

Other methods. No methods of preparing tissue slices for metabolic work have been as well evaluated and used satisfactorily in so many different circumstances as those just described, which cut by metal blades. It is valuable, however, to have alternative methods; among these is the use of a nylon thread (Frank, 1970). Methods which involve tissue disintegration followed by reassembly of chosen particles on a filter are described, for example as 'synaptosome-beds' in Chapter 10. Tissues may also be cut to slices, as well as being if necessary further comminuted, by the chopping now to be described.

PREPARATION OF TISSUE BY CHOPPING

The small tissue sections needed for metabolic work may be prepared not only by cutting but also by chopping. The distinction between the two processes lies in the motion of a cutting tool in relation to the tissue specimen. In cutting proper, the motion of the cutting blade can be resolved into movement towards the specimen, and movement across it. In chopping the movement of the blade is in one direction only, that is into the specimen (Fig. 6.5). A mechanical chopper (Fig. 6.6) has been built to carry out rapidly a series of chops at predetermined distances apart. This is done by moving the specimens between successive blows of the chopping blade, and while the blade is out of the tissue (Fig. 6.5, C). This yields a series of slices from a block of tissue: for example a

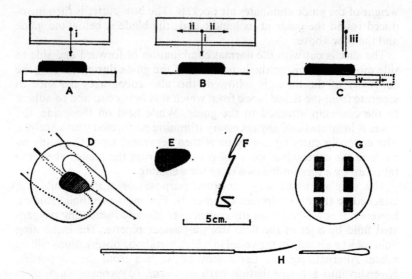

Fig. 6.5 Chopping fresh tissue for metabolic studies. *A:* Motion of the blade in chopping (i), contrasted with that in cutting: *B* (i) and (ii). *C:* Motions of the mechanical chopper: (iii) of the blade into the specimen and (iv) of the specimen, horizontal and intermittent.

D: Plan of the piece of tissue seen in elevation in *C*, showing the result of one series of chops between which the tissue was moved in direction (iv) yielding slices; *E:* The tissue after a second series of chops, also vertical but at right angles to the first series. *D* shows also the use of a spatula in transferring the chopped specimen. *F:* Rider of torsion balance for weighing small specimens before chopping. *G:* Arrangement of six small specimens for chopping together, mechanically. *H:* Mounted bristle or loop of nylon fibre for dispersing chopped specimen.

piece 3 mm thick and 10×10 mm in area, lying on its large face, would yield 30 slices $10 \times 3 \times 0\cdot33$ mm in size.

Moreover, the mechanical chopper makes it easy to subdivide the tissue still further, though this is necessary for certain purposes only. Thus if the specimen just described is chopped again in a direction at right angles to the first, it is converted to 900 square prisms (Fig. 6.5, *E*), $3 \times 0\cdot33 \times 0\cdot33$ mm in size.

Circumstances when chopping is advantageous

Chopping is an alternative to slicing which is sometimes advantageous and sometimes not. The following notes will assist the choice of method.

(i) The mechanical chopper was devised in order to prepare for metabolic experiments, small and irregularly shaped specimens of animal tissues. Small or irregular pieces are often the only specimens available from anatomically defined areas of the neural system of laboratory animals. They may also be all that is available from human biopsy. In both cases it can be important to use experimentally the whole of the specimen available, and not to leave a residual 'unsliceable' portion adhering to a filter paper on a cutting table, as is usual when tissue is sliced by the methods previously described.

(ii) These small specimens can be chopped completely and regularly to either slices or prisms, and equivalent samples distributed among several experimental vessels. The chopped specimens remains a cell-containing tissue, behaving metabolically in ways akin to the larger slices produced by the techniques already described, and radically differently from ground tissues. Nevertheless, finely chopped tissue can be handled in suspension in order to distribute samples among several vessels of a subsequent experiment.

(iii) Relatively uniform samples can be obtained also from portions of tissue which are known to be heterogeneous in structure. Thus the cerebellum or midbrain of the rat can be chopped and distributed among several experimental vessels. This has been applied also to the whole brain of newly-born guinea pigs or rats.

(iv) The tissue to be studied can be prepared as a block without addition of fluids, weighed at this stage, chopped with the addition of a small amount only of aqueous fluid, and transferred completely to the vessel of a subsequent experiment. Results are thus referred immediately to the fresh weight of the tissue sample, obtained directly and without the computation necessary when tissues have been sliced in aqueous fluids. This advantage is of course shared with tissue sliced with the bow cutter in absence of added fluid, but the mechanical chopper is the more easily used and more rapid instrument.

(v) Certain parts of the nervous system are difficult to prepare as thin sheets by ordinary slicing methods and are more easily prepared by chopping. These include areas with much white matter; the brain of newly born or foetal mammals; and some areas with modified or fibrous tissue formed under certain pathological or experimental conditions.

(vi) For preparing a considerable weight of slices the mechanical chopper is less tedious than ordinary slicing. The production of some grams of finely subdivided, cell-containing tissue is also of advantage in studying exchange between tissue and fluid, especially in relation to poorly diffusing substances.

A major situation in which slicing with blade and guide is preferable to chopping is in the production of relatively large sheets of tissue, of a few square centimetres in area. Such sheets offer the advantages of a lesser cut area in proportion to tissue weight; ease of transfer in aqueous fluids; and greater versatility in the electrical studies of Chapters 8 and 9. Also, it is easier during slicing to ensure the retention of a chosen neural structure, e.g., the lateral olfactory tract within a slice of the piriform cortex, as described in Chapter 12.

Apparatus

Most apparatus is based on the mechanical tissue chopper of Fig. 6.6. Construction and performance of the prototype were described by McIlwain & Buddle (1953) and McIlwain (1961), and the instrument has been developed by Mr. H. Mickle (The Mickle Laboratory Engineering Co., Mill Works, Gomshall, Surrey). Applications in a variety of experimental circumstances are described by Forda & McIlwain (1953), Leslie & Paul (1954), McIlwain (1954, 1956, 1959), Sproull (1956), Brierley & McIlwain (1956), Bell (1958), Abraham & Chaikoff (1959), Bassham *et al.* (1959), Paul (1959), Rinaldini (1959) and Mongar & Schild (1960).

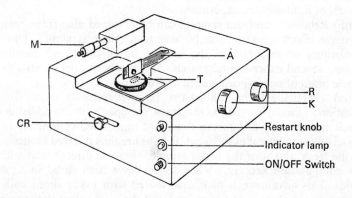

Fig. 6.6 Mechanical tissue chopper. The tissue is shown by the dotted outline on the cutting table (T). This travels mechanically from left to right (of an operator facing the instrument) between successive downward cuts of the chopping blade attached to arm (A). The table can also be moved manually after pulling out the carriage release (CR). The rheostat (knob R) controls the speed of operation. Knob K is at the end of the motor-driven shaft carrying a cam which raises arm A. With the motor switched off, knob K can be turned clockwise manually, and it then raises arm A clear of the specimen. The micrometer (M) adjusts the thickness of the slices cut. If the table is allowed to run to the full extent of its travel the motor is automatically switched off; the table must then be repositioned and the restart knob pressed, for cutting to begin. (Mr H. Mickle is thanked for contributing details of the current model of the machine. For other details, see the text and Fig. 6.5.)

The chopper operates as follows. Tissue on the table (T, Fig. 6.6) is cut by a blade clamped in the arm (A). The arm is raised by a motor whose speed is controlled by the rheostat (R). The arm and blade chop by being pulled downward into the tissue by a spring; tension on the spring is controlled by a knob at the left-hand side of the machine. The arm and blade are then raised by the motor and the carriage is caused to move to the right. The distance moved is controlled by a screw micrometer

(M), preset to give the desired slice thickness. The table then halts, and the blade is caused to make another chop at a point a little to the left of the previous cut. The cycle is repeated one or a few times each second until the specimen is completely sliced (Fig. 6.5).

Other apparatus and supplies needed when chopping tissue for metabolic experiments are: a torsion balance, the shallow dishes, spatula, and hardened filter paper of p. 110. A type of rider suitable for the torsion balance is illustrated in Fig. 6.5, F. A fine hair paint brush, with head about 3 mm in diameter and 1 cm long, is used for applying fluid to the tissue and blade. For dispersing the chopped tissue in fluid, a mounted nylon bristle or loop (Fig. 6.5, H) or a vibrator or a Pasteur pipette is used. The pipette should have its drawn-out end 1–1·2 mm in diameter and rounded at the tip, and is used with a teat or piston attachment. Ordinary safety-razor blades are used in the chopper; if the blades have been packed with grease or wax this should be removed with solvents. After chopping, if this has involved only a few hundred cuts, the blade can be washed, dried and returned to its wrapping for a subsequent experiment.

Procedure

All preparations which can be made ahead for the metabolic experiment in which the chopped tissue is to be used, are first made. A little fluid medium is then put in the shallow dish, three or four hardened filter paper circles 5·5 cm in diameter are quickly passed through the medium and excess fluid shaken off them. The filter papers are put on the table of the chopper and stroked with a spatula to remove any air locked between them. The table is brought to the left-hand side of the machine after releasing the control CR (carriage release). A blade is put loosely in position in the chopping arm and the arm brought to its lowest position, when it rests against a stop, by turning the knob K. With the arm and table in these positions the blade is allowed to rest on the pad of filter papers, along the whole length of its cutting edge. While still in this position, it is clamped by tightening the screw which grips it to the chopping arm. The machine is run for a few seconds at a few chops per second, and switched off with the blade raised. The blade should have just cut through, or just lightly scored, the top filter paper. If the top paper is cut through it is removed; otherwise it is left in position, and the papers now constitute the cutting pad on which specimens will be placed. The cutting table is now released (knob CR) and returned to the operator's left-hand side of the machine.

The specimen or specimens to be chopped are weighed and transferred with the rider of the torsion balance to the cutting pad so that they rest on their largest flat surface. A completely rounded specimen should be cut in half with a scalpel and put on the newly cut surfaces. If surfaces of the specimen differ in the proportion of blood vessels or fibrous tissue which they present, the surface with most vessels or fibres should be uppermost. On the upper surface of each specimen is placed a little fluid from the brush. The machine is switched on. The tissue should be cut smoothly and cleanly.

The specimens tend to spread a little during cutting: that is, to increase a little the area of paper occupied; but the newly formed slices remain adherent and collectively in almost the same shape as the original piece of tissue. If the tissue tends to stick to the blade and to leave the paper then (A) the filter paper pad may be too wet, and can be dried by pressing filter paper scraps against it; or (B) the specimen may be too dry, and need the application of a little more fluid to it or to the blade; or (C) the machine may have been run too fast. If several specimens are being cut together (as in G, Fig. 6.5) the blade should be watched carefully after it has passed the first group of specimens to ensure that small fragments of tissue are not being carried on it. Such fragments can result from the last cuts made by the blade when the tissue has been almost completely chopped. If such fragments are seen, the machine is stopped and the fragments are returned to the appropriate specimen with the tip of the hair brush, moist with medium. The machine is then switched on again to complete the cutting.

This yields a series of slices, and if further subdivision of the specimens is required, the chopper is first switched off with its blade out of the tissue. The carriage with its cutting table is then brought back to the left of the machine after releasing control CR. The table is rotated on its vertical axis through 90°, and chopping recommenced with the same precautions as during the first chopping. The chopping should proceed in the same fashion, and the tissue remain in much the same form as before chopping, but it now consists of a series of adherent prisms.

To transfer cut specimens of 20–200 mg, the cutting pad carrying them is held in the left hand with two or three fingers below it as shown in Fig. 6.5, D. The thin blade of the spatula, about 5–6 mm wide, is pressed against the pad behind the specimen and the pad, slightly curled over a finger, is moved away from the worker so that it leaves the specimen on the spatula. This transference can easily be quite complete. For larger specimens a larger spatula is used or the operation is repeated. The spatula carrying the specimen is placed below the fluid of an experimental vessel, avoiding contact with the sides of the vessel or with other dry surfaces. The tissue sticks to dry surfaces and when it is being moved it is easily smeared and damaged by a glancing contact with them. If accidental contact is made with the inside of a vessel, the adhering tissue can be washed into the liquid medium by liquid applied with the spatula used in transference, or with a pipette, or by tipping the vessel.

Dispersing chopped tissue in fluid

It is next necessary to suspend the chopped tissue in the fluid medium, and the method of choice depends on the size of specimen and the degree to which it has been subdivided in chopping.

A small sample chopped to a few slices, in a vessel which it is undesirable to shake strongly, can be separated by a mounted bristle or by a pipette. The spatula carrying the tissue is placed below the liquid, given a few jerks to detach the tissue, and removed. This probably will also partly separate the tissue to slices. The mounted bristle or loop (Fig. 6.5, *H*) is taken between thumb and forefinger, placed at a cut between adhering slices, and twirled. This detaches several slices and two or three repetitions should disperse the specimen. The most suitable pipette for separating the tissue is a small drawn-out tube with teat or piston, which will squirt about 1 ml of fluid through its capillary of some 0·6 mm diam. Fluid from the reaction mixture is drawn to the pipette and, keeping the tip of the pipette below the fluid, liquid is squirted at the tissue. Slices need not be completely separated if the reaction mixture is to be shaken, as in a manometric experiment, for partly separated tissue will then come apart to its slices.

Larger quantities of tissue, or tissue which has been chopped to many very small fragments, is best dispersed by mechanical movement or by a large pipette. The pipette should have a volume about 1/10 of the reaction mixture, and a smooth orifice appreciably larger than the particles. Fluid is moved in and out of the pipette while its tip is just above the tissue, so that clumps of prisms become detached from the bulk of tissue, and in the pipette or in the swirling fluid become further dispersed. Mechanical movement is often the best method of dispersing finely chopped tissue. It may be caused in a tube carrying fluid and specimen, by sharp blows with the knuckles; or the tube or other vessel may be held against the rubber-covered jaws of a mechanical flask-shaker. A Vibromix, giving small amplitude movement at electric-mains frequency, is also effective.

Analysis

The simplest procedure is to add a fixing or extracting agent, as trichloroacetic acid, to the whole reaction mixture. If there are doubts about the speed of penetration of the agent, it can be added to the mixture in a tube with a loosely fitting plunger, with which the tissue is disintegrated immediately after addition of the fixing agent.

Fluid without tissue may be sampled with small pipettes by putting their tip in the fluid or against a glass surface, away from the tissue; for larger samples a fragment of cotton wool may be put in the reaction mixture and the top of the pipette rested on the cotton while sampling. For repeated sampling, a wisp of fibres drawn from cotton can be wound round the pipette tip.

Tissue may be collected from the reaction mixture by sieving or centrifuging. Cloth, plastic, or metal gauze may be used in a funnel.

Also, for small specimens, a scoop of metal gauze mounted on a handle is useful. Silver gauze mounted in a holder of the type used for bacteriological wire loops can be used to scoop up tissue from a reaction mixture of a few millilitres, and to place it in a tube of fixing agent.

CELL SUSPENSIONS FROM NEURAL TISSUES

The term 'cell suspensions' in this connection has often represented an aim rather than an achievement. At different times in the metabolic study of neural tissues, preparations have been found valuable which are neither cell-containing organized tissues, as those already described, nor entirely cell-free, as are those of Chapter 10. There are, however, inherent difficulties in making satisfactory preparations of intermediate types. Typical neurones carry long and ramifying axons, often entangled with other cells or their processes. The cells obtained in suspension from an organ such as the brain are thus almost always damaged to varying degrees. The degree of damage and the extent to which intracellular substances and particles have been released by the different procedures, require specific evaluation by both metabolic and microscopic techniques.

Earlier methods

Grinding in isotonic media. The thorough grinding of neural tissues to obtain maximal release and separation of subcellular particles is the main subject of Chapter 10. Here are noted the effects of a lesser degree of grinding, carried out for example in a test tube homogenizer or in a blender. A test-tube homogenzier with a relatively loosely-fitting pestle, about 0·2–0·4 mm less in diameter than its tube, gives suspensions which have many of the properties of cell-containing tissues. The suspensions do in fact still contain cells or large parts of cells, damaged to varying degrees. Their metabolic properties have been described by Birmingham & Elliott (1951). For these studies whole rat brains were ground in 5–10 vol. of fluid media, which included Krebs-Ringer similar to that of Table 7.1. Isotonic phosphate, glucose and sucrose were also used. The suspensions obtained, when incubated in oxygenated glucose salines, respired at rates about 80 % of those of sliced tissues, and showed many characteristics of sliced tissue, including reciprocal relationships between respiration and glycolysis.

Crushed tissues and breis. Breis have been prepared by rubbing or crushing neural tissues as such, without, in the first place, any suspending fluid. Pigeon brain mashed with a bone spatula has been used by Peters *et al.* (1935), yielding a paste from which uniform samples could be taken and weighed. These were suspended in phosphate salines with substrates and other additions, and measurement made of respiration and substrate utilization. Suspension of the already-crushed tissue was brought about with a flat-ended glass rod. Factors affecting the properties of the suspensions included the temperature at which the brei was prepared; greater effects of added thiamine were seen when tissue was crushed at 38° than when it was prepared at lower temperatures. A brei has also been made from rat brain by placing it in a 5 ml glass syringe and squeezing it through the nozzle with the usual glass plunger (Aldridge & Cremer, 1955). Brei (1·2 g) was squeezed directly into 6 ml of 0·1 M-sodium phosphate buffer and then suspended in it by repeatedly sucking in and out of the syringe until after some six or eight times a smooth dispersion resulted. This is analogous to the preparations of Peters et al., in that the tissue is first disrupted by shearing and crushing forces without added fluid, and subsequently suspended in fluid. Aldridge & Cremer (1955) also describe a brain suspension made by grinding in a mortar cooled in ice and salt, until the resulting paste froze, after which it was suspended in fluid.

Neuronal perikarya and glial clumps, by manual dissection or collection

It is to be emphasized that the perikarya which these methods yield are likely to be a minority only of the volume of the cells from which they derive; they have inevitably suffered damage in losing cell-processes, but they retain the cell nucleus. Methods in which the dissection to separate each cell is carried out manually, are most applicable to large cells in specified parts of the brain, and may be carried out with frozen-dried tissue or with fresh tissue.

Use of *frozen-dried* tissues is exemplified by assay of enzymes in cell-bodies from the spinal cord, ganglia and brain of rabbits (Kato & Lowry, 1973). The selected samples were dissected free, frozen, cut by microtome at $-24°$ to sections 8–10 μm thick, and dried. While viewed microscopically, the chosen cell bodies were removed with a mounted needle. They were weighed with a quartz-fibre balance (typical wt, 1–10 ng) and transferred to reaction mixtures under light mineral oil.

Investigation of cell bodies dissected from *fresh* tissues is described by Giacobini (1969). More material may be obtained from the brain when the initial disintegration is by sieving. In this method (Roots & Johnstone, 1965, 1972) a chosen region, for example the lateral vestibular nucleus from ox brain, is placed with medium on a nylon sieve of diameter 5–6 cm. The sieve has apertures of 0·3–0·35 mm, and the medium may be isotonic sucrose or a Ringer type of saline. The nylon sieve rests on a polyethylene support above a Buchner flask. Pressure on the tissue with a flat-ended glass pestle and suction from the flask, are used to force the tissue through the sieve; it is collected from the other side of the sieve and suspended in the chosen medium at 4°C. This suspension is then subjected to a second process of sieving through monofilament nylon cloth with apertures of about 110 μm. After keeping this suspension at 4° for 10 min, samples are examined with a stereomicroscope at ×20 magnification. Individual identified cell bodies are collected with a nylon loop, 0·15–0·2 mm in diameter and made from 12–15 denier nylon thread mounted in a glass handle. Some 70 or more cells can be collected per min in this way, and transferred to a chosen medium and apparatus; for further details see Roots & Johnstone (1972).

Passage through sieves features also in a method for bulk separation of cell bodies from neural tissues, described below.

Tissue disintegration for bulk separation of material originating from different cell-types

These methods are distinguished from those just described, by the greater quantity of material treated or obtained in the specimens after separation.

Disintegration aided by enzymes. These methods employ initial chopping, with little or no added fluid, and subsequent incubation of the tissue while it is suspended in enzyme-containing media (McIlwain, 1954; Poduslo & Norton, 1972). The chopping may be done manually; for greater uniformity the mechanical tissue-chopper is recommended, set to cut at intervals of 0·1 mm. The selected tissue sample of 1 to 30 g was weighed, chopped and transferred to a flask together with, per g of tissue, 10 ml of freshly-prepared 'trypsinizing medium'. This contained 10 mg/ml of crystallized, salt-free trypsin (from beef pancreas) dissolved in an albumen–hexose phosphate solution, A-solution. The A solution contained 10mM-KH_2PO_4 brought to pH6 by NaOH, 5% w/v glucose, 5% w/v fructose and 1% w/v of bovine serum albumen fraction V of Cohn, supplied by Sigma Chemical Co.

The resulting mixture was then incubated at 37° with shaking for a chosen period which was usually between 30 and 90 min, depending on the material involved. Thus 30 min was most suitable for obtaining from rat brain, neurones still carrying many cell processes; but 90 min incubation gave more satisfactory preparations of astrocyte and neuronal perikarya. Incubation with the trypsin medium for 30 min was also used to obtain motor neurone perikarya from the grey matter of rat spinal cord. For cell bodies of oligodendroglia from 30 g batches of the white matter from calf brain, incubation was for 90 min with 0·1% trypsin in A-solution (Poduslo & Norton, 1972).

Separation after enzyme treatment. The trypsin treatment was halted at the chosen time by cooling the digest in ice and adding 0·2 vol of buffered calf serum (made with 9 parts of calf serum and 1 part of 10mM-KH_2PO_4–NaOH at pH 6). It was then transferred to centrifuge cups and most of the trypsin removed by depositing the tissue, centrifuging at 140g for 5 min. The supernatant was discarded and the tissue washed twice with, and then suspended in, A-solution. This suspension was drawn by suction through nylon bolting cloth of 100–150 mesh in a Buchner funnel, aiding the passage by rubbing the material on the cloth with a smooth glass surface and adding further A-solution. The resulting suspension was then passed similarly on the funnel through a stainless steel sieve of 200 mesh with apertures of 74 μm, and its final volume brought with A-solution to 20 ml/g original tissue.

The subsequent separation recommended (by Poduslo & Norton 1972, q.v.) is by density gradient centrifugation using sucrose in the A-solution. Centrifugation yielded fractions differentially enriched in myelin and cell bodies originating from different cell-types. Neuronal perikarya were obtained with little admixture of other components; glial fractions required further density gradient separation.

Other methods. For appraisal of initial sieving methods, see Cremer et al. (1968) and Poduslo & Norton (1972). Separation after homogenizing has yielded myelin fragments from the brain (Spohn & Davison, 1972) and cell-complexes derived from the glomeruli of the cerebellum (Balazs et al., 1974; Tapia et al., 1974).

Comment

The preparations of cellular material which have been described in this section are characterized by their containing cell nuclei plus some surrounding cytoplasmic constituents, within membranes believed to be mainly those of the cells of origin, though necessarily disrupted and in part reformed. More drastic mechanical disintegration is typically used to obtain subcellular entities and smaller cell-fragments, as is described in Chapter 10. The fragments so obtained include the synaptosomes, which derive from one of the cell-regions most drastically disrupted when nerve cells are separated. The synaptosomes carry part of the outer membranes of both of the cells which were formerly in synaptic relation. They thus carry some of the membrane removed from the perikarya whose preparation has just been described. The fragments also include myelin lamellae which can be prepared centrifugally as a distinct subfraction, and which represent material from a particular cell type, again difficult to obtain intact, for their biological role inherently concerns the investment of other cells.

REFERENCES

Abraham, S. & Chaikoff, I. L. (1959) *J. biol. Chem.* **234**, 2246.
Aldridge, W. N. & Cremer, J. E. (1955) *Biochem. J.* **61**, 406.
Balázs, R., Hajós, F., Johnson, A. L., Tapia, R. & Wilkin, G. (1974) *Biochem. Soc. Trans.* **3**, 682.
Bassham, J. A., Birt, L. M., Hems, R. & Loening, U. E. (1959) *Biochem. J.* **73**, 491.
Bell, J. L. (1958). *J. Neurochem.* **2**, 265.
Birmingham, M. K. & Elliott, K. A. C. (1951) *J. biol. Chem.* **189**, 73.
Brierley, J. B. & McIlwain, H. (1956) *J. Neurochem.* **1**, 109.
Cremer, J. E., Johnston, P. V., Roots, B. I. & Trevor, A. J. (1968) *J. Neurochem.* **15**, 1361.
Forda, O. & McIlwain, H. (1953). *Br. J. Pharmacol.* **8**, 225.

Franck, G. (1970) *Sur la composition ionique des tranches de cerveau de rat*. Liège: Université faculté de médicine.
Giacobini, E. (1969) *Handbook of Neurochem.* **2,** 195.
Kato, T. & Lowry, O. (1973) *J. Neurochem.* **20,** 151.
Leslie, I. & Paul, J. (1954) *J. Endocrinol.* **11,** 110.
McIlwain, H. (1954) *Proc. Univ. Otago Med. School* **32,** 17.
McIlwain, H. (1956) *Biochem. J.* **63,** 257.
McIlwain, H. (1959) *Biochem. J.* **71,** 281.
McIlwain, H. (1961) *Biochem. J.* **78,** 213.
McIlwain, H. & Buddle, H. L. (1953) *Biochem. J.* **53,** 412.
McIlwain, H. & Snyder, S. H. (1970) *J. Neurochem.* **17,** 521.
Majno, G. & Bunker, W. E. (1958) *J. Neurochem.* **2,** 11.
Mongar, J. L. & Schild, H. O. (1960) *J. Physiol.* **150,** 546.
Paul, J. (1959) *Cell and Tissue Culture*. Edinburgh: Livingstone.
Peters, R. A., Rydin, H. & Thompson, R. H. S. (1935) *Biochem. J.* **29,** 63.
Poduslo, S. E. & Norton, W. T. (1972) *Res. Meth. Neurochem.* **1,** 19.
Pull, I. & McIlwain, H. (1972) *Biochem. J.* **126,** 965; **130,** 975.
Rinaldini, L. M. (1959) *Exp. Cell Res.* **16,** 477.
Rodnight, R. & McIlwain, H. (1954) *Biochem. J.* **57,** 649.
Roots, B. I. & Johnston, P. V. (1965) *Biochem. J.* **94,** 61.
Roots, B. I. & Johnston, P. V. (1972) *Res. Meth. Neurochem.* **1,** 3.
Spohn, M. & Davison, A. N. (1972). *Res. Meth. Neurochem.* **1,** 33.
Sproull, D. H. (1956) *Biochem. J.* **62,** 372.
Tapia, R., Hajós, F., Wilkin, G., Johnson, A. L. & Balázs, R. (1974) *Brain Res.* **70,** 285.
Umbreit, W. W., Burris, R. H. & Stauffer, J. F. (1964) *Manometric Methods*. Minneapolis: Burgess.
Varon, S. & McIlwain, H. (1961) *J. Neurochem.* **8,** 262.

7. Metabolic Experiments with Neural Tissues

H. McILWAIN

I. Salines and other aqueous suspending media	133
Bicarbonate saline	134
Alternative buffers	135
Oxidizable substrates	137
Other organic constituents of fluid media	138
II. Gas supply and measurement	140
Respiratory measurements in aqueous media	140
Anaerobic experiments	141
Respiratory experiments with minimal fluid	142
Non-aqueous fluid	143
III. Incubation, sampling and analysis of tissues and metabolic mixtures	144
Quick transfer holders	145
Thermostat-bath and assembly	147
Procedure	148
Illustrative experiments; related apparatus	149
IV. Superfusion of isolated tissues	151
Superfusion method for multiple tissue samples	151
Procedure; illustrative examples	154
Notes on other perfusion and superfusion methods	156
References	156

Neural tissues can be examined by most of the methods which are applied to the metabolic study of tissues from other organs of the body. Such general biochemical methods need only a brief notice here, and workers are referred to texts or monograph series including those of Glick (1971), Lajtha (1969–1972) and Umbreit, Burris & Stauffer (1972) for many details. Most space below will be given to methods which have received specific elaboration in connection with neural tissues. We comment successively on: (1) the aqueous media normally used in metabolic studies; (2) the gases supplied and measured; (3) incubation, sampling and analysis of neural tissues; and (4) superfusion methods.

I. SALINES AND OTHER AQUEOUS SUSPENDING MEDIA

The present description is of media suitable for mammalian tissues; for data concerning other classes of animals see Altman & Dittmer (1972). With neural as with other mammalian tissues, many current investigations employ Krebs–Ringer solutions or others similar to them.

Practical details of their preparation are given by Umbreit et al. (1972) and need not be repeated here. A summarized comparison of their composition with that of cerebrospinal fluid and of blood plasma is quoted in Table 7.1, and various enriched media are listed in Table 7.2. In the notes which follow, it is assumed that glucose is supplied as oxidizable substrate unless specified otherwise, and that experimental arrangements will include in a given experiment, tissues under some chosen and relatively normal condition as well as others under the different conditions being investigated.

Bicarbonate saline

Of the media of Table 7.1, the bicarbonate saline is closest in composition to blood plasma or cerebrospinal fluid. In particular it employs the bicarbonate–CO_2 system, a major buffer of body fluids, as its main buffer. Consequently, bicarbonate saline is often the medium of first choice in exploring new properties of neural tissues. Thus it was employed in seeking evidence for the maintenance of resting potentials in cerebral tissues (Chapter 9). It must however be used in the knowledge that it lacks very many constituents present in blood or cerebrospinal fluid; see, with respect to cerebrospinal fluid, Lowenthal (1972).

Table 7.1 Components of blood sera, of cerebrospinal fluid and of Krebs'–Ringer solutions modified for use with mammalian cerebral tissues

Component	Concentration (mM) in:				
	Rat serum	Human CSF	Human serum	Bicarbonate saline	Phosphate saline
Na^+	134	141	142	150	134
K^+	5·1	2·5	5	6·2	6·7
Ca^{2+}	3	2·5	2·5	2·4	2·6
Mg^{2+}	1·3	1·3	1·5	1·2	1·3
Cl^-	102	101	103	134	105
PO_4^{3-}	2·4	2	1·1	1·2	20
SO_4^{2-}	—	—	0·5	1·2	1·3
HCO_3^-	22	27	27	26	0
Glucose	6·1	2·8	4·5	12	12

For sources of data and further comments on the significance of the comparisons, see Krebs (1950), Keesey, Wallgren & McIlwain (1965) and McIlwain & Bachelard (1971).

The bicarbonate–CO_2 system is an excellent buffer, and the reasons for choosing other buffers in the alternative salines which follow, are usually (i) that it is necessary to maintain a CO_2 atmosphere to bicarbonate buffers; or (ii) that the alternative buffers facilitate measurement, for example of respiration. Respiration is a valuable and simply-measured criterion of the normal behaviour of a tissue sample and in bicarbonate salines it must either be measured by an indirect method or replaced by some other criterion of the tissue's behaviour. For measurement of respiration in bicarbonate salines, see Umbreit *et al.* (1972). With cerebral tissues, alternative criteria of the tissue's normality which are recommended are the measurement of phosphocreatine or of potassium content. These substances are chosen because so many other of the tissue's chemical and physiological activities are dependent on energy-rich phosphates and phosphocreatine is the substance of this category which in cerebral tissues has proved most susceptible to adverse conditions; or they depend on membrane potential which requires a normal content of K^+. Phosphocreatine and K are readily determined in 50–100 mg of cerebral tissue (Chapter 3). An instance of the use of these criteria in guiding the choice of metabolic conditions is given by McIlwain & Snyder (1970). Here tissues were being incubated in superfusion apparatus which precluded manometric measurements. Phosphocreatine level has also been used as a criterion of adequate glucose and oxygen supply to tissues maintained at the surface of experimental salines for electrical measurements. Further suggested criteria are the membrane potentials observable in tissue samples, and their rate of glycolysis.

Alternative buffers

Inorganic orthophosphate has in the past been the buffer most commonly used in place of bicarbonate. The 25–33 mM-phosphate usually employed represents a 20-fold increase of the usual blood level (Table 7.1). Neural tissues have proved relatively insensitive to many changes in the anions of their environment.

Orthophosphate is however an important tissue metabolite and this often conditions its use as a buffer. Thus tissue levels of both inorganic phosphate and of phosphocreatine are increased by increase in the inorganic phosphate of incubating media. Further, when inorganic phosphate is to be determined in tissues which have been incubated in media at all high in phosphate, the phosphate of the medium must in some way be allowed for or eliminated. The orthophosphate of cerebral tissues is 3–4 μmol/g, or about 1/10 the concentration of phosphate-buffered media. It is often possible to diminish greatly the carry-over of a constituent by brief rinsing in a saline lacking the particular constituent, but such a step must be demonstrated to be innocuous before being adopted. The simplest procedure, therefore, is to choose some buffer other than phosphate in such experiments. A further situation requiring alternative buffers was encountered in seeking to incorporate added isotopic phosphate as rapidly as possible into cerebral tissues (Heald ,1956). Here, carrier-free isotopic orthophosphate was added to tissues in phosphate-free media.

Glycylglycine has proved valuable as an alternative buffer for use with cerebral tissues; it can be regarded as simulating the buffering capacity contributed by the plasma proteins to the blood. A 30–40 mM solution gives adequate buffering in a typical experiment of 1–2 h with about 20 mg tissue/ml, and there appears no reason for not increasing this concentration at least twofold if necessary, with appropriate diminution of NaCl. Blood plasma contains some 6% by weight of protein constituents; these are of high molecular weight, and for special purposes their colloid osmotic pressure has been replaced by other materials of high molecular weight (Marks & McIlwain, 1959). No advantage accrued in this instance.

Glycylglycine has been in use as a buffer in these laboratories since its introduction for metabolic experiments involving phosphate determination (Buchel & McIlwain, 1950). An 0·3 M stock solution is prepared from 3·96 g by dissolving in about 60 ml of water, making to pH 7·4 by N-NaOH (about 5 ml) and diluting to 100 ml. Situations have not been encountered in which its use is undesirable, though in studying peptide metabolism caution would appear to be necessary.

Tris buffer (2-amino-2-hydroxymethylpropane-1, 3-diol hydrochloride) has also been extensively used with neural tissues, as solutions of 25 or 30 mM. It is characteristically different from glycylglycine in being not a zwitterion but a dissociated salt of a strong base akin rather to choline chloride. It can indeed be used to replace completely the Na or K salts of media in special situations (Hillman & McIlwain, 1960). More frequently, however, it has been used in place of phosphate in situations where tissue phosphate is being investigated. Tris of highest purity has been used to prepare an 0·5 M stock solution by dissolving 15·1 g in about 200 ml of water, taking the pH to 7·4 by N-HCl (about 40 ml) and diluting to 250 ml.

Other ions. It will be recalled that Ringer's initial contribution to balanced physiological salines was the addition of calcium salts to NaCl; the solutions of Table 7.1 were not developed primarily in relation to neural tissues and though they have been found satisfactory, specific demonstrations of their virtues are rare. The need for Na, K, and Ca salts in approximately the concentrations stated, is undoubted; they have major effects on the excitability of isolated cerebral tissues (see McIlwain, 1967; McIlwain & Bachelard, 1971). Excess K salts depolarize the tissue (see Chapter 9). Statements that K salts 'stimulate' the tissue in the fashion of electrical pulses are however incorrect; the many differences between the actions of electrical pulses and high concentrations of potassium salts are probably more interesting than the similarities.

An instance showing the value of the Mg of Krebs–Ringer salines is found in Dickens' (1946) demonstration of the more satisfactory retention of respiration in cerebral cortex on exposure to oxygen at pressures above 1 atmosphere. The further addition of salts of Mn and Co also improved performance under these conditions. The brain contains many other inorganic substances in small amounts (Stitch, 1957; Tower, 1969; Pfeiffer, 1972).

Oxidizable substrates

A major incentive to developing methods of handling neural tissues in chemically defined media has been the understanding of relationships between the tissue and organic constituents of its environment. Its relationship to oxidizable substrates is the most thoroughly investigated aspect of this and results have been incorporated in general neurochemical writings (McIlwain & Bachelard, 1971; Lajtha, 1969–72) and a few practical points only, are described here.

Glucose is clearly the main energy-yielding substrate of the brain under normal conditions and is the substance most frequently provided in experimental mixtures; only for special reasons are other substances substituted. Glucose at 5 or 10 mM with tissue at about 20 mg/ml gives an excess throughout an experiment of 1–2 h and at levels approximating to those of the blood (usually 4·5 mM in man). Oxidation of glucose by cerebral tissues is accompanied by its glycolysis with accumulation of lactic acid. Media thus become acid during incubation and the choice of buffers and of their concentration is partly governed by the accumulation which is anticipated. Situations in which accumulation is unusually great may require additional buffer, and it is advisable to measure roughly the pH of samples of all reaction mixtures at the end of incubation. An average glycolytic rate of cerebral tissues of 10–20 μmol/g.h is increased by a variety of agents, electrical and chemical, or by anoxia, to up to 150 μmol/g.h.

Lactate and pyruvate. Lactate itself acts as an energy-yielding substrate with many neural tissues, and in some situations will simulate glucose when provided in relatively high concentrations (20–100 mM). The same is true of pyruvate even at relatively low concentrations (2–10 mM). It should however be noted: (i) that when these acids are provided as substrates in initially neutral solutions, their oxidation yields alkaline solutions; (ii) pyruvate probably represents the normal oxidizable substrate of cerebral mitochondria, and lactate readily yields pyruvate; however, (iii) neither yields energy by a process akin to glycolysis, which is the second major energy-yielding reaction of the brain. Although the yield by glycolysis is relatively small under ordinary conditions cerebral tissues contain sufficiently potent glycolytic systems to yield for brief periods about the same quantity of energy-rich phosphate by the anaerobic process of glycolysis as by respiration.

Acetoacetate and 3-hydroxybutyrate. Realization that these ketone bodies are major cerebral substrates in man and the rat during starvation is relatively recent (Cahill *et al.*, 1968; Itoh & Quastel, 1970; Pull & McIlwain, 1971; Hawkins *et al.*, 1971). The capacity of cerebral tissues for their utilization, *in vivo* and *in vitro*, is independent of the nutritional state of the animals; but their utilization increases when the concentration of the ketone bodies increases, as occurs in starving animals. The increased utilization was shown between 0·2 and 1mM acetoacetate, and 0·2 and 3mM 3-hydroxybutyrate, with both suckling and with adult rats. Acetoacetate could suppress the oxidative utilization of pyruvate, but the formation of lactate from glucose continued in the presence of acetoacetate. Tissue levels of adenosine triphosphate, maintained by glucose, were reported to be unchanged when 5mM-acetoacetate was present together with 5mM-glucose. Part of the ^{14}C supplied to cerebral tissues as [$3^{14}C$] acetoacetate was subsequently found in several of the amino acids of the tissues.

Other substances which potentially act as substrates for cerebral tissues, including acids alternative to pyruvate, and carbohydrates alternative to glucose, are described elsewhere (McIlwain & Bachelard, 1971; Lajtha 1969–72).

Additional oxidizable substrates incorporated in media already containing glucose were used by Krebs (1950) in the experiments which are included in Table 7.2. In the medium with pyruvate and glutamate, respiration of cerebral and other tissues took place at greater rates than when glucose was the only substrate. The rates were however less stable and with cerebral tissues were accompanied by lower levels of phosphocreatine: results which emphasize that respiratory rate is one criterion only of the many by which the satisfactoriness of experimental media may be judged. Succinate also affords higher rates of oxygen uptake.

Other organic constituents of fluid media

Of the multiplicity of tissue constituents, several have been studied sufficiently for it to be known whether they are maintained in the tissue during typical experiments *in vitro*, or whether specific additions to suspending fluids are needed to achieve this. Of such constituents some, as nicotinamide nucleotides, are well maintained in ordinary oxygenated glucose salines. The nicotinamide nucleotides are however gradually lost anaerobically; nicotinamide or the nucleotides themselves do not then increase the tissue level. Other constituents, as phosphocreatine or adenosine triphosphate are in part lost in ordinary glucose salines, aerobically, and need specific additions for their maintenance at normal level. Table 7.2 summarizes several observations of this sort.

In the instances in which it occurs, reassimilation of a tissue constituent may be relatively slow, requiring an hour or more even when the appropriate substance or its precursor is provided at concentrations equal to or greater than those of the blood or cerebrospinal fluid. This is consistent with the operation in the isolated tissue of specific processes of assimilation which, though slow, are adequate to maintain normal equilibria between tissues and body fluids. *In vitro* experiments understandably make unusually great demands on such mechanisms, and

Table 7.2 Further additions to buffered glucose salines used with mammalian cerebral cortex

Substances added	Basis on which the addition appears advantageous, and other notes
Mn^{2+}; Co^{2+}, 20–50 μM (Dickens, 1946)	Maintenance of respiration with oxygen pressures above 1 atm; addition of some 30 other substances examined
Amino acids (Lajtha, 1967; Piccoli et al., 1971; Jones & McIlwain, 1971b)	Normal amino acid content of tissue maintained or modified
Glutamate, 1–10 mM (Terner et al., 1950; Pappius & Elliott, 1956; Pull et al., 1970)	Increased tissue K^+, which is however associated with increased water content, decrease in phosphocreatine and fall in resting membrane potential
Pyruvate, fumarate and glutamate, approx. 5 mM (Krebs, 1950)	Increased respiratory rate, which is however less stable and accompanied by lower tissue phosphocreatine
Creatine, mM (Thomas, 1956)	Tissue creatine and phosphocreatine restored to in vivo values
Adenosine, guanosine and creatine, each 1 mM (Thomas, 1957, Kakiuchi & Rall 1970; McIlwain, 1972b, c)	Tissue adenosine and guanine nucleotides in-increased towards in vivo values; adenosine 3′, 5′-cyclic monophosphate increased
Noradrenaline, 50 μM; histamine, 100 μM (Kakiuchi et al., 1969)	Adenosine 3′, 5′-cyclic monophosphate increased
Ascorbate, 0·03–1 mM (McIlwain, Thomas & Bell, 1957; McIlwain & Snyder, 1970)	Tissue ascorbate could be adjusted to in vivo values; noradrenaline oxidation diminished
Glutathione, 0·3 mM (Martin & McIlwain, 1959)	Maintained tissue glutathione close to in vivo values; reduced form most effective
Gangliosides and plasma proteins (Marks & McIlwain, 1959; McIlwain, 1961)	Tissue excitability restored after loss on keeping in cold media

thus may offer good opportunities for their study; this is especially so when superfusion methods are used (see below). Examples concerning amino acids and neurotransmitters are given by McIlwain & Bachelard (1971); Jones & McIlwain (1971a, b) and Pull et al. (1972).

Present knowledge does not permit the recommendation of some one medium combining the various additions of Table 7.2. The Table does, rather, present a set of possibilities which should be examined in any investigation of neural tissues in vitro, before it is concluded that isolated tissues do not reproduce some chosen in vivo phenomenon. Extension of the type of experiment represented by the Table appears very desirable. It remains valuable, also, to include body fluids in the media being examined in particular experimental situations. Addition of cerebrospinal fluid or blood plasma to bicarbonate salines presents no problems. If, however, the additions are to be made to media lacking bicarbonate and without a CO_2-containing atmosphere, the native bicarbonate should be removed from such fluids for example as described by Field (1948); an example of the use of this technique is given by Brierley & McIlwain (1956).

II. GAS SUPPLY AND MEASUREMENT

Experiments with isolated tissues are carried out in defined gas atmospheres, and for reasons outlined previously (p. 134) O_2–CO_2 mixtures are the gases most frequently used. The chosen gas is typically supplied in one of the following two ways. In the first, now to be described, the tissue, media and gas are sealed for the duration of an experiment; consumption of O_2 or output of CO_2 can then be used to follow the progress of these aspects of metabolism, as in manometric experiments measuring tissue respiration. In the second group of methods, described subsequently (p. 145), persistence of the chosen gas atmosphere is ensured by a continuous gas flow. The flow displaces any entering air, avoids the need for a sealed system, and allows much more manipulation or sampling of tissue or incubation fluid.

Respiratory measurements in aqueous media

There is little which is specific to neural tissues in the use of ordinary manometric apparatus, or of related apparatus (Gilson, 1963), for following tissue metabolism by measurement of gaseous exchange. Cerebral cortex and retina were included in Warburg's experiments of the 1920s while the now conventional manometric vessels and associated apparatus were being devised (Warburg, 1930). Most departures from the conventional conical vessel have served to emphasize the virtues of that shape. Typical experiments employ 10–100 mg fresh wt of tissue in 1–5 ml of fluid in a conical vessel of total volume 10–25 ml. When tissue or a medium constituent is limiting, vessels of some 3 ml volume are valuable, being used with about 0·6 ml of fluid and 2–10 mg tissue (see for example Greengard & McIlwain, 1955). When more tissue is to be incubated in order to isolate a metabolite or tissue-constituent, vessels of 100–150 ml have been used (Wolfe & McIlwain, 1961). We have preferred the vessels of Fig. 7.1 of Krebs & Eggleston (1945), but made with shorter centre-chambers. They accommodate up to 2 g of tissue in 25 ml fluid. Both the large and the small vessel can be used with ordinary manometers.

Typical rates of respiration of mammalian cerebral cortex are 55–80 μmol O_2/g wet wt/h under ordinary conditions, and 100–200 μmol/g.h with electrical or other stimulation. (For relationship of wet wt to fresh wt, see Chapter 6.) The values for white matter are approximately half those of the cortex. These rates result in pressure changes of about 50–150 mm/h in the ordinary manometric apparatus quoted above. Thus readings at 5 min intervals differ by about 5–10 mm. A typical experiment with tissue from one experimental animal would comprise six to nine vessels and an example of their arrangement is quoted in Table 7.3. In such an experiment the vessels are filled and

Table 7.3 A metabolic experiment with slices of cerebral cortex

Vessels in order of manipulation:	Thermo-barometer	(1)	(2)	(3)	(4)	(5)	(6)
Added agent:	0	0	a	b	b	a	0
Tissue: hemisphere and slice in order of cutting:	0	First hemisphere: 1st 2nd 3rd			Second hemisphere: 1st 2nd 3rd		

Solutions common to all vessels were added first: 0·12ml 5 N-NaOH to a centre well, and the saline to the main compartments. The added agents and tissues followed in that order.

manipulated, and tissues are cut and transferred, in the order (1) to (6) in which they appear in the table. The arrangement of duplicate vessels within the experiment is chosen in order to reveal any effects which may follow from the order of manipulation, and also any which may be due to the anatomical site from which the tissue is derived. In a replicate experiment, with another animal on another occasion, the reagents of vessels (3) and (4) would be exchanged with those of vessels (2) and (5), or (1) and (6).

It is to be emphasized that there is no fixed routine in gasometric measurements, which can be introduced at various stages of an experiment: after pretreatment of tissue, or before its transfer to other conditions or removal for analysis. Apart from making additions from side arms, gasometric experiments can be interrupted and the complete medium exchanged (Marks & McIlwain, 1959; McIlwain, 1956; 1959b). Electrical stimulation during manometric experiments is described in a subsequent section and in Chapter 8. For polarographic measurement of oxygen, see p. 237.

Anaerobic experiments

As conventional manometric apparatus is designed for good gaseous exchange, it is very simply used for anaerobic experiments whether or not manometric measurements are being made. Usually the most informative measurement which can be made anaerobically with neural tissues is of acid formation in bicarbonate buffers, revealed by displacement of CO_2.

In such experiments, fluid media are usually equilibrated with a commercial mixture of N_2–5% CO_2. Small sticks of yellow phosphorus about 2–3 mm in diameter and 10–14 mm long are used in the centre well. These are prepared by melting yellow phosphorus under hot water and sucking it into glass tubing of the chosen bore. After cooling, the phosphorus is cut to length and kept under water in a stoppered bottle. It is handled with forceps, transferring rapidly and with minimal exposure to the air; thus, the manometric apparatus in which it is to be placed is already receiving a stream of N_2 before the phosphorus is taken. The phosphorus is used repeatedly, scraping a fresh surface on the stick when necessary.

Formation of lactic acid from glucose by cerebral tissues proceeds anaerobically at up to 200 μmol/g.h. For measurement, tissues may be prepared aerobically in the usual way. They are then placed into bicarbonate media in manometric vessels. A stick of yellow phosphorus is placed in the centre well of one vessel and this immediately attached to its manometer, through which N_2–CO_2 is already flowing. A wisp of smoke may appear in the vessel and be carried out by the gas; it does not contribute appreciable phosphate to the medium. Equilibration and manometric measurement follows the same course as in aerobic experiments, after which the vessels are, one at a time, removed from their manometers and the yellow phosphorus immediately removed and placed below water in a bottle. For examples of such measurement, see McIlwain (1956). Experiments may have successive aerobic and anaerobic phases.

Respiratory experiments with minimal fluid

Preparation of tissues for such experiments has been described in Chapter 6; the tissue is obtained as a thin sheet weighed on a coverslip. If the tissue is not more than 0·2 mm thick, it plus coverslip may be used in some types of subsequent experiment. A typical tissue slice 0·35 mm thick needs to be oxygenated from both sides, and may be transferred to a gauze support sufficiently small to be inserted in an experimental vessel (Fig. 6.4, supports D or E; and Fig. 7.1). For transfer, the coverslip carrying the slice is inverted over the gauze support, and the narrow spatula of Fig. 6.1 is used to separate the slice from the coverslip; this separation may be facilitated by adding a small quantity of aqueous fluid to the spatula, if such fluid is being used in the experiment for which the slice is being prepared. Alternatively, to avoid transfer, a piece of nylon gauze about 1·2 × 2 cm may be used in place of the coverslip of Fig. 6.4.

Substrates and other additions can be made in small volumes of aqueous fluid during the preparation of the tissue, or immediately before placing it in manometric vessels. These vessels then receive NaOH and a paper wick in the centre wells and are equilibrated with oxygen which has passed through water or isotonic NaCl solution. The respiration of cerebral tissues under these conditions, with or without added glucose, is initially of 40–100 μmol O_2/g.h (Rodnight & McIlwain, 1954).

Respiratory experiments in non-aqueous fluids

Electrophysiological work with neural tissues *in vitro* and *in vivo* has frequently involved covering the exposed cerebral cortex or spinal cord with liquid paraffin; this allows access to the tissues while preventing their drying, and the paraffin does not short-circuit electrodes as would aqueous fluids. Excitability is maintained under these conditions; thus the liquid paraffin is not deleterious and such fluids are available for use also with isolated tissues. They then give opportunities of adding substrates or inhibitors during an experiment, without addition of water.

Non-aqueous fluids have been used to supply materials in two ways: in suspension and in solution. The most important substance supplied in solution is oxygen. The relatively high solubility of oxygen in non-aqueous solvents is often not realized; it is valuable in connection with all metabolic studies in non-aqueous fluids. Table 7.4 lists the solvents

Table 7.4 Non-aqueous fluids used in metabolic experiments

Fluid	Solubility of oxygen (α-O_2, 38°)
Liquid paraffin, sp. gr. 0·835	0·098
Olive oil, pharmaceutical grade	0·102
Silicone, viscosity 5 centistokes	0·24
Silicone, viscosity 1 centistoke	0·30
Water or Ringer's solution	0·024

The solubilities are expressed as Bunsen coefficients: the volume of gas, calculated to 0° and 760 mmHg, which is dissolved by 1 vol. of fluid at the temperature of the experiment (Rodnight & McIlwain, 1954).

employed, and the extent to which they dissolve oxygen at 38°. Nitrogen also is much more soluble in these fluids than in water, and this necessitates special arrangements for gaseous equilibration; apparatus devised for the purpose is shown in Fig. 7.1.

Methods. Using liquid paraffin, sodium pyruvate and glucose have been supplied as suspensions prepared from the solids and paraffin in an ordinary glass homogenizer; atropine, phenobarbitone, butobarbitone and 2:4-dinitrophenol have been dissolved in the paraffin and placed in side arms of manometric vessels. An outline of the procedure follows; for further details see Rodnight & McIlwain (1954) and Hosein *et al.* (1962).

The oil is first saturated with water and oxygen at 38° in vessels *A* and *C* (Fig. 7.1). Manometric flasks are prepared and any necessary reagents added to side arms and centre wells. Tissue slices are cut with

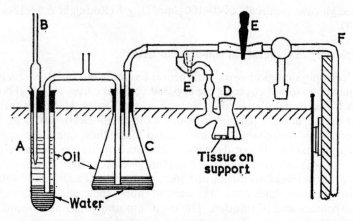

Fig. 7.1 Apparatus and equilibration arrangements for manometric experiments with non-aqueous solvents (Rodnight & McIlwain, 1954). *A:* Tube in which the oil is saturated with water and oxygen before transferring it by pipette (*B*) to the manometric vessel (*D*). This vessel receives through its sidearm stopper, oxygen saturated with water and with the vapour of the non-aqueous solvent, from flask (*C*). Clip (*E*) is used to direct the flow of oxygen from the vessel sidearm to the manometer (*F*), after the vessel *D* has been placed on its manometer. The position of the 3-way tap on manometer *F* differs from its position on conventional manometers, but is necessary for this apparatus and is advantageous also in other studies of tissue metabolism.

the bow cutter (q.v.), transferred to supports or electrodes, weighed and placed in the flasks. The oil is added to the flasks in a fashion which minimizes its contact with the air: a flask is clamped in the position in which it is shown in Fig. 7.1, oxygen passed through the moist oil to the flask, and oil then pipetted from tube *A* to the flask. The flask is then immediately transferred to its manometer, which is adjacent, and the passage of oxygen continued, with shaking, for a further 1·5 min. In the meanwhile, successive vessels receive the oil; with two workers an experiment involving an oil thermobarometer and four vessels with tissue can be assembled so that the first readings are taken 30 min after obtaining the tissue.

III INCUBATION, SAMPLING AND ANALYSIS OF TISSUES AND METABOLIC MIXTURES

This section concerns the handling of isolated tissues and of incubation fluids for analysis; the analytical methods have been described in Chapters 2 to 5. A feature of experiments with isolated tissues which is outstandingly advantageous, is that the necessarily traumatic removal of the tissue from an animal does not intervene between the establishing of chosen experimental conditions, and tissue analysis. Procedures such as rapid freezing in liquid nitrogen before the use of an extracting agent are therefore not necessary for the preservation of labile constituents, though they may be employed for other reasons.

Changes in composition necessarily occur in tissues during their preparation for metabolic experiments, and one objective in the choice of incubating conditions is that they should as far as possible restore and maintain important tissue components. It is feasible to use the tissue content of phosphocreatine and of potassium for judging tissue normality, as noted above (p. 135). Also, an investigator will usually also analyse the tissues for further constituents more specifically related to the aspect of tissue functioning which is being studied.

Quick transfer holders

Tissue incubation in appropriate conditions must be followed by prompt transfer of the tissue samples to chosen fixing or extracting agents. Several agents, chemical or electrical, alter the metabolism and composition of neural tissues in a few seconds. Specific apparatus has been devised for sampling tissue in brief periods after applying such agents (Fig. 7.2). The apparatus is also advantageous when the fluid surrounding the tissue is to be sampled repeatedly, either during brief periods or during the course of a more ordinary experimental period of an hour or so.

The apparatus in effect puts the tissue on a handle in a vessel from which it can be rapidly removed. The quick-transfer holder (Fig. 7.2) is a plastic frame which carries the tissue at its lower end and within a grid of wires or fibres. By the holder the tissue is picked up after its preparation, transferred and held during incubation, and rapidly removed for fixation. The plastic frame is arranged to be easily opened from its upper 'handle' end; the lower end which takes the tissue keeps it fully extended in 4–7 ml of saline. The saline is in a small beaker in a thermostat and is circulated and oxygenated by a stream of O_2 or of O_2–CO_2. An oxygen atmosphere is maintained above the fluid by a sponge-rubber barrier, attached to the holder and well above the level of the fluid; the rubber contributes to retaining the gas supplied and also keeps the tissue-holder approximately in position. For experiments in a superfusion system (see below, p. 152) there has been devised a vessel which is more completely sealed and in which the relative positions of the tissue and fluid are more defined; this vessel may replace that of Fig. 7.2.

The holder of Fig. 7.2 is made in two parts, from strips of transparent plastic (Perspex; Lucite) $1 \cdot 5 \times 22$ mm in cross-section and bent as shown. Part a is shaped at the top to take the thumb; at the bottom it carries a grid on which the tissue lies horizontally. Part b has an additional piece c of plastic strip cemented across its upper portion and this is shaped to take two fingers; below it carries a grid which closes on that of a so that the tissue is held between the two grids as between two

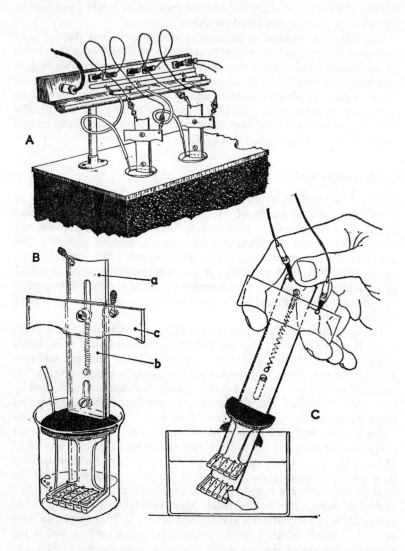

Fig. 7.2 Quick-transfer apparatus for metabolic experiments with tissue-slices (McIlwain, 1960, 1972a). *A*. Part of a thermostat with arrangements for incubating 6 to 9 quick-transfer holders, and carrying distributors for supplying gas mixtures and electrical pulses to the tissues. *B:* An individual holder in its beaker. *C:* Releasing a slice from the holder to a dish of fluid, e.g. ice-cold isotonic sucrose. A similar dish, with incubation media at 38°C, may be used for mounting the tissue in electrodes at the beginning of an experiment.(Drawn by Mr R. J. Woodman. Scale: *A*, 1/10; *B*, 1/1·5; *C*, 1/2.)

jaws. Most of the slice is however freely in contact with the saline; only at defined points at its edges is it gripped between a wire of the grid of one part of the holder, and the frame of the other part. Part *b* also carries bolts which slide in grooves cut in *a* so that the two parts are held together, but the jaws can be opened and closed. In use they are normally held closed by a spring but a pressure of 250–300 g between the thumb and finger opens the holder to insert or release the slice (see Fig. 7.2).

The parts of *a* and *b* which enter the beaker are cut away as shown in Fig. 7.2, *B* so that medium does not creep between them by capillary attraction. The greater part of the plastic in the horizontal portions is also cut away so that rectangular frames are left, across which wires or nylon nets are fixed. Further constructional details are given by McIlwain (1961) and the apparatus is supplied by Specialized Laboratory Equipment Ltd, 15 Campbell Road, Croydon, Surrey CR02SQ, U.K.

Thermostat-bath and assembly

A commercially-available thermostat with a tank of water about 20×55 cm and 16 cm deep (these dimensions are not critical) is readily adapted to take a series of quick-transfer holders. The thermostat is already provided with a good stirrer and an electrically-operated relay which can be set to keep the temperature at $37 \cdot 5 \pm 1°$. A transparent plastic sheet (Perspex; Lucite) is made to cover the top of the thermostat and is kept in position by flanges on its lower surface. The sheet is 8 mm thick and provided with holes to take 6 to 9 of the beakers used with the holders. The beakers are 3·2 cm external diameter and 4·8 cm high, flanged but without spouts, and of nominal capacity 30 ml. The sheet also carries uprights on which are mounted gas distributors, and distributors of electrical pulses when these are to be applied.

The gas, usually oxygen or oxygen–5% carbon dioxide, comes from a cylinder through a sintered-glass bubbler which contains water and is immersed in the bath. The gas then goes to a tube which carries 8 fine nipples on which can be slipped thin flexible plastic tubing (vinyl tubing, elastic grade, bore 1 mm, external diameter 2 mm; no. 2E of Portland Plastics, Hythe, Kent). Each tube passes under a screw clamp and ends in a narrower and stiffer piece of tubing 6 cm long, detachable and cut obliquely at each end as is a hypodermic needle (polythene tubing, surgical quality, bore 0·5 mm, external diameter 1 mm; no. 48 of Portland Plastics). This is the portion which enters the beaker and during experiments it is held in position by the sponge-rubber seal which presses it against the wall of the beaker. A chosen position is readily maintained; no further fixing has been found necessary. The most effective size and position for the bubbler-tube has been determined by observing the motion of particles of decolorizing charcoal suspended in fluid in the assembled apparatus. The position should be such that the bubbling end is at the level of the slice, and between it and the wall of the beaker, to either the right or left of the holder when viewed as in Fig. 7.2, *B*. Placing the tube behind the holder is especially to be avoided. The oblique ends of the polythene tube make it easy to insert into the wider tubing and unlikely to be blocked by contact with other parts of the apparatus. Fine tubes are preferable in giving a continuous stream of bubbles without too great a gas flow, and also because less individual adjustment is needed to maintain bubbling in all the beakers of the set. This is presumably because the narrow pieces of tubing, which are all similar, offer the main resistance to gas-flow.

When electrical pulses are to be applied to the tissue, they are supplied from apparatus to be described in Chapter 8 to output sockets mounted above the quick-transfer holders. To the output sockets, flexible insulated wire connections are plugged; these carry at their other end clips which grip terminal lugs of the quick-transfer holders. These lugs form the upper ends of insulated wires carried in grooves up the handle of the holder from the grid below which surrounds the tissue.

Procedure

The thermostat is brought to temperature and apparatus for preparing the tissue is assembled. A small auxiliary thermostat (bath WB/100, Luckham Ltd, Burgess Hill, Sussex) may be used to maintain a dish of incubating fluid at 38° for mounting the tissue in the holders (Fig. 7.2, *C*), and this also is brought to temperature. Reagents are prepared for subsequent fixing and analysis of tissue or fluid. The medium for tissue incubation is prepared and distributed in the beakers together with any additions whose effect is to be examined. The beakers, usually 6 or 8, with a specific numbered transfer-holder adjacent to each, are left in sequence at room temperature or in a cold-bath.

The chosen tissue is prepared and weighed; two people should work together at this stage of the experiment to minize delay. As successive tissue samples are weighed, they are placed in the dish of incubating medium and picked from the dish by taking the transfer-holder in one hand, opening its jaws, and resting the lower jaw on the bottom of the dish. With a spatula in the other hand, the tissue is manoeuvred, fully extended in the saline, above the jaw of the holder and with it is picked from the saline; the jaw is then closed and the holder put loosely in its beaker. The beakers with their holders and tissues are taken to the thermostat, the gassing-tubes inserted and the gas turned on. In the first beaker the gassing tube is placed in position and gripped by pushing the rubber of the holder into the top of the beaker as shown in Fig. 7.2,*B*. This is then done with the following beakers. Any unused gas outlets are closed by their screw-clips and a gas-flow of 1 to 1·5 l/min is needed for six beakers. If, rarely, it is necessary to equalize gas flow in the beakers the other screw clips are adjusted. This stage in the experiment should be reached in 12–20 min after the tissue has been removed from an animal. A chosen period of preincubation is allowed for tissue composition to reach a stable *in vitro* value. This is typically 20–40 min, depending on the constituents being studied, and after this the tissues receive any further treatment which is part of the experiment: for example the addition of a reagent or the application of electrical pulses. Small volumes of reagents can be added by pipette or hypodermic syringe without removing the holder. The beakers are then sampled in the order in which they were prepared.

Tissue may be sampled as follows. (i) By immediate transfer to a measured quantity of fixing agent in a beaker about 3·5 cm in diameter and 3·5 cm deep. These beakers are wider and squatter than the beakers used in incubation, and permit the holder to be opened and jerked backwards in about 2 ml of fluid, so allowing the slice to float free. If release is into an acid reagent such as trichloroacetic acid, then the holder immediately after releasing the slice is placed into a nearby beaker of $NaHCO_3$ (about 300 ml of 2% solution), for acid damages the plastic. (ii) It has been found advantageous in some cases to release the tissue first to an ice-cold isotonic solution and after 1–3s only, to pick the slice from this with a mounted, bent wire for immediate transfer to a fixing agent. The fixing agent can then be in a smaller, more convenient vessel, for example in a centrifuge tube of 3–5 ml, containing 1 ml of fixing solution and provided with a loosely-fitting pestle or plunger by which the tissue can be promptly disintegrated in the solution. (iii) When speed of transfer is not critical, the tissue may be released from its holder to the batch of fluid in which it has been incubating, and then picked from the fluid with a mounted bent wire and placed in a test-tube homogenizer containing a fixing or extracting solution.

Fluid is sampled by pipette or syringe. (i) Some analyses permit taking small samples of 0·05–0·5 ml by capillary pipette or by syringe on several occasions during the course of an experiment. This can be done by pushing the pipette or needle between the sponge rubber and the wall of the beaker, at the side opposite to that which has received the gas bubbler. (ii) If larger samples are taken on more than one occasion it is advisable to make good the volume removed, by adding fresh medium. The removal and replacement of medium should be carried out in all beakers of an experiment if they are to remain comparable. (iii) If no other measure of normality of tissue metabolism is being obtained, lactate formed from glucose and accumulating in the medium may be determined in the fluid remaining after the slice has been removed at the end of the experiment. This can indicate the adequacy of oxygenation.

Illustrative experiments; related apparatus

1. Breakdown of the phosphocreatine of cerebral tissues by electrical

pulses, and its inhibition by cocaine (Bollard & McIlwain, 1959; McIlwain et al., 1969). Guinea pig cerebral cortex was used in glycylglycine-buffered glucose salines, with oxygen as gas phase; six tissue samples from one animal may be examined in one experiment, preparing and handling them in sequence in the fashion illustrated earlier (Table 7.3). Three slices of about 110 mg were cut successively from the first hemisphere and used in holders *1, 2* and *3* respectively; those from the second hemisphere were used in sequence in holders *4, 5* and *6*. Cocaine (20 μM) was present in the medium of the even-numbered beakers, and all tissues were preincubated for 30 min. Tissue *1* was then released into ice-cold trichloroacetic acid and the holder put into bicarbonate. Immediately afterwards, electrical pulses were applied to tissue *2* for 10 s; during the tenth second the holder, still with the tissue moist with fluid and receiving pulses, was removed from the vessel and plunged to an adjacent beaker of ice-cold trichloroacetic acid. This beaker was returned to ice, the holder released from its electrical supply and put into bicarbonate. Tissues *3, 4* and *5* were successively treated as tissue *2* and tissue *6* as tissue *1*.

The tissue content of phosphocreatine (μmol/g wet wt) was *1*, 1·5; *6*, 1·4. Those with pulses and cocaine had not greatly altered: *2*, 1·4; *4*, 1·3. Those with pulses and no cocaine lost about half their phosphocreatine: *3*, 0·7; *5*, 0·7. In a replicate experiment with the same electrodes and vessels in the same order cocaine was placed in the odd-numbered beakers. Corresponding experiments with different periods of stimulation showed breakdown of phosphocreatine to occur at over 1000 μmol/g.h and to be inhibited 94–97% by 20 μM cocaine.

2. Ion movements. Cummins & McIlwain (1961) and McIlwain et al. (1969) describe in detail the employment of the apparatus for following loss and gain of Na^+ and K^+, showing the effect of electrical pulses on each movement and on net changes in K^+. This involved the use of radioisotopes, frequent sampling of medium and tissue, and brief rinsing of the tissue in cold salt-free solutions before extraction. Sensitivity of the electrically-stimulated entry of Na^+ to inhibition by 20 nM-tetrodotoxin was demonstrated.

3. Metabolic effects of neurotransmitters. Unexpectedly large changes in the cyclic AMP of incubated neocortical tissues were found to follow electrical excitation in the presence of noradrenaline or histamine (Kakiuchi, Rall & McIlwain, 1969). After exposure to the agents for chosen periods, tissues were released from their holders into the incubating fluid and from there transferred to a test-tube homogenizer containing 30 mM-HCl which was at 0° and which already contained a standard amount of [^{3}H]cyclic 3′,5′-adenosine monophosphate. The transfer required about 10 s and the tissues' cyclic AMP was demonstrated to be stable for about 2 min under the conditions of transfer. The tissue sample was ground in the HCl and kept at 0°, under which conditions it could be assayed within 8 days by column chromatographic

separation and determination of the specific activity of the isolated nucleotide.

IV. SUPERFUSION OF ISOLATED TISSUES

Superfusion methods cause incubating fluid to flow over and around tissue samples, so that the tissues are constantly supplied with new substrates and their metabolic products are removed. Thus in superfusion methods a chosen experimental fluid replaces the blood stream more directly than in the other metabolic procedures so far described. This is the case even though the capillary bed is not used in superfusion. Addition of a substance to the inflowing fluid can be made as a brief 'pulse', or can be sustained for the duration of an experiment. The effluent fluid, like venous blood, can be collected at chosen intervals for analysis and then gives a measure of the progress of tissue metabolism. The intervals at which it is appropriate to make additions to superfusion fluids, or to take samples from them, are conditioned by the volume and flow rate of the fluid; and these variables are dependent on the choice of apparatus and experimental conditions now to be described.

Superfusion method for multiple tissue samples

This method (McIlwain & Snyder, 1970; McIlwain, 1972a) enables four to seven samples of tissue to be studied simultaneously. Each sample is in a separate vessel in which it may remain for the whole of an experiment involving preincubation, incorporation of labelled reagents, removal of excess reagents, application of specified chemical, pharmacological or electrical agents or conditions, and removal and measurement of metabolic products. The tissue, finally, can be quickly removed from its superfusion vessel for extraction or further examination.

Apparatus. Tissues are held in quick-transfer holders, of the type shown in Fig. 7.3, which carry the following components. (i) Inflow and outflow tubes for incubation media. (ii) An inflow tube for gas, which is supplied from cylinders through a bubbler containing water and a circuit with a Rotameter for measuring gas-flow. (iii) Tissue-holding jaws which are usually also electrodes, and are held in position by a rod and slotted tube. (iv) A flanged lid with a neoprene gasket which joins components i–iii to an incubating beaker. The holder is, however, readily removable from the beaker and the flange ensures that it can be returned to its original position after removal. The flanged lid is accurately machined and it and the gasket replace the rubber seal of the quick-transfer holder described in Fig. 7.2; the structures which support and release the tissue-holding jaws have also been modified as is noted in Fig. 7.3.

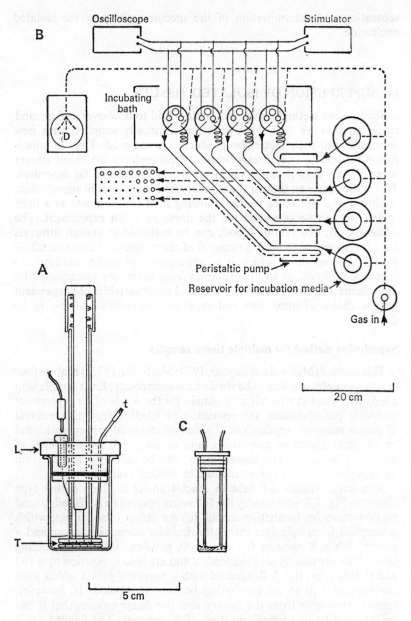

Fig. 7.3 Apparatus for superfusion of isolated tissues. *A:* Tissue in transfer-holder, by which it is held in a fixed position in an incubating beaker. The flanged lid (*L*) carries the inflow tubes (not shown) for gas and incubation medium, and the outflow tube (*t*) which conditions by its position, the volume of medium surrounding the tissue. The lid (*L*) also carries two electrical connections, of which

one is shown, leading to grids of electrodes around the tissue (T) in the fashion shown in Fig. 7.2. *B:* Plan of superfusion arrangements for four incubation beakers, each with a tissue sample in a transfer-holder, and each receiving incubation media from separate reservoirs through a peristaltic pump and warming coils. Gas mixtures, usually O_2–5% CO_2, are also supplied to each beaker, to the reservoirs and to a crystallizing dish (D). The dish contains about 15 ml of incubation medium and is used for mounting the tissue samples in their transfer-holders; it is covered by a spring-loaded lid which can readily be opened when a tissue is being mounted, and is kept at 38° by a small auxiliary bath (see text). *C:* Primer which is used prior to superfusion, in order to fill the pump-tubing, warming coils and connection tubes with incubation media. Each flow-circuit from the pump is provided with a primer, which is a 5 ml sample tube with a plastic stopper, and is clipped to the incubating bath. The inflow and outflow tubes which serve a particular beaker pass through the stopper to the bottom of a primer, and at the beginning of an experiment the pump is run for a few minutes with the tubes in this position until a continuous flow reaches the collecting vessels. The pump is then halted, the stopper removed from the primer, and its tubes attached to those of the corresponding incubating beaker for the remainder of the experiment. The stopper links the inflow and outflow tubes serving a given beaker and it is convenient to leave it in position on the tubes throughout a number of experiments; between experiments, the stopper and its flow-tubes are returned to the primer.

Flow of incubating fluid is caused by a multi-channel peristaltic pump. The flow rates of such pumps are conditioned by the diameter of the tubing which they compress, 1–5 mm internal diam. being typical. In addition, the pump should have facilities for changing from one predetermined speed to another at chosen times during superfusion. Matched sets of pump-tubing, as received from manufacturers (Tygon: U.S.Stoneware, Akron, Ohio, U.S.A.) differ slightly in rate of flow. It is possible to calibrate the flow rates given by individual tubes in the pump and to select closely-matching pairs for input and output at a given beaker, so that the volume of fluid in the beaker remains constant Except during brief superfusion experiments, this procedure is less satisfactory than that by which the volume of fluid in the beaker is conditioned by the position of the outflow tube (t, Fig. 7.3). The pump-tube connected to the outflow tube t should then be chosen as one which gives a slightly greater flow-rate than the inflow tube of the same beaker. With this arrangement, the rate of superfusion corresponds to that of inflow; the outflow tube carries incubation media plus occasional gas bubbles.

The incubating fluid comes from a reservoir bottle (or bottles, if fluids of different composition are used) of 0·2–2 l. The reservoirs have plastic stoppers perforated to take gassing tubes and one or more tubes leading to the pump. These and other connecting tubes are usually of 0·75 mm bore polythene tubing (Portland Plastics Ltd, Hythe, Kent) with ends cut obliquely and joined by pushing them into short sections of Tygon vinyl tubing of bore slightly smaller than the outer diam. of the polythene tubing. From the pump the fluid is impelled to a glass heating coil (Technicon) in the incubating bath and from there to the incubating beaker. From the beaker it returns to another channel of the pump. A 12-channel peristaltic pump is thus needed for six beakers; the flow of fluid in the different channels is kept separate even if several inflow tubes take fluid from a single reservoir.

The outflow-tubes from the pump run to sample-receivers, usually test tubes in racks. The outflow is transferred to successive tubes at chosen intervals, for example of 0·5 to 5 min. A simple plastic transfer-stick is convenient for the transfer: it holds the four or six output tubes at a spacing equivalent to that between the rows of tubes in the racks, so that all output-tubes can be moved simultaneously. The sample-collection has not been further mechanized because during this part of an experiment, the investigator is necessarily at the apparatus in order to make other adjustments.

Evaluation of superfusion and perfusion systems has received comment from Bernstein (1971) and Pull, Jones & McIlwain (1972). Using the apparatus described, and a flow rate of 3·5 ml/min, a change of fluid at the intake tube reached the incubating beaker about 30 s and the collecting tube about 50 s later. The time taken for a change of input fluid to produce a new steady state in composition of the output fluid, depends also on the volume of fluid in the incubating beaker, and an example is quoted below. Between the components of the superfusion system, the tubing used is narrow-bore and of the grade employed in automatic analytical systems; little mixing occurs in the tubing. Though the superfusion fluid is gassed while in its reservoir, gas-flow to the beakers remains necessary: this is to aid mixing in the beaker, and also to equilibrate with the chosen gas at the temperature of the incubation bath.

Procedure; illustrative example

Major applications of superfusion techniques have been in the investigation of tissue responses to electrical excitation, and such use is described in the following chapter. Here, procedures are illustrated first by an examination of how hypoxia affects the metabolism and output of amino acids and of lactate from superfused tissues, measuring molar amounts and products from an isotopically-labelled precursor (Jones & McIlwain, 1971a, b; Pull, Jones & McIlwain, 1972).

Glucose–bicarbonate incubation media were prepared and 5ml placed in stoppered incubation-beakers; in some experiments the beakers also contained [^{3}H]leucine. Most of the incubation medium was kept in two reservoirs, the larger batch in equilibrium with O_2–5% CO_2 and the other with N_2–5% CO_2; incubation medium with O_2–CO_2 was also placed in the auxiliary tissue-mounting bath and it and the main bath were brought to 38°C. The main bath carried two gas manifolds, supplying respectively O_2–CO_2 and N_2–CO_2 to the incubation beakers.

Racks of collecting tubes were prepared to receive the superfusates, and ice-cold homogenizer-tubes with trichloroacetic acid were prepared to receive the tissues at the end of the experiment.

Two investigators, working together at this point, cut guinea pig neocortical tissues, weighed them still attached to the cover-glass insert used in cutting (Fig. 6.4), and washed them from the cover glass to the mounting-dish of warm, oxygenated medium. Here each in succession was immediately mounted in a quick-transfer holder, the holder placed in its beaker, connected to the gassing-tube carrying O_2–CO_2, and placed in the incubation bath. An initial incubation period of 30 min followed, during which the superfusion system was primed with incubation medium as described in Fig. 7.3, the pump switched off and the inflow and outflow tubes from each primer connected to those of the corresponding incubating beaker.

Superfusion was then commenced at a chosen speed, e.g. of 3–4 ml/min. This flow first removed excess [^{3}H]leucine, when this had been added, and lactate which had accumulated during the initial incubation. Collection of superfusion fluid for analysis usually commenced after 10–16 min flow, and was made each 1–4 min. Hypoxia was imposed by replacing the reservoirs gassed with O_2–CO_2 with those gassed with N_2–CO_2, and replacing the tubes carrying O_2–CO_2 to the incubating beakers by tubes carrying N_2–CO_2. These replacements were shown in preliminary experiments to bring the O_2 tension of the fluid in the beakers to 0·05 atm in 3 to 5 min. Collection of effluent samples was continued for periods of up to 2 h in all. Separate measured portions of the effluents were taken for determination of lactate, and of ^{3}H by scintillation counting. The distribution of radioactivity in individual amino acids was carried out with pooled samples of effluent media from 10 to 20 min flow; these were desalted using Dowex columns, and the amino acids separated chromatographically.

After superfusion of the tissues they were promptly and successively released from their holders to the fluid of the incubating beakers, and with a mounted, bent wire were immediately transferred to trichloroacetic acid at O°C in homogenizer tubes, and dispersed in the acid. After keeping at O° and centrifuging, measured portions of the supernatants were taken for determination of ^{3}H, K and phosphocreatine.

These experiments showed lactate formation by the superfused tissues to increase promptly from its normal value of some 20 μmoles/g.h, when oxygen tension fell to 0·2 atmos. The rate at 0·05 atmos. O_2 was over 200μmoles lactate/g.h, which is greater than obtains in non-flowing systems. Phosphocreatine in tissues at 0·05 atm O_2 was about 50% of its normal value, and tissue K, about 80% of that in 95% O_2. Of ^{3}H which had been supplied as [^{3}H]leucine, most remained in this form in tissues and superfusates, but appreciable proportions were found as glutamate and glutamine. A notable net output of amino acids took place from the tissues under normal conditions, and this increased during hypoxia. The hypoxia however, diminished the output of ^{3}H from tissues which had assimilated [^{3}H]leucine during an initial period.

Notes on other perfusion and superfusion methods

Perfusion of the brain *in situ* by its normal vasculature is an important experimental approach for which other publications must be consulted (Sacks, 1969; Diczfaluzy, 1971; Ross, 1972). Control of the cerebral blood vessels presents many problems in using such methods. Local perfusion of the brain has been carried out by push-pull cannulae inserted so that they take minute samples of the extracellular fluid of small, chosen parts of the brain (De Feudis *et al.*, 1970). Output of amino acids from the amygdala, hypothalamus and other parts of the monkey brain was measured in this way. Part of the cortical surface of anaesthetized animals, exposed at operation, can be enclosed by sealing to it a small cylinder; flow of fluid in the cup so formed gives a superfusion system in which output of K^+, of amino acids and of neurotransmitters has been observed (Van Harreveld & Kooiman, 1965). Ventricular perfusion methods have also been greatly used in the study of neurotransmitters, transport and metabolism within the brain; recommendations are made by Fenstermacher (1972) regarding cannulae, perfusion fluids, and flow rates.

Superfusion of isolated cerebral tissues for metabolic studies in parallel with electrical observations, have employed a tissue-chamber described in a following chapter (Fig. 9.2; Heller & McIlwain, 1973). Superfusion of chopped fragments of cerebral tissue is described by Srinivasan *et al.* (1969). Apparatus specifically devised for superfusion and stimulation of ganglia from small animals involved tissue of about 1 mg and flow-rates of 0·3–0·8 ml/h (Larrabee, 1958); respiratory measurements were made with oxygen-sensitive electrodes. Other metabolic and electrical measurements in such ganglia were carried out in small vessels without superfusion (Larrabee *et al.*, 1963; Halstead & Larrabee, 1972). Extraction of sympathetic ganglia for chemical analysis, after superfusion and stimulation, is described by McAfee *et al.* (1971).

REFERENCES

Altman, P. L. & Dittmer, D. S. (1972) *Biology Data Book* **1**, 444. Bethesda, Md: Fed. Am. Soc. Exp. Biol.
Bernstein, E. F. (1971) *Karolinska Sympos. Res. Meth. Reprod. Endocrinol.* **4**, 44.
Bollard, B. M. & McIlwain, H. (1959) *Biochem. Pharmacol.* **2**, 81.
Brierley, J. B. & McIlwain, H. (1956) *J. Neurochem.* **1**, 109.
Buchel, L. & McIlwain, H. (1950) *Br. J. Pharmacol.* **5**, 465.
Cahill, G. F., Owen, O. E. & Morgan, A. P. (1968) *Advances in Enzyme Regulation* **6**, 143.
Cummins, J. T. & McIlwain, H. (1961) *Biochem. J.* **79**, 330.
De Feudis, F. V., Delgado, J. M. R. & Roth, R. H. (1970) *Brain Res.* **18**, 15.
Dickens, F. (1946) *Biochem. J.* **40**, 145.

Diczfaluzy, E. (1971) *Edit. Karolinska Sympos. Res. Meth. Endocrinol.* **4**.
Fenstermacher, J. D. (1972) *Res. Meth. Neurochem.* **1**, 165.
Field, J. (1948) *Meth. Med. Res.* **1**, 289.
Gilson, W. E. (1963) *Science, N.Y.* **141**, 531.
Glick, D. (Ed.) (1971) *Methods biochem. Res.* **20**.
Gore, M. B. R. & McIlwain, H. (1952) *J. Physiol.* **117**, 471.
Greengard, P. & McIlwain, H. (1955) *Proc. Internat. Neurochem. Sympos.* **1**, 251.
Halstead, D. C. & Larrabee, M. G. (1972) In *Immunosympathectomy*, ed. Steiner, G. & Schönbaum, E., p. 221. Amsterdam: Elsevier.
Hawkins, R. A., Williamson, D. H. & Krebs, H. A. (1971) *Biochem. J.* **122**, 13.
Heald, P. J. (1954) *Biochem. J.* **57**, 673.
Heald, P. J. (1956) *Biochem. J.* **63**, 242.
Heller, I. H. & McIlwain, H. (1973) *Brain Res.* **53**, 105.
Hillman, H. H. & McIlwain, H. (1960) *J. Physiol.* **152**, 59P.
Horowicz, P. & Larrabee, M. G. (1958) *J. Neurochem.* **2**, 102.
Hosein, E. A., Emblem, M., Rochon, S. & Morch, S. E. (1962) *Archs Biochem. Biophys.* **99**, 414.
Ingvar, D. H. (1961) *Proc. int. Neurochem. Symp.* **4**, 118.
Itoh, T. & Quastel, J. H. (1970) *Biochem. J.* **116**, 641.
Jones, D. A. & McIlwain, H. (1971a) *J. Neurochem.* **18**, 41.
Jones, D. A. & McIlwain, H. (1971b) *J. Neurobiol.* **2**, 311.
Kakiuchi, S., Rall, T. W. & McIlwain, H. (1969) *J. Neurochem.* **16**, 485.
Kakiuchi, S. & Rall, T. W. (1970) *Molec. Pharmacol.* **4**, 367–379.
Keesey, J. C., Wallgren, H. & McIlwain, H. (1965) *Biochem. J.* **94**, 289.
Krebs, H. A. (1950) *Biochim. biophys. Acta* **4**, 249.
Krebs, H. A. & Eggleston, L. V. (1945) *Biochem. J.* **39**, 408.
Lajtha, A. (1967) *Prog. Brain Res.* **29**, 201.
Lajtha, A. (1969–72) *Handbook of Neurochemistry.* New York: Plenum.
Larrabee, M. G., Klingman, J. D. & Leicht, W. S. (1963) *J. Neurochem.* **10**, 549.
Larrabee, M. G. (1958) *J. Neurochem.* **2**, 81.
Lowenthal, A. (1972) *Handb. Neurochem.* **7**, 429.
Marks, N. & McIlwain, H. (1959) *Biochem. J.* **73**, 401.
Martin, H. & McIlwain, H. (1959) *Biochem. J.* **71**, 275.
McAfee, D. A., Schorderet, M. & Greengard, P. (1971) *Science, N.Y.* **171**, 1156.
McIlwain, H. (1956) *Biochem. J.* **63**, 257.
McIlwain, H. (1959b) *Biochem. J.* **73**, 514.
McIlwain, H. (1960) *J. Neurochem.* **6**, 244.
McIlwain, H. (1961) *Biochem. J.* **78**, 24.
McIlwain, H. (1967) *Prog. Brain Res.* **29**, 270.
McIlwain, H. (1972a) In *Experimental Models of Epilepsy*, ed. Purpura, Penry, Tower, Walter & Woodbury. New York: Raven Press.
McIlwain, H. (1972b) In *Effects of Drugs on Cellular Control Mechanisms*, ed. Rabin & Freedman. London: MacMillan.
McIlwain, H. (1972c) *Biochem. Soc. Symp.* **36**, 69.
McIlwain, H. & Bachelard, H. S. (1971) *Biochemistry and the Central Nervous System.* Edinburgh: Churchill Livingstone.
McIlwain, H., Harvey, J. A. & Rodriguez, G. (1969) *J. Neurochem.* **16**, 363.
McIlwain, H. & Snyder, S. H. (1970) *J. Neurochem.* **17**, 521.
McIlwain, H., Thomas, J. & Bell, J. L. (1957) *Biochem. J.* **64**, 332.
Pappius, H. M. & Elliott, K. A. C. (1956) *Can. J. Biochem. Physiol.* **34**, 1007, 1053.
Pfeiffer, C. C. (1972) *Int. Rev. Neurobiol.* Suppl. 1.
Piccoli, F., Grynbaum, A. & Lajtha, A. (1971) *J. Neurochem.* **18**, 1135.
Pull, I., Jones, D. A. & McIlwain, H. (1972) *J. Neurobiol.* **3**, 311.
Pull, I. & McIlwain, H. (1971) *J. Neurochem.* **18**, 1163.
Pull, I., McIlwain, H. & Ramsay, R. L. (1970) *Biochem. J.* **116**, 181.

Rodnight, R. & McIlwain, H. (1954) *Biochem, J.* **57**, 649.
Ross, B. D. (1972) *Perfusion Techniques in Biochemistry.* Oxford Univ. Press.
Sacks, W. (1969) *Handb. Neurochem.* **1**, 301.
Srinivasan, V., Neal, M. J. & Mitchell, J. F. (1969) *J. Neurochem.* **16**, 1235.
Stitch, S. R. (1957) *Biochem. J.* **67**, 97.
Terner, C., Eggleston, L. V. & Krebs, H. A. (1950) *Biochem. J.* **47**, 139.
Thomas, J. (1956) *Biochem. J.* **64**, 335.
Thomas, J. (1957) *Biochem. J.* **66**, 655.
Thomas, J. & McIlwain, H. (1956) *J. Neurochem.* **1**, 1.
Tower, D. (1969) *Handb. Neurochem.* **1**, 1.
Umbreit, W. W., Burris, R. H. & Stauffer, J. F. (1972) *Manometric Techniques.* Minneapolis: Burgess.
Van Harreveld, A. & Kooiman, M. (1965) *J. Neurochem.* **12**, 431.
Warburg, O. (1930) *The Metabolism of Tumours*, trans. Dickens, F., London: Constable.
Wolfe, L. S. & McIlwain, H. (1961) *Biochem. J.* **78**, 33.

8. Electrical Stimulation of the Metabolism of Isolated Neural Tissues

H. MCILWAIN

Electrode materials	160
Silver	160
Gold/Platinum/Other electrode materials	162
Electrode systems	165
Concentric electrodes	165
Grid electrodes	168
Rapid-transfer electrodes	171
Electrodes for localized application of pulses	171
Experiments appraising electrode systems	172
Sources of electrical pulses	173
Condenser pulses	173
Square-wave pulses and sine-wave current	175
Metabolic effects of electrical stimulation	177
Stimulation and inhibition of cation movements and glycolysis	178
Respiratory response and inhibition by chlorpromazine	179
Noradrenaline output and glycolysis, in a superfusion system	182
Summarizing notes and tables	185
References	188

In its normal state *in vivo* an animal's nervous system continually receives electrical impulses, at varying frequencies and levels of intensity, and either from receptors in contact with the environment or from other parts of the animal. Many *in vitro* experiments with neural tissues must therefore include some appropriate counterpart of the tissue's normal electrical environment as well as of its chemical environment.

The facility with which the level of activity of neural tissues can be altered by electrical means, indeed offers very great experimental opportunities. Advantage has been taken of this for nearly two centuries in more physiological investigations. Metabolic studies of isolated peripheral nerves at different levels of electrically-induced activity commenced about 40 years ago. Only in the last 20 years have corresponding studies been made with that part of the nervous system about which most chemical data are available, namely the mammalian central nervous system. It is with such studies that the greater part of the present chapter is concerned. Their significance can be judged from the knowledge that applied electrical pulses very markedly alter the composition and metabolism of cerebral tissues: in 5s their level of phosphocreatine can be halved, their lactic acid commences to accumulate and their respiration soon reaches double its previous rate.

Applying electrical stimuli to neural tissues involves a selection of electrode materials, electrode systems, and sources of electrical pulses, which are described successively below; metabolic effects of the pulses are then illustrated and summarized. For a general account of electrophysiological methods in biology, see Bureš, Petráň & Zachar (1967).

ELECTRODE MATERIALS

Careful consideration should be given to the selection of materials for use as electrodes when they are to dip into solutions in which metabolic change is being measured, or to come directly into contact with the neural preparation which is subsequently to be analysed. The dangers inherent in these situations will be evident; however, wide experience has now been gained with a variety of electrodes and several clear recommendations about the virtues and disadvantages of different materials can be given. Silver, gold, and platinum, which have been most widely applied, are described first.

Silver

Silver electrodes have a long tradition of use in electrophysiological experiments, both for recording potentials (which does not concern the present section) and for applying stimuli. This employment has largely concerned intact organs or organ-parts, for example peripheral nerve with its sheath intact. Deleterious effects of the silver do not appear to have been noted under these circumstances. It has also proved satisfactory with cell-containing, sliced tissues, as is described below. In cell-free systems its use must be regarded with suspicion, for a number of extraneous effects, due specifically to the silver, have been observed (see Chappell & Greville, 1954; Narayanaswami & McIlwain, 1954). Further, some grades of silver carry additional hazards through the addition of other metals; thus copper is added in 'sterling silver' to give a more hard-wearing metal, but it adds also a further potentially toxic agent. It is therefore necessary to specify pure silver for the preparation of electrodes; this is available commercially in a wide variety of forms (Johnson, Matthey & Co. Ltd, Hatton Garden, London).

Effects of currents at silver electrodes. Currents applied to silver electrodes in aqueous solution rapidly cause polarization. Thus electrodes H used with cerebral slices (see below) are comparable to two silver wires a few centimetres in length and 2 mm apart. If these are placed in 0·15 M-NaCl or one of the salines of Table 7.1, and momentarily connected to a source of potential of a few volts, the current flowing may at first correspond to a resistance of about 20 ohms at the electrodes. In a second or so, however, this has risen to some hundreds of ohms. A subsequent contact with the same source of potential with the same polarity, registers only the higher resistance. However, if the polarity is reversed, low resistance is again momentarily recorded. These relationships persist through very large numbers of reversals in polarity, and are the basis for the use of alternating rather than unidirectional series of pulses. When polarization has caused a high resistance at the electrode, only part of the potential difference applied to the electrodes is exerted in the fluid between the electrodes, the remaining fall in potential occurring at the metal–solution interface.

Silver electrodes undergo also, more permanent and extensive changes when pulses (for example, alternating pulses of 0·3 ms duration of 5 V peak potential at 100/s) are applied to them in ordinary physiological salines. These lead to 'chloriding', the formation of a surface film of presumably AgCl which under ordinary laboratory conditions in the presence of light or of tissue or both, becomes black. This may be Ag, Ag_2O or Ag_2S; the process is however limited to a surface layer. It is accompanied by relatively little change in impedance with pulses of the characteristics quoted.

The silver electrode-systems described below have been carefully tested under the conditions of typical metabolic experiments to see whether pulses induced changes likely to simulate or modify metabolism. The inhibitory phenomena found in cell-free systems were not exhibited with tissues. The situations examined included the use of oxygenated glucose salines, variously buffered, when alternating condenser pulses of up to 12 V, time-constant 1 ms, at 50Hz caused no change in oxygen or glucose. Oxygen uptake is however catalysed by such pulses at silver electrodes when the salines contain certain easily oxidized substances: as adrenaline, ascorbic acid or cysteine (Lewis & McIlwain, 1954; McIlwain, unpublished). Appropriate control experiments (see below) must therefore be carried out with any unusual reaction-mixtures.

Other properties; cleaning. Silver is of too low a melting-point to be fused into glass. It is too reactive to receive more than the briefest exposure to acids usually employed in cleaning glassware. Consequently silver electrodes have been made detachable from glassware and have been cleaned mechanically. When electrodes *H* or the quick-transfer electrodes have been used in the experiments described below, they acquire the black surface film. After each experiment, tissues are removed promptly and the electrodes washed briefly in running tap water and then in three changes of distilled water during 30 min. They are then shaken or blotted from adhering water and left loosely covered to dry in air at room temperature. Just before the subsequent experiment, the parts of the silver which constitute the electrodes and their attachments (i.e. the parts not enamelled) are cleaned until they show as uniformly bright silver. This is carried out at the electrode portions by scraping with a scalpel, using the minimal pressure compatible with removal of the black film. The sockets of the electrodes are cleaned internally, lightly with a roughened dissecting needle or a rat-tail file.

After a group of experiments, or when it is seen to be necessary, the electrodes are also treated with solvent and acid to remove grease and calcium salts. The solvent must be chosen in relation to the grease used, for example at ground joints, and with which the electrodes may have been in contact; and also in relation to the properties of any plastic or enamelled portions of the electrodes. Diethyl ether is suitable for the electrodes described below. After drying the electrodes as described in the preceding paragraph, they are swirled momentarily in a small beaker with ether, then ethanol, and then with distilled water. The water is then replaced with 20 mM-HCl for 2 min while the electrodes are swirled in the beaker; the acid is removed and the electrodes washed in 3 changes of distilled water, being left in the last for 30 min.

Gold

Gold has been found suitable as a material for electrodes which are to be used in metabolic experiments. It is sufficiently acid-resistant to be cleaned with the glassware, but of too low a melting-point to be fused into glass apparatus. Consequently, platinum (q.v.) which is itself not suitable for electrodes but can be fused in glass, has been gold-plated; practical details follow.

Electrodeposition of gold. A cyanide plating solution is prepared by dissolving gold chloride equivalent to 0·34 g Au, and 1·9 g KCN, in 100 ml of water. The vessel with the electrodes which are to be plated, is cleaned by standing for 1 h or by heating for a few minutes, in concentrated nitric acid and then rinsed free from acid with water. It is filled with the cyanide plating solution and placed in a water-bath at 70° in a fume cupboard. Connections are made from the cathode of a 2 V accumulator to the electrodes which are to be plated, preferably at points outside the electrode vessel (e.g. at the electrode arms of the vessels of Figs 8.1 or 8.2). The anode of the accumulator is taken through an ammeter to a heavy gold wire. This is held in a clamp carried above the electrode vessel by a rack and pinion, so that the wire can be lowered gradually into the plating solution. Plating is commenced by lowering the wire and observing the current registered on the ammeter, until a current is shown which corresponds to 6–10 mA/cm^2 of electrode being plated. Plating should be adequate in 20 min. It is stopped while the surface is still relatively smooth and bright, and before crystal-like growth has commenced. The plating should be entirely adherent even with rough mechanical handling.

Plating is stopped by removing the gold anode, and the plating solution must then be tipped immediately from the electrode vessel and the vessel rinsed with water. This is because the deposited gold will redissolve in the cyanide solution. For this reason, also, all electrodes which are to be plated within a vessel should be plated at the same time by connecting them together to the cathode of the accumulator. If plating is found to be incomplete, the vessel is again cleaned in acid and the plating repeated. Wire of 26-gauge (0·46 mm diam.), which is recommended for vessels A and E, has a surface area of 14·5mm^2/cm length.

Cleaning gold-electrode vessels. Gold is resistant to ordinary nitric or chromic-sulphuric cleaning mixtures provided these are made from pure chemicals and not contaminated. As ordinarily made and handled, however, these mixtures may be found to dissolve gold and this can probably be attributed to the presence of chlorine in them. It is therefore recommended that for cleaning gold-electrode vessels the cleaning baths be made from laboratory- and not commercial-grade chemicals, and that additional care is taken that all apparatus placed in them is rinsed free from salts. Vessels should be left in the acid for a chosen time, as 0·5 h, and not overnight.

Use in metabolic experiments. No deleterious effects have been found on applying sine-wave alternating currents of a few volts, or brief alternating condenser pulses (up to 15 V, time constant 0·5 ms, at 100/s), to gold electrodes in ordinary oxygenated glucose-salines. A number of other substrates, and tissues slices, may be present without interactions due to the gold itself being evident; this has been examined from the point of view of oxygen uptake, substrate utilization, acid formation, and the continuance of normal tissue metabolism after a period of electrical stimulation.

In solutions containing gelatin, pulses applied to gold electrodes have been found to yield in about 30 min a pink-coloured gold sol. Although this was not observed to be associated with change in tissue metabolism, use of gold electrodes in such solutions is obviously to be avoided; gold alloys should not be used.

Platinum

Platinum can be used as an electrode material in neurochemical work only with great caution; in many situations its use is most undesirable. The significance of several investigations which have used platinum electrodes is open to doubt, for changes which have been ascribed to tissue metabolism could have resulted from catalysis at charged platinum surfaces.

Platinum in finely divided form, as platinum-black or spongy platinum, is well known as a catalyst in many situations including aqueous solutions at room temperatures, but the potency of ordinary bright metallic platinum has often been overlooked. The metal is particularly active as a catalyst when it is electrically charged. Thus the stimulating pulses or currents normally used in electrophysiological experiments are sufficient to catalyse oxidation of glucose in ordinary oxygenated media, when the pulses or currents are applied at platinum electrodes. Thus small platinum electrodes 1 mm^2 in area, when conducting 8 mA sine wave alternating current of 50 Hz at 3 V through ordinary glucose-phosphate or bicarbonate salines, catalysed the absorption of 10 μmol O_2/h, and the formation of acid. If such a change took place in a vessel in which the respiration of 50 mg of neural tissue was being measured, it would represent a change equivalent to 150–400% of the tissue's normal respiration (the magnitude of this percentage depending on the respiratory rate of the tissue).

Situations in which platinum electrodes may be used. The properties of platinum which have been outlined indicate the obligation of an investigator to appraise carefully the use of platinum in his own experimental situation. In some, indicated below, it is innocuous. The virtues of platinum which have led to its widespread use, are well known. It can be fused in glass (though seals in borosilicate glass are not always satisfactory), cleaned by strong acids or heat, readily made to a variety of shapes, and in the absence of applied potentials it is not associated with reactions among ordinary tissue constituents.

When carrying quite small currents, as in recording electrodes, there is no indication that platinum is deleterious. The use of platinum stimulating electrodes outside the experimental chamber proper, is also in most cases satisfactory. In these situations the stimulus is carried as a nerve impulse to the preparation being studied. The stimulus applied at the electrodes is usually chosen as being above the value needed for full response by the preparation. An attempt to determine change in threshold of stimuli applied to platinum electrodes should however be regarded with suspicion unless adequate control experiments are recorded, for the stimuli are likely to bring about chemical change in fluid with which they are in contact and this, for example a change in pH, may alter threshold. Acid formation has been noted above: oxidation may yield hydrogen peroxide in reactions catalysed by platinum.

Other electrode materials

Several samples of stainless steel, as well as other ferrous metals examined, were found unsuitable as stimulating electrodes in aqueous reaction mixtures. Stainless steel, after being inert for a period, could then develop patches of rusting which spread rapidly and on some occasions were associated with uptake of oxygen at appreciable rates; this was especially so if the steel had been adequately cleaned by scraping or by acid. Stainless steel has however been used for clips in electrodes *H* (Fig. 8.2); the clips do not carry currents.

Tungsten and molybdenum can be sealed through glass and have not been found to produce artifacts in typical metabolic experiments. Tungsten is difficult to work by mechanical means except by grinding, but molybdenum wire can readily be shaped to produce the electrodes of vessels *E*, below. Its use in this way is described by Narayanaswami & McIlwain (1954). Molybdenum must be given some protection from oxygen during glass-blowing with high melting-point glass, by introducing nitrogen to the vessel and working quickly.

ELECTRICAL STIMULATION OF TISSUE METABOLISM 165

ELECTRODE SYSTEMS

Instances in which electrical stimuli are applied to a nerve outside a reaction-vessel, and conducted into the vessel by existing neural connections, raise few problems which are specific to neurochemistry. An example of such experiments is given in Chapter 12. The present section is therefore concerned with electrode systems which stimulate by establishing electrical potential gradients at the metabolizing tissue itself. In these situations both electrical and metabolic factors must be taken into account in designing electrodes and vessels. In particular, the presence around the tissue of several times its volume of aqueous fluid carrying oxygen, salts and substrates must usually be accepted.

Electrode systems designed on these principles are described below, commencing with those in which gaseous exchange is measured. Conical manometric vessels have formed the most suitable basis for these, despite experiments with a variety of other forms. To the conical vessels have been fitted either permanent concentric electrodes, or removable grids. For analysis of tissue, or for repeated sampling of the fluid which surrounds it, quick-transfer electrodes and superfusion experiments are subsequently described.

Concentric-electrode vessels

These are illustrated in Fig. 8.1. Electrode vessel E has been based on the conventional conical manometric vessel of 15–20 ml total volume, taking about 3·5 ml of saline. Electrode vessel F is of 4–5 ml total volume and takes 0·7–1 ml of saline. The following characteristics are important in their use. (1) The electrodes are disposed so that all or most of the reaction mixture is between them. (2) The tissue can thus be floating freely within the fluid, and the vessels are suitable for quite small and irregular pieces of tissue, or for tissue-suspensions produced by chopping. (3) The voltage-gradient between the electrodes is not uniform; measurements showing a two fold variation are quoted in Fig. 8.1. (4) The electrodes are permanently fixed in the vessel and have therefore been made of gold-plated platinum, or of molybdenum, so that they can be cleaned *in situ*.

Preparation of the vessels. These are received from the glassblower (Messrs. A. W. Dixon & Co., 30 Anerley Station Road, London, S.E. 20) with their electrode wires made approximately to the size and shape of Fig. 8.1. The electrodes are usually of 26-gauge platinum, welded to tungsten at the points where they leave the vessel, and the welding sealed completely in the glass of the electrode arms. The tungsten may then be taken completely up the electrode arm, or left as a stub. In the laboratory the vessels are then prepared for use as follows (see McIlwain, 1961a).(1) They are tested for any leak by filling with water and applying a pressure of about 10lb/in². External connections are then made to the tungsten by cleaning it with sodium nitrite,

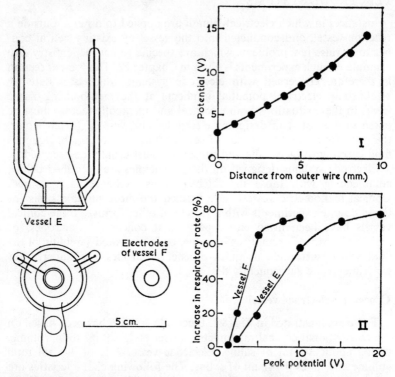

Fig. 8.1 Elevation and plan of concentric-electrode vessel E, and plan only of the electrodes of vessel F which is of similar but smaller design. The side-arm, to take a valve stopper, is shown in plan only.
I: Potential gradient established in saline between the electrodes of a vessel E when condenser pulses of peak potential 18 V, 0·4 ms time-constant, were applied at 100 Hz. The electrodes were of gold-plated platinum wire of 0·46 mm in diameter. The points record potentials observed with a probe electrode; the line is drawn according to theoretical expectation (Ayres & McIlwain, 1953).
II: Respiratory response of guinea-pig cerebral cortex in vessels E and F to condenser pulses of varying voltage, other characteristics being as in I. When response in the two vessels was expressed in terms of peak potential gradient, the two sets of points coincided (McIlwain, 1954a).

copper-plating it, and tinning with solder. If the tungsten reaches beyond the electrode-arm the tinned end forms a terminal, to receive a clip from the pulse-generator. If the tungsten ends as a stub in the electrode-arm, there is soldered to it a length of tinned copper wire which extends just beyond the electrode arm, which is then closed by cement. (2) Any excess electrode-wire inside the vessel is cut off, and if the volume of the vessel is to be determined using mercury, the calibration is carried out at this stage. (3) The wires of the electrodes are bent free from any contact with each other or with the walls of the vessel, the vessel cleaned with nitric acid and washed well. The wires are then gold-plated as described above, and bent to their final positions. This is done with forceps which have pieces of plastic tubing slipped over their jaws to ensure that the plating will not be damaged.

Electrical characteristics and metabolic response. The impedance of the conical electrode vessels under the conditions in which they are used in metabolic experiments, is equivalent to resistances of 40–50 Ω (vessel E) and 35–40 Ω (vessel F), determined by substitution as described below. Stimulation of tissue in the vessels involves applying to the electrodes brief pulses of 1–18 V in peak potential (Fig. 8.1). Electrical events in the fluid during such application have been determined by measuring the potentials attained at a probe wire placed by a travelling microscope at different, measured, positions between the electrodes. An example of the findings is quoted in Fig. 8.1, I; this shows (1) that an abrupt potential change occurs between the electrodes and the solution; and (2) that in the solution there is a gradual change in potential which is reasonably in agreement with theoretical expectation. The voltage gradient in the solution is given by the slope of the line of Fig. 8.1, I; this is not constant, but a vessel designed for greater uniformity in potential gradient proved less satisfactory in other respects. The potential gradients established in vessels E and F on applying pulses of several voltages are recorded by Ayres & McIlwain (1953), together with the abrupt change at the electrodes.

Metabolic change on stimulating tissue in concentric-electrode vessels has been correlated with electrical characteristics of the applied pulses, as is illustrated with respect to tissue respiration in Fig. 8.1, II. This is a voltage-response curve, other characteristics of the stimuli being kept constant; above a threshold voltage, respiration is seen to increase with increase in applied potential until it reaches a maximum about 85% above its unstimulated value. For a given change it is necessary to apply greater potentials to the larger vessel E than to vessel F, but when expressed in terms of voltage gradient the results in the two vessels are found to be equivalent, a peak potential gradient of about 0·6 V/mm being necessary in each to give an increase in respiration of 50% (McIlwain, 1954a). Response has also been correlated with time-characteristics of the applied pulses, generated and measured as described below. The effect of pulses applied at 50 Hz increased with increase in time-constant up to 0·3 or 0·4 ms. However, when response was plotted against the product of pulse-duration and pulse-frequency, pulses at 3 or 10/s were found most effective in causing respiratory response.

Other concentric-electrode systems. In place of gold-plated platinum, the most suitable metal found for use in electrode vessels E is molybdenum. Gradient measurements in vessels made according to Fig. 8.1, I but with two molybdenum electrodes showed a rather greater fall in

M

potential at the metal-fluid interface, but otherwise relationships were similar to those at gold electrodes (Ayres & McIlwain, 1953). Detachable silver electrodes have been used by Wallgren & Kulonen (1960).

Electrodes approximating in size to those of vessel F have been made from gold, silver or molybdenum wire and mounted on a glass rod (Ayres & McIlwain, 1953). The rod is held by a stopper in a test-tube and the assembly can be shaken in a thermostat. In this way it is easy to demonstrate a glycolytic response of chopped cerebral tissue to electrical pulses. The arrangement has the virtue of being easy to assemble cheaply for an occasional demonstration, but is not as satisfactory or versatile as the quick-transfer apparatus described subsequently.

Grid electrodes

Grid electrodes (H) are shown in Fig. 8.2 in a conical manometric vessel (A) of conventional size and fitted with electrode arms. The following characteristics explain the design and use of the electrodes and vessels. (1) The electrodes are disposed around a relatively small volume of fluid (0·05 ml) within which a tissue slice is held. During an experiment the slice is in contact with a much larger volume of fluid, for the electrode holds the slice only loosely, and fully extended, in 3–5 ml of fluid which moves over the slice when the vessel is shaken in the usual way. (2) The slice is held at 14 points near its edges but is relatively little damaged, the tissue being displaced rather than crushed at the points of contact. A voltage gradient can then be established in a definite direction and position in relation to the tissue. (3) Voltage gradients between the electrodes have been measured (Fig. 8.2); the wires of the electrodes are only 2 mm apart and the voltages used are smaller than with vessels E. The impedance of the electrodes, observed with pulses of the characteristics causing stimulation, and measured by substitution as described subsequently, is equivalent to 40–50 ohms without the tissue, and about three times this value with a slice in position. (4) As it is necessary to manoeuvre a slice into an individual electrode, the electrodes have been made removable from the vessels. The electrodes are thus cleaned and prepared separately from the glassware, and this enables silver wire and a plastic frame to be used in constructing them.

Vessels A. These vessels, designed to take the grid electrodes, have two specific features. They carry electrode arms from which project, into the vessel, stubs of 26-gauge platinum wire; and the reagent-well which is normally fused to the centre of the base of a manometric vessel, is displaced towards the outer wall. In making the vessels (Dixon & Co., see above) the reagent well, electrode arms, and side-arm for reagents should be in the relative positions shown in Fig. 8.2. After glass-blowing, the vessels are prepared for use in a fashion similar to that described for vessels E: they are tested for any leakage; external

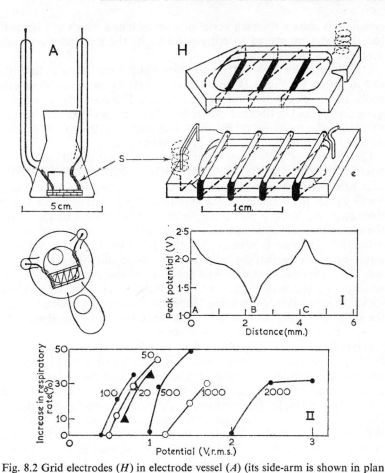

Fig. 8.2 Grid electrodes (*H*) in electrode vessel (*A*) (its side-arm is shown in plan only). On the right the two halves of the electrode are shown separated. Each half is wound with a continuous length of enamelled wire, but for clarity in the figure the wire is shown in full only when it is serving as an electrode. Black areas: enamel intact. *S*: connecting spirals; *e*, end held when inserting the electrode into its vessel. A larger version of electrodes *H* is described in the text.
I: Potential gradients in saline between wires *A*, *B* and *C* of part of an electrode *H*, when sine-wave current of 3·15 V peak potential at 50 Hz was applied.
II: Voltage-response curves of guinea-pig cerebral cortex in electrodes *H*, to sine-wave a.c. of the frequencies (Hz) quoted in the diagram.

connections are made to their electrode stubs and the stubs trimmed, bent into the positions of Fig. 8.2 and gold-plated.

Grid electrodes H. Constructional details of the electrodes are given by Ayres & McIlwain (1953) and McIlwain (1961a); they are manufactured and supplied with their silver wire completely enamelled. An electrode is chosen for each vessel of a set of 6 or 8 which will be used together, and numbered to correspond to it; the electrodes are then prepared for use as follows.

(1) The wires of each of the two grids which constitute an electrode end in spirals (S, Fig. 8.2) which form sockets for attaching the electrodes to the electrode stubs of the vessels. These spirals, which are quite springy and flexible but can be bent into new positions, are bent to shapes similar to those of Fig. 8.2. The length of the spirals is also adjusted if necessary, by cutting off or stretching.

(2) The end of each spiral is now shaped so that it will make firm electrical contact with the stubs. The shaping is carried out with the tip of a needle-file, which comes to a point and is chosen to be of a size which equals the diameter of the stub (about 0·5 mm) at about 4 mm from its point. About 8 mm of the file are inserted in the wire spiral and the end of the spiral held against the file with a finger of the right hand. While the electrode is held in the other hand the file is rotated in a direction which tightens the spiral until its inside takes the shape of the file. The file is now rotated against the inside of the spiral sufficiently to remove enamel and some silver and to leave about half the inside of the socket as bare silver. After being shaped in this way the electrodes grip the stubs when they are pressed on to them, and retain good metallic contact during experiments of some hours.

(3) The electrodes are put into their vessels. An electrode is gripped firmly by forceps at its 'back' end (e, Fig. 8.2) and inserted into the vessel to reach the position shown; the flexibility of the electrode arms allows them to bend when passing the neck of the vessel and regain their positions inside. The front end of the electrode reaches the base of the vessel first and the other end is then also guided down by the forceps. With the forceps, the two spirals in turn are pressed home on the stubs. Note that the spirals are bent to facilitate insertion in this manner: their ends should point away from the worker when he is inserting the electrodes into the vessels. Any small adjustments needed in the length or position of the electrode spirals will now be evident, and can be made when the electrodes are removed from the vessels; the removal also is done with forceps.

(4) The parts of the enamelled silver wire which are to act as electrodes are now bared by scraping. There are four transverse pieces of wire on the lower and three on the upper electrode. They are scraped with a scalpel and observed with a $\times 8$ magnifier to ensure that all enamel is removed from the parts shown; unless the wires are examined closely it is easy to overlook patches of enamel.

An experiment illustrating the use of the electrodes is given in a subsequent section (Table 8.1). A larger model of electrodes H is advantageous and has been used extensively; it carries four electrodes on its upper frame and five on its lower, spaced as illustrated, each frame being 10×21 mm.

Rapid-transfer electrodes

A major feature of the electrical stimulation of neural tissues is the speed of their metabolic response: change in tissue composition may go to completion during the application of pulses for 5 or 30 s only. The manometric apparatus which has been described is suitable for measuring sustained changes in metabolism, and may be used to observe changes in tissue composition during intervals of a minute or more. For observing more rapid changes, the quick-transfer apparatus of Fig. 7.2 and 7.3 was devised.

When this apparatus is used with quick-transfer electrodes, the jaws of the quick-transfer holders are wound with the enamelled silver wire used with electrodes H, again as grids 2 mm apart. Holders so prepared are supplied (Specialized Laboratory Equipment Ltd; see p. 147) with the enamel of the wires intact. For use, the portions of the wires which are to act as electrodes are bared by scraping with a scalpel as previously described. The potentials established in salines between such electrodes when pulses are applied to them, has been found similar to that recorded in Fig. 8.2 at electrodes H. The impedance of the quick-transfer electrodes to alternating condenser pulses of 10 V peak potential and 0·4 ms time-constant, at 50 Hz, was found by substitution to be equivalent to about 20 Ω.

Several experiments illustrating the use of the quick transfer electrodes are given in Chapter 7 and later in the present chapter.

Electrodes for localized application of pulses

The several types of electrodes described above were designed to give maximum metabolic response and this was found to be brought about by placing the whole of a tissue-sample within an appreciable voltage-gradient (McIlwain, 1951a). In this situation, any effects depending on transmission within the tissue are not likely to be shown, and to exhibit any such effects the electrodes of the present section have been used. These are based on the grid electrodes H or the quick-transfer electrodes, and produced simply by baring limited portions only of their enamelled wires. To do this in a reproducible manner, templates have been made into which the electrodes can be fitted (McIlwain, 1961a).

The template used in preparing the quick-transfer electrodes is shown in Fig. 8.3. The template is fitted over one of the jaws of the holder, and 1·5 mm portions only of the wires are then accessible through a slit. A chosen wire is scraped with a small tool, which is made from a lancet. The same template is then fitted over the other jaw, and 1·5 mm of an opposing wire similarly bared. Templates have also been made on the same principle for exposing limited portions of the grid electrodes H. Metabolic findings with the electrodes are described by Bollard & McIlwain (1959), Kurokawa (1960) and Yamamoto & McIlwain (1966).

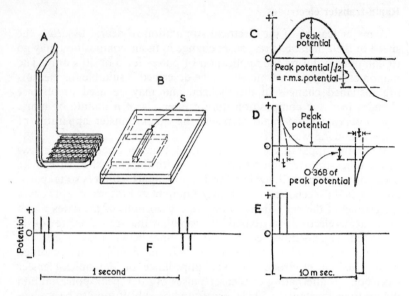

Fig. 8.3 A. The lower jaw of a quick-transfer holder, equipped with a grid electrode and nylon gauze. For the complete holder, with its beaker, thermostat and distributors for stimuli and oxygen, see Fig. 7.2.
B: A template for baring a limited portion of the grid electrode wires. In use it is placed over the electrode and a wire scraped free from a portion of its enamel, through the slot s of the template. The same template serves also for the upper electrode.
C to E: Time-voltage relationships: in (C), sine-wave a.c; (D), condenser pulses (t = time constant: see text), and (E), square waves. F: Time-voltage relationships in a series of pulses at the output of the pulse-interruptor.

Experiments appraising electrode systems

Some appropriate control experiments may be noted, although this can be done in general terms only and the control experiments necessary will depend greatly on the systems being studied. (i) The electrode in the experimental salines, with and without pulses of the maximal intended characteristics should not affect the process being measured: e.g. cause gas change in a manometric experiment, or change in glucose if this is being measured. This can be examined in the absence of the tissue preparation. (ii) However, additional factors contributed by the tissue must be evaluated. The tissue alters the medium by adding or removing metabolites, and also adds some of its own substance. Consequently, medium in which tissue has been incubated under the chosen experimental conditions, and from which tissue has been removed, should also be exposed to pulses with concomitant measurements as described above. (iii) The possible production at the charged electrodes of substances modifying tissue metabolism, is also to be evaluated. Their production may not be accompanied by changes in the characteristics measured. Thus it is necessary to expose media to charged electrodes as in (i) and (ii) and subsequently to use the exposed media in the chosen experimental system: observing whether, for example, it changes tissue phosphates. (iv) If the electrodes enclose only a small volume of the fluid in which tissue is being studied, it is also possible to appraise extraneous effects by comparing the result of applying pulses while tissue is within the electrodes, with the result of applying them while the tissue is outside the electrodes.

These suggestions are based on experiments of McIlwain (1951a), Narayanaswami & McIlwain (1954), and Lewis & McIlwain (1954) The controls which they offer are related purely to changes observed by chemical means. In addition, the electrophysiological status of the tissue before and after stimulation can be appraised as described in the following chapter.

SOURCES OF ELECTRICAL PULSES

In the electrode systems described, the electrode vessels or holders show impedance equivalent to 30–150 Ω measured with the quantities of tissue and fluid typical of metabolic experiments. Six such electrodes used in parallel in an experiment thus offer an impedance of only 5–25 Ω, and a source of pulses is required to establish the stimulating potential across such a resistance. The potential gradients found necessary for stimulation are of 0·3–3 V/mm, requiring in the different electrode systems the application of 1 to 12 V peak potential during brief periods. These factors condition the choice of instruments used for electrical stimulation.

Condenser pulses

Electrical conditions most effective for stimulating many excitable tissues involve the rapid establishing of the chosen potential for a brief period, followed by rapid fall, and with the peak potential maintained for only a small proportion of the total time. A simple way of producing such pulses of current is by the charge and discharge of condensers. The pulses then have the exponential time-voltage relationships shown in Fig. 8.3; by using both charge and discharge the pulses alternate in polarity and electrode polarization is minimized.

When a capacitance c of 4 μF is charged to a potential e_0 through an electrode vessel such as those described, which with saline and tissue are of impedance r equivalent to some 100 ohms, the formula:

$$\log_\epsilon \frac{e^0}{e} = -\frac{t}{cr}$$

gives the time-constant as 0·4 ms. This is the time t required for the potential to fall by a fraction $(\epsilon-1)/\epsilon$ or 0·632 of its peak value, and equivalent also to the extrapolation shown in Fig. 8.3. Such pulses, applied at 20Hz and at potentials which give gradients of a few volts per millimetre, ordinarily induce maximum metabolic response by isolated cerebral tissues. It will be noted that in the electrode systems which have been described the electrodes are from 2–10 mm apart, and that applied potentials of 5–12 V are required to give the voltage gradients associated with maximal response.

These characteristics are embodied in the condenser-pulse generators made commercially, for example by Specialized Laboratory Equipment, Croydon, Surrey. Principles of construction are described by Bureš *et al.* (1967). Such instruments should supply up to six electrode vessels. Condensers of appropriate capacity can be brought into use and charged to potentials of 0·5–30 V. In this apparatus, as in the examples just quoted, the impedance of the electrode vessels themselves conditions the time-constant of the discharge. This can however be brought to chosen values by the controls of the instrument. The controls allow (Fig. 8.4): continuous variation in output potential; stepwise and continuous variation in pulse frequency. The pulses are required to be uniformly spaced in time and alternating in potential, and for most metabolic work it is adequate for them to be supplied at frequencies between 1 and 100 Hz (i.e., 2 to 200 pulses/s). The output when supplying a circuit of impedance equivalent to 5 Ω should be capable of adjustment between 0 and 12 V, giving pulses with time-constant between 0·02 and 1 ms.

In using the generator, leads are taken from it to the vessels to be stimulated and to an oscilloscope, as shown in Fig. 8.4. The instruments are switched on at the mains and when they come into operation the oscilloscope is set to display the pulse which it is intended to use. The pulse generator is set to give minimal voltage and duration and switched into connection with the vessels and the oscilloscope. The duration and voltage of the pulses are now increased gradually while observing their values on the oscilloscope, until they reach the chosen values. The oscilloscope is left running, with the pulse displayed, for the remainder of the experiment. A suitable oscilloscope for use with the generator is model D54 of Telequipment, London (also available through Tektronix, Oregon, U.S.A.). Further details of the use of the generator are given in the illustrative experiment at the end of this section.

The voltage and time characteristics of the pulse from a given setting of the generator can be observed by the oscilloscope to depend on the impedance of the circuit to which its output is connected. With minimum resistance in the output circuit of the instrument connection of 20 Ω to the output can cause a 30% fall in potential. (At other settings of the instrument, it can be made less sensitive to an attached resistance.) The impedance of electrode vessels can be measured approximately in this way, under the conditions in which they are to be used. This is done by adding the appropriate amount of fluid and connecting the vessel to the generator, oscilloscope and resistance box as in Fig. 8.4. Pulses of the characteristics used with the vessel are displayed on the oscilloscope, the vessel switched into circuit, and the resulting change of potential noted. The vessel is now switched off, the resistance box switched into circuit and the resistance of the box adjusted until it brings about the same change as was caused by the vessel. Results obtained by this method of substitution have been quoted in describing the individual electrode systems.

If it is attempted to measure the resistance of the vessels to direct current, using a moving-coil instrument, much higher values are obtained. These are not however relevant to the use of the vessels with alternating pulses because of polarization phenomena at the electrode-fluid interfaces.

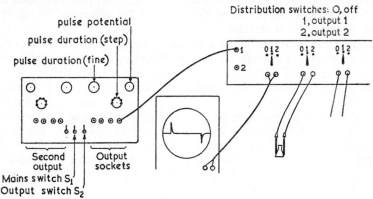

Fig. 8.4 Connections between pulse generator, oscilloscope and distributor supplying several electrode-vessels in parallel. The distributor is mounted immediately above the electrode vessel, e.g. on the thermostat of a quick-transfer apparatus (Fig. 7.2) or on the tank of a Warburg thermostat. A resistance-box may be connected in the place of a vessel. For clarity in the diagram, connections from the second output of the generator to the distributor are not shown.

Square-wave pulses and sine-wave current

The following apparatus is needed to supply and measure alternating rectangular pulses of appropriate characteristics. (i) A square-wave generator giving alternating pulses of duration 0·02 to a few ms, at intervals of 1–100 ms. It should be capable of supplying pulses which are alternating in polarity; alternatively, two stimulators are needed which can be coupled and caused to deliver the alternating sequence (see McIlwain & Joanny, 1963; Heller & McIlwain, 1973). (ii) A power amplifier with an output of 5 watts or more, and with minimal distortion. Such instruments are made for amplification of sound. The amplifier is needed because the current from typical square-wave stimulators is inadequate to supply multiple electrodes immersed in aqueous salt solutions. In normal physiological use such stimulators supply electrodes which are smaller than those used in the present work, and which are placed in contact with tissues of relatively high impedance. If the output from a square-wave stimulator is connected directly to, for example, the four electrode vessels of Table 8.1, the change in potential which is caused at the electrodes is much smaller than that of the nominal output and is distorted in time-voltage relationship. The output from the stimulator is therefore supplied to the power amplifier, and the amplifier supplies the electrodes; for an example of the controls and connections involved, see McIlwain & Joanny (1963) and Bureš et al., (1967). (iii) An oscilloscope is required; this will be evident from the foregoing statement. The instrument is connected in parallel with the group of electrode vessels, and is used to monitor the pulses supplied to them throughout the period of stimulation. A Telequipment (Tektronix) oscilloscope D54 is suitable.

To stimulate neural tissues with sine-wave a.c. requires a relatively large expenditure of electrical energy. Therefore such currents are not usually the stimuli of choice, but have been used in two situations. (i) In comparing the relative effectiveness of stimuli of a variety of time voltage relationships (see, in relation to cerebral cortex, McIlwain, 1954a). (ii) In examining modified tissues, for example in the presence of drugs, when their normal excitability is altered and specific effects of a.c. currents of particular frequencies have been found (Greengard & McIlwain, 1955a). It is to be noted that currents applied as sine-wave a.c. do not necessarily retain these voltage-time relationships between electrodes in aqueous solutions (see Ayres & McIlwain, 1953). Mains alternating current of 50 or 60 Hz is of a suitable frequency for stimulation of neural tissues and with a transformer is easily supplied at an appropriate range of potentials, for example between 0·5 and 3·5 V r.m.s. or virtual voltage measured with ordinary moving-coil instruments: Fig. 8.3.

Sine-wave generators are available commercially which give current of a wide range of frequencies. The frequencies of interest in relation to cerebral tissues lie between 10 and 2000 Hz (Fig. 8.2). In order that a number of electrode vessels be suppied simultaneously, it is again necessary to use a power amplifier between the oscillator and the vessel, and to observe with an oscilloscope the time-voltage relationships of the current supplied while the vessels are connected to the amplifier. The response of cerebral tissues to a.c. supplied to grid electrodes H is shown in Fig. 8.2. Two characteristics of the current are concerned: frequency and voltage, and response-voltage curves are shown at six frequencies. By interpolation between these and other results, the frequency at which a given degree of stimulation was obtained with minimal current or voltage, was concluded to be about 80 Hz; the threshold was then about 0·4 V r.m.s.

The instruments described above give continuous series of pulses of chosen types. Specific effects have however been obtained by applying bursts of pulses separated by intervals during which no pulses are applied and which last between 0·5–5s; for these and the apparatus used for interrupting the series of pulses, see McIlwain (1954a, 1961). With pulses patterned in this way, evidence relating to persistence of the effect of pulses, or to a refractory period, has been obtained; the tissue also shows unusual relationships to certain added agents.

METABOLIC EFFECTS OF ELECTRICAL STIMULATION

Electrical excitation causes a multiplicity of chemical changes in responsive tissues, and the changes are summarized in Tables 8.4 and 8.5 at the end of this chapter; observation of associated electrical changes are the subject of the following chapter. The changes produced on excitation can be considered in relation to the following sequence. (i) The stimulating pulses momentarily alter the electrical charge at polarized membranes of component cells of the tissue, which (ii) respond by a transitory increase in permeability to Na^+. This allows greater Na^+ entry and accentuates the depolarization caused by the initial permeability change; the depolarization may also be conducted to contiguous regions. (iii) Depolarization promptly causes increased permeability to K^+ which is lost from cellular regions (where it is at high concentration), so causing them to regain their negative charge. (iv) More slowly, the Na^+ movements are reversed with the utilization of ATP and its conversion to ADP and inorganic phosphate. (v) These products and ATP are in equilibrium with phosphocreatine and creatine, and the less phosphorylated components are capable of accelerating the oxidative and glycolytic processes of phosphorylation which resynthesize ATP. (vi) The depolarization of certain regions, especially of synapses, causes release of some of their constituents to extracellular fluids; some of the compounds released function as neurotransmitters. (vii) Numerous secondary processes of translocation and metabolism occur in the compounds involved in (v) and (vi).

Three experimental arrangements will now be described which offer characteristically different advantages in the measurement of changes in this sequence. (I) First, the use of quick-transfer electrodes for measuring changes in cations and in glycolysis. (II) Second, manometric experiments for determining respiratory changes and for imposing anoxia. (III) Third, a superfusion system for measuring neurotransmitter output and carbohydrate metabolism.

I Stimulation and inhibition of cation movements and glycolysis

This experiment exhibits the specific action of tetrodotoxin, which must be handled with caution as a scheduled poison (see below). Six tissue samples are incubated aerobically in quick-transfer holders fitted with silver-grid electrodes, using the apparatus of Fig. 7.2 and 8.3; two investigators work together in preparing and stimulating the tissues. Beakers and electrodes are arranged and numbered as in Table 8.1 and the bared portions of the silver electrode-wires freshly scraped clean. The solutions specified are pipetted to the beakers. Isotonic sucrose in a crystallizing dish and 6% trichloroacetic acid in homogenizer tubes are put in crushed ice.

The brain of one or more guinea pigs is removed for preparation of tissue slices as described in Chapter 7. The weighed tissues are successively mounted in the transfer electrodes, the electrodes placed in their beakers, the flow of O_2–CO_2 begun and continued throughout the incubation period, and the beaker placed in the incubating bath at 38°C. After about 20 min incubation, the stimulator and oscilloscope are switched on, run at about 0·5 V peak potential with pulses of time-constant 0·4 ms, and the continuity of connections is tested to each of

Table 8.1 Stimulation and inhibition of cation movements and glycolysis

A

Components	Beaker contents in order of manipulation					
	(1)	(2)	(3)	(4)	(5)	(6)
Quick transfer electrodes						
Glucose-bicarbonate saline (ml)	5	5	4·5	4·5	5	5
Tetrodotoxin solution, 0·75 μM	0	0	0·5	0·5	0	0
Tissues in order of cutting	(weight recorded, and time of commencing incubation)					
Stimulation, S	0	S	S	S	S	0

B

Substances determined	Values in individual beakers					
Lactate accumulation in incubation fluid (μmol/g tissue/expt.)	36	68	40	37	62	31
K^+ content of tissue, (μ equiv./g)	75	52	70	68	50	73
Na^+ content of tissue (μ equiv./g)	75	92	77	86	97	84
'Non-inulin' Na^+ of tissue, (μ equiv./g)	19	50	20	24	58	22

The incubation medium was equilibrated with 5% CO_2 in O_2 and contained 1% of inulin when the 'non-inulin' Na was to be measured. For further details, see Chapter 2; McIlwain & Joanny, 1963; Keesey *et al.*, 1965; and McIlwain *et al.*, 1969. Section *B* of the Table gives typical findings from an experiment in which the tissue samples were incubated for 57–63 min and stimulated during the terminal 15 min of that period.

the electrodes which are to receive pulses. This is done by displaying the time-voltage relationship of the pulses on the oscilloscope, and momentarily switching each vessel in turn into circuit; continuity is indicated by a fall in voltage comparable to that seen when a resistance of about 30 Ω is connected between the output terminals.

Stimulation: at a chosen time 30–40 min after incubation was commenced, the vessels which are to receive pulses are switched into connection with the stimulator. This is now run at a potential sufficient for stimulation, e.g. at 8 V peak potential, 0·4 ms time-constant, and 60 pulses/s (30 Hz). The output from the stimulator will require adjustment as each vessel is connected or disconnected, and the time-voltage relationships of the pulses are observed on the oscilloscope throughout the experiment to ensure the supply of pulses of the characteristics chosen. Stimulation is terminated by releasing the tissues successively from their quick-transfer electrodes into the dish of ice-cold sucrose. This is done after a chosen period of stimulation, and within 5 s of its release the tissue is picked from the dish with a mounted, bent wire, drained by drawing it up the side of the dish, and dropped to the cold trichloroacetic acid. Here it is promptly dispersed, and samples are later taken for determining Na, K and inulin. Samples of incubation fluids from the beakers are taken for determination of lactic acid.

Results are exemplified by the data of Table 8.1. Their rate of lactate formation in absence of excitation averages 33 μmol/g.h, and in those stimulated in absence of tetrodotoxin this has become 65: that is, it has increased by 32 μmol as a result of 15 min excitation. Assuming, as is likely, that this increase occurs during the stimulation, glycolysis during this period is occurring at 33 + (4 × 32), or 161 μmol/g.h. This gives a value nearly 5 times the resting rate of glycolysis; such increase is a typical response to maximal excitation. Values for non-inulin Na are calculated and appraised as described in Chapter 2 and with appropriate reservations are regarded as a measure of intracellular Na content. Excitation has increased their average value from 21 to 54 μequiv/g tissue, while K has fallen by a nearly equal amount.

Tetrodotoxin has almost completely inhibited those effects of stimulation which have been measured in the present experiments. The compound (see Kao, 1966; McIlwain et al., 1969; Pull & McIlwain, 1973) is highly toxic, is purchased and handled with admixed citrate buffer; aqueous solutions are kept at 0° for not more than 3 days and pipetted by syringe. The experiment as quoted in Table 8.1 is one of two or four which would be carried out to determine the action of one concentration of an added agent such as tetrodotoxin. In the parallel experiments, the tetrodotoxin and the stimulation would be applied to different vessels and tissues of the sequence: for example, to numbers 1, 2, 5 and 6.

II Respiratory response and inhibition by chlorpromazine

The experimental arrangement outlined in Table 8.2 employs grid electrodes in conical manometric vessels of total volume about 15 ml; alternative arrangements are noted below.

The six electrodes are already prepared to fit their vessels; an ordinary Warburg vessel is used as thermobarometer. Thermostat, gassing arrangements, saline and the preparation of the tissue are those usual for the measurement of tissue respiration. The blade and glass guide, cutting block and shallow porcelain dishes of Chapter 6 are used for slicing the tissue.

Table 8.2 Respiratory and glycolytic responses to electrical pulses, and their inhibition by chlorpromazine

Vessel	Thermo-barometer	Electrode vessels A with grid electrodes H					
		(1)	(2)	(3)	(4)	(5)	(6)
Added agent	0	0	0	C*	C*	0	0
Tissue slice		First hemisphere			Second hemisphere		
	0	1st	2nd	3rd	1st	2nd	3rd
Respiratory rates during successive 30 min periods (μmol/g h)	1st	61	60	62	60	62	60
	2nd;	62	P114	P74	P78	P111	58
	3rd;	60	P112	P75	P72	P113	59
Lactic acid accumulation (μmol/vessel)		32	72	42	41	69	32

To all vessels were added 3·5 ml phosphate-buffered glucose saline (Table 7.1) which when asterisked * contained 10 μM chlorpromazine; and, in the reagent-well, 0·12 ml 5 N-NaOH. Electrical pulses were applied to vessels (2), (3), (4) and (5) during the second and third periods. For other details see text.

Immediately before the experiment the electrodes are scraped clean and solutions pipetted to the vessels according to Table 8.2. The cerebral hemispheres are taken from a guinea pig, slices cut by one worker and successively trimmed to size and weighed by another. As they are weighed the first worker fits them to the electrodes: a slice is floated, fully extended and not folded, in an auxiliary bath of saline at 38°C. The lower grid of an electrode is taken by its wire spiral in the left hand, and with a spatula (g, Fig. 6.1) the floating slice is manoeuvred over the electrode and with the electrode is lifted out of the dish. The upper part of the electrode is now taken in the right hand, its pointed end inserted under the wire loop of the lower electrode, and the two parts then brought together at their other ends, when they are guided into position and held firmly together by their clip. The assembled electrode with its tissue is then taken firmly by forceps at the clip end and inserted, the other end foremost, into the vessel so that it rests in the position shown in Fig. 8.2. With the same forceps, the two electrode sockets are pushed home on the electrode stubs of the vessel. Successive slices are mounted similarly and their electrodes connected in the vessels; papers are then placed in the alkali-wells of the vessels and these placed on their manometers, equilibrated with oxygen and placed in the thermostat as in an ordinary manometric experiment. Mounting the tissue takes little time and the six vessels should be in the thermostat 25–30 min after exsanguinating the guinea pig.

Manometric readings of gas pressure or, in Gilson apparatus, micrometer measurements of gas volume, are then taken each 5 min. During the first few minutes the pulse-generator and oscilloscope are switched on at the mains and connected to the distributor (Fig. 8.4). Leads from the distributor are clipped to the electrode arms of the vessels, but pulses not yet switched to the vessels. Sufficient readings (probably six) are taken to give accurate values for the initial respiratory rates. The pulse generator is then adjusted to give a small discharge, for example of 0·5 V, 0·4 ms time-constant, and the resulting pulse displayed on the oscilloscope. Output from the generator is now switched momentarily to the first vessel only. This should cause the voltage and duration of the pulse on the oscilloscope screen to fall to an extent similar to that caused by a resistance of about 100 ohms. The generator is similarly switched, briefly, to each vessel in turn to confirm that electrical contacts are in order. An electrical fault is rare; if detected, lead wires, plugs and switches should be examined and can be replaced, but a break of contact inside the vessel can only be noted for attention later. The checking of contacts takes only a minute.

Electrical stimulation of all the vessels is now commenced by switching them together to the generator and adjusting its output to pulses of 10 V peak potential and 0·4 ms time-constant at a chosen frequency, as 10–50 Hz. This is done by setting the voltage and time base of the oscilloscope to the appropriate ranges, and gradually increasing both the voltage and duration controls of the generator until the required pulse-type is obtained; the vessels remain connected to the generator during this adjustment. The pulses being delivered are displayed on the oscilloscope throughout stimulation; the oscilloscope picture should remain unchanged. Manometric or other readings are taken each 5 min throughout the following hour. After the last reading the stimulating voltage is brought to a value of a volt or less, all the vessels switched off, the oscilloscope trace noted, and the effect again observed of switching in each vessel individually. The change seen on the oscilloscope should be similar to that seen at the beginning of the experiment, confirming that connections have remained satisfactory in each individual vessel.

Leads from the electrical distributor are then unclipped from the vessels, these are removed from the manometers and the centre-well papers taken from the vessels. The electrodes are then removed by unclipping their two spiral arms from the stubs with forceps and withdrawing the electrode by one spiral arm. If the tissue is to be investigated further, the electrode carrying it is put in a dish of saline, opened, and the tissue floated free. Otherwise the electrode is unclipped and the tissue washed away under the tap; the electrode with others is put into a beaker for washing as described above. From the vessels, 0·5 ml of incubation fluid is immediately taken for determination of lactic acid.

Results exemplified by those quoted in Table 8.2 show in vessels (1), (2), (5) and (6) the typical respiratory and glycolytic response of cerebral tissues to maximal stimulation by electrical pulses. The respiratory response is seen to be sustained for an hour; it would continue for longer. If pulses were not applied for the third 30 min period, the rates during this would be similar to those of the first period. This arrangement is not suggested in the illustrative experiment as the glycolytic response would then be much less evident for it depends on lactate accumulation. The experimental arrangement is partly conditioned by the possibility that slices cut from different depths in the cortex might behave differently. No major effect of this is demonstrated; in other experiments of the same investigation chlorpromazine would be included in vessels (1) and (6), or in (2) and (5), in place of (3) and (4). Results of such a study, in which the concentration of chlorpromazine and the potential of the applied pulses were also varied, are given by McIlwain & Greengard (1957). Studies of these variables enabled voltage-responses curve to be constructed in the presence and absence of the drug, and concentration-action curves to be constructed for chlorpromazine and a related phenothiazine.

In the illustrative experiment, chlorpromazine is seen to be without action during the first, unstimulated, period but to inhibit when pulses are applied. The tissue is then highly susceptible to chlorpromazine, and this constitutes the most sensitive *in vitro* system to the drug. Associated experiments showed chlorpromazine to be without action at concentrations up to 1 mM, unless pulses were applied. This exemplifies many situations in which electrically stimulated tissues become highly sensitive to added agents which have little or no effects in the absence of pulses. Further instances are given in Table 8.4 together with other situations in which the response of the tissue has been modified. Respiratory, glycolytic, or phosphate measurements in such systems serve as indexes of the tissue's response and do not imply direct actions of the agents on the metabolic process measured. Respiration increased by succinate, 2,4-dinitrophenol, or increased concentrations of potassium salts is not comparably sensitive.

III Noradrenaline output and glycolysis, in a superfusion system

The present illustrative experiment (McIlwain & Snyder, 1970) uses the superfusion system of Fig. 7.3. Four tissue samples receive [^{3}H] noradrenaline under circumstances in which it is actively taken up by the tissues. They are then superfused and for part of this time are electrically stimulated. Frequent collections of effluent fluids are taken to show the progress of tissue glycolysis and of the output of ^{3}H derivatives (Fig. 8.5).

Procedure. Four quick-transfer electrodes of the type illustrated in Fig. 7.3 are prepared, and the incubating baths of the superfusion apparatus are brought to 38°C. Glucose bicarbonate saline is prepared and placed in a reservoir of the superfusion apparatus and also in the bath used for mounting the tissue. The incubating beakers receive 5 ml of this medium to which has been added nialamide and ascorbate to prevent oxidation of noradrenaline. Racks of collecting tubes are arranged to receive supernatants, and ice-cold perchloric acid is placed in homogenizer test-tubes to receive the tissues at the end of incubation.

Two workers cut, weigh, and mount tissue sections from the brain of a guinea pig. Successively, the sections in their electrodes are placed in the incubation beakers, which are connected to their supply of O_2–CO_2 and placed in the thermostat bath. When all the beakers with their electrodes and tissues have been placed in the bath, measured additions of [^{3}H]noradrenaline or other chosen substrate are made to each and incubation continued for a chosen period, usually of 20–30 min and termed 'preincubation'. During this time (i) the flow system is filled with superfusion fluid by briefly running the pump with its output tubes connected to the primer (Fig. 7.3). With the pump halted, the stopper carrying these tubes is then removed from the primer and the tubes are attached to the supply tubes of the quick-transfer electrodes, so that superfusion fluid can now be circulated through the incubation beakers. (ii) Also during preincubation, the stimulator and oscilloscope are switched on and the output from the stimulator is adjusted to deliver stimuli of approximately the characteristics of those which will subsequently be applied to the tissues. The continuity of connections from the stimulator to the leads of the individual vessels are then tested by momentarily connecting in place of each vessel, a resistance of about 10 ohms; a characteristic small displacement should be seen on the oscilloscope screen when connection is made.

After the chosen period of preincubation, superfusion is commenced; initially, this removes the incubation fluid which was first placed in the beakers and which now carries accumulated tissue metabolites and excess reagents. A brief and relatively rapid flow at 5 to 10 ml/min may be used for this purpose, and the fluid flowing during the first few minutes discarded. The flow-rate is then brought to a selected value of, e.g., 2 to 5 ml/min and the effluent collected each 1 to 5 min subsequently (Fig. 8.5). During this period reagents or stimuli are applied. In the experiment of Fig. 8.5, some vessels were stimulated electrically for 1 min, and this was sufficient to increase 3-fold the output of ^{3}H from tissues preincubated with [³H]noradrenaline, but not from others which had been preincubated with [³H]glycine. Stimulation may also be continued for longer periods in order to study the maintenance of steady-state outputs of the labelled derivatives. The quick-transfer electrodes allow the tissues to be released for analysis during or after a period of stimulation. They are typically released to the incubating fluid of the beakers after halting the pump, and from the beakers are taken with a mounted, bent wire and transferred to the ice-cold fixing solution.

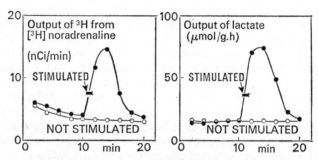

Fig. 8.5 Output of ³H added as [³H] noradrenaline, and of lactate, from guinea pig neocortical tissues incubated and superfused in the apparatus of Fig. 7.3. Data were obtained by analysis of samples collected each 2 min; the tissues indicated were stimulated for 1 min with pulses of peak potential 10V, time-constant 0.4ms, alternating polarity and 25 Hz (McIlwain & Snyder, 1970 and unpublished).

Expression of results; associated experiments. In Fig. 8.5 the output of lactate is quoted in μmol/g tissue.h, calculated from the quantity of lactate found in each 2 min collection of superfusion fluid. Radioactivity measurements give most directly the nCi of material collected during the 2-min periods, but better comparison is obtained between data from different tissue samples or different experiments by calculating the proportion of the tissue's radioactivity which is being released in a given interval. This is obtained by measuring the nCi of ³H in the tissue at the end of the experiment, as well as in each fluid sample. The proportion is then given by the ratio: (nCi in sample s)/(0·5 nCi in s + nCi in subsequent samples + nCi in tissue).

In the work quoted in Fig. 8.5 (McIlwain & Snyder, 1970) associated findings were cited to show that nialamide did not inhibit uptake of noradrenaline to the tissues. Of the [³H]noradrenaline taken up by the tissue, more than 80% remained as the catecholamine, and it was present in the tissue at concentrations 4-5 times greater than in surrounding fluids. Also, the ³H was preferentially localized in synaptosomal fractions. Such associated experiments are usually necessary in appraising the status of findings such as those of Fig. 8.5.

NOTES AND TABLES SUMMARIZING METABOLIC FINDINGS WITH ELECTRICALLY STIMULATED TISSUES

Application of the techniques described in this chapter is briefly illustrated by results obtained with tissues from the mammalian brain.

It is seen from Table 8.3 that most of the constituents of typical incubation media need to be present, in order that the tissues should respond to electrical excitation by increase in respiratory rate. Similar but not identical requirements are found when other metabolic responses to excitation are measured. For example, neocortical tissues under hypoxic conditions showed little or no glycolytic response to excitation, but still retained noradrenaline and serotonin and remained capable of responding by increased release of the two compounds on excitation (Pull, Jones & McIlwain, 1972). Pulses also bring about changes in tissues which are incubated completely anaerobically (McIlwain, 1956a) and for such experiments arrangements similar to those of Table 8.2 may be employed, but using yellow phosphorus to maintain anaerobiosis as described in Chapter 7. Change in tissue composition with electrical pulses can be followed in the apparatus of Fig. 8.2 provided that rapid changes are not involved; this procedure has the advantage of permitting respiratory measurements with the same tissue samples as are later analysed. In other situations the rapid transfer apparatus is employed.

An arrangement similar to that of Table 8.2 can be used to show that cerebral tissues require glucose or a similar substrate for respiratory response to electrical stimulation. Slices were prepared in media lacking glucose, and these media were placed in all the vessels. Vessels (2) and (5) then also received glucose, and (3) and (4) succinate: e.g. 0·035 ml of molar solutions. Respiratory rates in the first period without pulses were: no substrate, 37; glucose, 60; succinate, 90. With pulses these became 40, 115 and 85 μmol O_2/g.h respectively. Summarized results of such experiments are included in Table 8.3.

The metabolic responses to excitation which have been observed in isolated cerebral tissues are numerous and are exemplified by the many categories of Table 8.4. Changes in efflux or influx rates of substances not necessarily undergoing chemical change are included, as well as modified rates of chemical conversion. Responses in these different categories have been changed further by the application of other agents or by disease. Many model systems are thus offered for study of the development of the brain and of alterations occurring in illness.

Techniques of electrical excitation have been applied to many parts of the brain, and special attention may be directed to those parts which have been found suitable for joint study by metabolic and electrophysiological means. These may be described as subsystems of the brain, and are listed in Table 8.5. Their electrophysiological study is described

Table 8.3 Requirements for respiratory response to electrical pulses by mammalian cerebral cortex

Substance omitted from medium	Effect
K^+	Response lost; regained with K^+, 1–2 mM.
Na^+	Response lost; regained with Na^+, 10–100 mM.
Ca^{2+}	Resting rate increased; response diminished
Mg^2 or Cl^- or SO_4^{2-}	Response not greatly altered
HPO_4^{2-}	Response lowered
Glucose	Response lost; regained with glucose, 0·3–2 mM; fructose, 10 mM; lactate, 5–10 mM; pyruvate, 5 mM; or oxaloacetate, 5 mM, and partly with glutamate. Not regained with succinate, fumarate, citrate, ketoglutarate, aspartate, asparagine or γ-aminobutyric acid
Oxygen	Pulses in absence of oxygen diminish response when, subsequently, oxygen is supplied.

The complete media were based on those of Table 7.1. Appropriate osmotic adjustment was made when media were altered. The experimental arrangement was similar to that of Table 8.2. Under these conditions the tissue retains, concentrates, or excludes many of its native substances and media constituents. Data from Gore & McIlwain, 1952; McIlwain & Gore, 1953; McIlwain, 1953a, b; McIlwain, 1956a, b; Kratzing, 1953, 1956; Cummins & McIlwain, 1961; Woodman & McIlwain, 1961; Keesey et al., 1965; Jones & McIlwain, 1971.

Table 8.4 Metabolic responses to electrical pulses, observed in cerebral tissues; and circumstances modifying them

Type of experiment	Variables studied, and references
A. Response by normal tissue	1. Ion movement. Ca^{2+}: Lolley, 1963. Na^+ and K^+: Cummins & McIlwain, 1961; Hillman et al., 1963; McIlwain & Joanny, 1963; McIlwain, 1967; McIlwain et al., 1969; Harvey & McIlwain, 1969.
	2. Substrate utilization: see Table 8.3, and McIlwain, 1956a, b, 1959.
	3. Change in tissue composition: McIlwain, 1952b, 1959; Rowsell, 1954; Heald, 1954, 1958, 1959; McIlwain & Tresize, 1956; McIlwain, Thomas & Bell, 1956.
	4. Change in output of: Acetylcholine: Rowsell, 1954; Katz & Chase, 1970. Adenine derivatives: Pull & McIlwain, 1972; McIlwain, 1972b; Heller & McIlwain, 1973. Amino acids: Srinivasan et al., 1969; Hammerstad & Cutler, 1972; Pull et al., 1972. Noradrenaline, serotonin: Baldessarini & Kopin, 1967; Chase et al., 1969; McIlwain & Snyder, 1970; Pull, Jones & McIlwain, 1972.
	5. Changed turnover or accumulation of: phosphoprotein-phosphorus: Heald, 1958; Rodnight, 1971; Jones & Rodnight, 1971; Reddington et al., 1973.

	Phospholipids: Pumphrey, 1969. Protein-amino acids: Jones & Banks, 1970; Jones & McIlwain, 1971. Cyclic AMP: Kakiuchi, Rall & McIlwain, 1969.
	6. Species and tissue: Kratzing, 1951; McIlwain, 1952a, 1953b, 1954a, b; Greengard & McIlwain, 1955b; Bollard & McIlwain, 1957.
B. Tissue modified *in vivo* by treatment of animal or by disease	1. Hypoglycaemia: Setchell, 1959. 2. Cold: Brierley & McIlwain, 1956. 3. Neoplasms; other conditions: McIlwain, 1952a, 1953b, 1954b; Katz *et al.*, 1969.
C. Tissue modified by treatment *in vitro*	1. Cold media: Marks & McIlwain, 1959; McIlwain, 1959, 1961b. 2. Pulses in absence of substrate or oxygen: McIlwain & Gore, 1953; McIlwain, 1956a, b; Pull *et al.*, 1972. 3. Minimal aqueous media: Rodnight & McIlwain, 1954. 4. Alkylating agents, neuraminidase: Evans & McIlwain, 1967.
D. Addition of substances during *in vitro* experiments	1. General depressants; barbiturates, ethanol: McIlwain, 1953c; Wallgren & Kulonen, 1960; Hillman *et al.*, 1963. 2. Analgesics: Bell, 1958. 3. Anticonvulsants: Forda & McIlwain, 1953; Greengard & McIlwain, 1955a. 4. Chlorpromazine, reserpine, yohimbine: McIlwain & Greengard, 1957; Hillman *et al.*, 1963. 5. Convulsants: Anguiano & McIlwain, 1951; McIlwain & Greengard, 1957. 6. Atropine, hyoscine, eserine: McIlwain, 1951b. 7. Tetrodotoxin, cocaine, procaine: Bollard & McIlwain, 1959; McIlwain *et al.*, 1969; Reddington *et al.*, 1973. 8. Ergot derivatives, mescaline, dibenamine: Lewis & McIlwain, 1954. 9. Protoveratrines: Wollenberger, 1955. 10. Azide, cyanide, malonate, fluoride: Heald, 1953; Kratzing, 1956. Copper salts, Epstein & McIlwain, 1966. 11. Basic proteins: McIlwain, 1961; Hillman *et al.*, 1963.

in Chapters 9 and 12. The anatomical structure of the samples examined is important in investigating these parts of the brain, especially when they are being examined by electrodes for local stimulation of a defined fibre tract (see Chapter 12).

Table 8.5 Subsystems of the mammalian brain which have been observed to respond in vitro to electrical excitation, by metabolic change

Part of the brain	Response observed: i, increase; d, decrease
Corpus callosum; subcortical white matter	Respiration i, glycolysis i, inorganic phosphate i, phosphocreatine d
Lateral olfactory tract	Respiration i, glycolysis i
Lateral olfactory tract–piriform cortex	Respiration i, glycolysis i, inorganic phosphate i, K content d, phosphocreatine d, output of noradrenaline, serotonin and adenine derivatives i
Optic tract–superior colliculus	Serotonin output i
Medulla	Respiration i, glycolysis i, inorganic phosphate i, phosphocreatine d
Hypothalamus	Serotonin and catecholamine output i
Corpus striatum	Output of glutamate, γ-aminobutyrate, noradrenaline and serotonin i
Neocortex	Respiration i, glycolysis i, inorganic phosphate i, cyclic AMP i, ATP d, phosphocreatine d, K content d, Na content i, Ca flux i Phospholipid turnover i. Output of acetylcholine, adenosine, γ-aminobutyrate, amino acids collectively, noradrenaline, serotonin i

For collected references, see Table 8.4; Katz & Chase, 1970; McIlwain, 1972a, b; and for subsequent work: Pull *et al.*, 1972; Heller & McIlwain, 1973; Pull & McIlwain, 1975.

REFERENCES

Anguiano, G. & McIlwain, H. (1951) *Br. J. Pharmacol.* **6**, 448.
Ayres, P. J. W. & McIlwain, H. (1953) *Biochem. J.* **55**, 607.
Baldessarini, R. J. & Kopin, I. J. (1967) *J. Pharmacol.* **156**, 31.
Bell, J. L. (1958) *J. Neurochem.* **2**, 265.
Bollard, B. M. & McIlwain, H. (1957) *Biochem. J.* **66**, 651.
Bollard, B. M. & McIlwain, H. (1959) *Biochem. Pharmacol.* **2**, 81.
Brierley, J. B. & McIlwain, H. (1956) *J. Neurochem.* **1**, 109.
Bureš, J., Petráň, M. & Zachar, J. (1967) *Electrophysiological Methods in Biological Research*. Academia: Prague.
Chappell, J. B. & Greville, G. D. (1954) *Nature, Lond.*, **174**, 930.
Chase, T. N., Katz, R. I. & Kopin, I. J. (1969) *J. Neurochem.* **16**, 607.
Cummins, J. T. & McIlwain, H. (1961) *Biochem. J.* **79**, 330.
Epstein, P. S. & McIlwain, H. (1966) *Proc. R. Soc.* B **166**, 295.
Evans, W. H. & McIlwain, H. (1967) *J. Neurochem.* **14**, 35.
Forda, O. & McIlwain, H. (1953) *Br. J. Pharmacol.* **8**, 225.
Gore, M. B. R. & McIlwain, H. (1952) *J. Physiol.* **117**, 471.
Greengard, O. & McIlwain, H. (1955a) *Biochem. J.* **61**, 61.
Greengard, P. & McIlwain, H. (1955b) *Proc. int. Neurochem. Symp.* **1**, 251.
Hammerstad, J. P. & Cutler, R. W. P. (1972) *Eur. J. Pharmacol.* **20**, 118.
Harvey, J. A. & McIlwain, H. (1969) *Biochem. J.* **108**, 269.
Heald, P. J. (1953) *Biochem. J.* **55**, 625.
Heald, P. J. (1954) *Biochem. J.* **57**, 673.
Heald, P. J. (1958) *Biochem. J.* **68**, 580.
Heald, P. J. (1959) *Biochem. J.* **73**, 132.

Heller, I. & McIlwain, H. (1973) *Brain Res.* **53,** 105.
Hillman, H. H., Campbell, W. J. & McIlwain, H. (1963) *J. Neurochem.* **10,** 325.
Jones, C. T. & Banks, P. (1970) *Biochem. J.* **118,** 791.
Jones, D. A. & McIlwain, H. (1971) *J. Neurochem.* **18,** 41.
Jones, D. A. & Rodnight, R. (1971) *Biochem. J.* **121,** 597.
Kakiuchi, S., Rall, T. W. & McIlwain, H. (1969) *J. Neurochem.* **16,** 485.
Kao, C. Y. (1966). *Pharmacol. Rev.* **18,** 997.
Katz, R. I. & Chase, T. N. (1970) *Adv. Pharmacol. Chemother.* **8,** 1.
Katz, R. I., Goodwin, J. S. & Kopin, I. J. (1969) *Life Sciences* **8,** 561.
Keesey, J. C., Wallgren, H. & McIlwain, H. (1965) *Biochem. J.* **95,** 289.
Kratzing, C. C. (1951) *Biochem. J.* **50,** 253.
Kratzing, C. C. (1953) *Biochem. J.* **54,** 313.
Kratzing, C. C. (1956) *Biochem. J.* **62,** 127.
Kurokawa, M. (1960) *J. Neurochem.* **5,** 283.
Lewis, J. L. & McIlwain, H. (1954) *Biochem. J.* **57,** 680.
Lolley, R. N. (1963) *J. Neurochem.* **10,** 665.
Marks, N. & McIlwain, H. (1959) *Biochem. J.* **73,** 401.
McIlwain, H. (1951a) *Biochem. J.* **49,** 382.
McIlwain, H. (1951b) *Br. J. Pharmacol.* **6,** 531.
McIlwain, H. (1952a) *J. ment. Sci.* **98,** 265.
McIlwain, H. (1952b) *Biochem. Soc. Symp.* **8,** 27.
McIlwain, H. (1953a) *Biochem. J.* **55,** 618.
McIlwain, H. (1953b) *J. Neurol. Neurosurg. Psychiat.* **16,** 257.
McIlwain, H. (1953c) *Biochem. J.* **53,** 403.
McIlwain, H. (1954a) *J. Physiol.* **124,** 117.
McIlwain, H. (1954b) *Archis Neurol. Psychiat., Chicago* **71,** 488.
McIlwain, H. (1956a) *Biochem. J.* **63,** 257.
McIlwain, H. (1956b) *Physiol. Rev.* **36,** 355.
McIlwain, H. (1959) *Biochem. J.* **73,** 514.
McIlwain, H. (1961a) *J. Neurochem.* **6,** 244.
McIlwain, H. (1961b) *Biochem. J.* **78,** 24.
McIlwain, H. (1967) *Progress in Brain Research.* Vol. 29: *Brain Barrier Systems,* ed. Lajtha & Ford, p. 273. Amsterdam: Elsevier.
McIlwain, H. (1972a) *Experimental Models of Epilepsy,* ed. Purpura, Perry, Tower, Walter & Woodbury, p. 270. New York: Raven Press.
McIlwain, H. (1972b) *Effects of Drugs on Cellular Control Mechanism,* ed. Rabin & Freedman, p. 281. London: Macmillan.
McIlwain, H. & Gore, M. B. R. (1953) *Biochem. J.* **54,** 305.
McIlwain, H. & Greengard, O. (1957) *J. Neurochem.* **1,** 348.
McIlwain, H., Harvey, J. A. & Rodriguez, G. (1969) *J. Neurochem.* **16,** 363.
McIlwain, H. & Joanny, P. (1963) *J. Neurochem.* **10,** 313.
McIlwain, H. & Snyder, S. H. (1970) *J. Neurochem.* **17,** 521.
McIlwain, H., Thomas, J. & Bell, J. L. (1956) *Biochem. J.* **64,** 332.
McIlwain, H. & Tresize, M. A. (1956) *Biochem. J.* **63,** 250.
Narayanaswami, A. & McIlwain, H. (1954) *Biochem. J.* **57,** 663.
Phillipa, A. (1970) *Bayer-Symposium* **11,** p. 258. Berlin: Springer, Verlag.
Pull, I., Jones, D. A. & McIlwain, H. (1972) *J. Neurobiol.* **3,** 311.
Pull, I. & McIlwain, H. (1972) *Biochem. J.* **126,** 965; **130,** 975.
Pull, I. & McIlwain, H. (1973) *Biochem. J.* **136,** 893.
Pull, I. & McIlwain, H. (1975) *J. Neurochem.* **24,** 695, in press.
Pumphrey, A. M. (1969) *Biochem. J.* **112,** 61.
Reddington, M., Rodnight, R. & Williams, M. (1973) *Biochem. J.* **132,** 475.
Rodnight, R. (1971) *Handb. Neurochem.* **5,** 141.
Rodnight, R. & McIlwain, H. (1954) *Biochem. J.* **57,** 649.
Rowsell, E. V. (1954) *Biochem. J.* **57,** 666.
Setchell, B. P. (1959) *Biochem. J.* **72,** 265, 275.

Srinivasan, V., Neal, M. J. & Mitchell, J. F. (1969) *J. Neurochem.* **16**, 1235.
Wallgren, H. & Kulonen, E. (1960) *Biochem. J.* **75**, 150.
Wollenberger, A. (1955) *Biochem. J.* **61**, 77.
Woodman, R. J. & McIlwain, H. (1961) *Biochem. J.* **81**, 83.
Yamamoto, C. & McIlwain, H. (1966) *J. Neurochem.* **13**, 1333.

9. Maintenance of Isolated Parts of the Brain for Electrical Measurements

H. MCILWAIN

Apparatus	191
Tissue chamber	192
Thermostat and auxiliary apparatus	193
Electrodes for stimulating and extracellular recording	196
Micropipette electrodes and intracellular recording	197
Procedure for measuring intracellular potentials	198
Cell-firing observed intracellularly	201
Procedure for extracellular observation of cell-discharge	203
Illustrative findings	206
References	206

Measurement of electrical and chemical events in the same neural specimen can give most valuable information about its functioning. Many of the methods for such studies have developed as part of electrophysiology and find adequate description within that subject. Maintenance of isolated mammalian cerebral tissues and dissected subsystems for electrical measurements is however described here, because such specimens are more exacting chemically than are the peripheral neural structures which have been most frequently employed in electrical studies. Also, mammalian cerebral tissues represent the neural system about which most chemical information is available, and which inevitably remains the most accessible source of neural material, in bulk, for chemical study. Further, the maintenance of cerebral tissues for electrical measurements involves handling them in a distinctive fashion, which itself gives new information and further experimental opportunities. It has guided the choice and handling of specified subsystems excised from the mammalian brain, as is described more fully in Chapter 12.

APPARATUS

Apparatus which has been successfully used in observing cell-membrane potentials in subsystems from the brain of laboratory animals and man, is shown in Fig. 9.1 and 9.2 and comprises a chamber for maintaining the tissue; a thermostat and other auxiliary apparatus; and stimulating and recording electrodes, their mountings and connections to amplifying and recording apparatus. These will now be described individually.

Tissue chamber

The chamber of Fig. 9.1 was designed to allow the following manipulations. (i) An initial incubation of tissue immersed in well-oxygenated nutrient media, under conditions which approximate to those of metabolic experiments. Such preincubation has been found necessary for reestablishing the tissue content of labile metabolites and also (see below) its membrane potentials. (ii) After preincubation, the chamber allows the tissue to be supported at the surface of the incubation medium, so that electrical observations can be made without excess aqueous fluid and with the tissue in a fixed position in relation to electrodes. (iii) The extensive outer ducting of the chamber allows the tissue to be adequately oxygenated while remaining accessible to electrodes from above; the electrodes are allowed considerable movement.

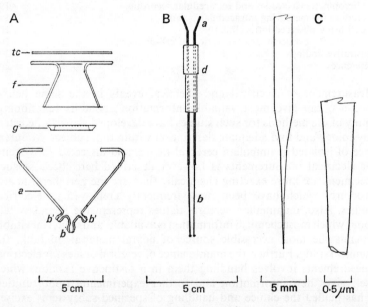

Fig. 9.1 Tissue chamber and electrodes (Li & McIlwain, 1957; McIlwain, 1972).
A: The central tissue chamber, conical, of glass, and shown in vertical section with its parts separated: a, the main vessel; b, b', nipples to take fine tubing for fluid and an earthing electrode (see Fig. 9.2); f, inner baffle; g, grid on which the tissue rests; tc, temporary cover used during preincubation and removed for inserting electrodes.
B: Silver, ball-tipped stimulating electrodes. The stout silver wires a carry fine silver wires b with fused spherical tips. The plastic mounting d fits to a micromanipulator. A single such ball-tipped wire, similarly mounted, serves for extracellular recordings at the surface of a tissue.
C: Glass micropipette electrode for intracellular recording. The drawing of the tip is based on an electron micrograph (Zachar, 1967) and shows a 1 in 12 taper. Similar electrodes with wider tips are suitable for extracellular recordings.

The mounted tissue-chamber and associated apparatus is shown in Fig. 9.2; this development of the apparatus of Li & McIlwain (1957) is described by Gibson & McIlwain (1965). The two parts a and f of the chamber are held a few millimeters apart and through the resulting aperture gas enters to the annular space between a and f. Their shape then guides the gas over the surface of the grid g which carries the tissue, and the gas leaves by the wide central aperture of f. This inner, detachable portion of the chamber can easily be lifted out for assembling or cleaning the apparatus and acts as a baffle, forming with the main chamber a wide duct for entering gas. The gas leaves by the wide central upper opening through which electrodes and pipettes are inserted, and experiments have shown that when 95% O_2–5% CO_2 was supplied at 100 ml/min, the gas sampled immediately above the tissue (above g, Fig. 9.2) contained 90% of O_2. At the beginning of experiments and before electrodes are inserted, the central opening of the chamber may be covered with a loosely-fitting glass plate.

The grid g which supports slices for electrical measurements rests at the apex of the chamber with approx. 1·5 ml of fluid below it, as a layer of average depth about 5 mm. During preincubation a further 2–4 ml of fluid may be added, allowing the slice to be immersed, and the gas then bubbles at the surface of the fluid, gently agitating it and the slice. The grid consists of glass fibres or a nylon net attached with a plastic cement to a rim turned from Perspex (Lucite) sheet. The rim is shaped to fit the side of the vessel and the grid then keeps its position without further fixing, but can be lifted out with forceps for cleaning. The characteristics of the nylon net are not critical, but one found suitable was of yarn denier 45(15)Z, mesh size 1·84 × 1·39 mm.

Three tubules lead from the apex of the chamber below the level of the fluid; these are of narrow bore and short to minimize breakage, and take flexible polythene tubing about 1 mm in internal diameter. One of these is connected to a syringe to supply, change, or adjust the level of fluid in the chamber. The other carries the earthing electrode of chlorided silver. The chamber is supplied by Messrs Dixon (p. 165).

Other designs of tissue-chamber, made in plastic and in some cases without arrangements for maintaining a defined gas atmosphere, have been used by Kawai & Yamamoto (1969), Yamamoto (1972) and Doré & Richards (1974). It is recommended that the performance of a tissue-chamber be validated by analyses of gas tension and of the composition of tissues which have been incubated in the chamber (see below).

Thermostat and auxiliary apparatus

A compact, robustly-made glass thermostat filled with water surrounds the tissue-chamber. It maintains the chamber, the incoming incubation fluid and the entering gas at the working temperature, usually 38°. Also, the entering gas mixture passes through a sintered-glass bubbler

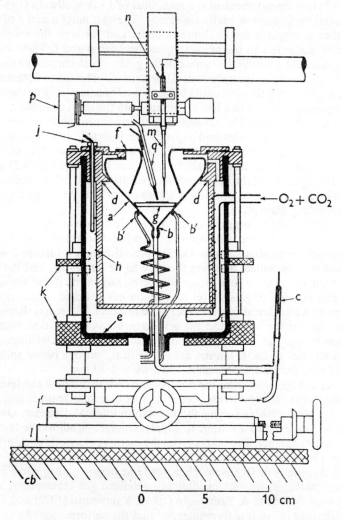

Fig. 9.2 Tissue chamber assembly: thermostat, mounting and moving parts (Gibson & McIlwain, 1965; Yamamoto & McIlwain, 1966; McIlwain, 1972). The central, conical tissue chamber a is shown separately in Fig. 9.1. The nipples b, b' carry connections for incubating fluid which is warmed by passage through the helix, and for the indifferent electrode c. A Perspex frame (d) is bolted to the top of the glass thermostat bath e, shown black in cross-section. Heating is by a coil (h) on a Perspex support and is controlled by the thermistor j and an external circuit. The Perspex and steel mounting k for the chamber is carried on the lathe slides l, l', the lower of which is bolted to a steel plate. For intracellula recordings this usually requires mounting on a heavy concrete block (cb).

into the water of the thermostat and thus arrives at the tissue-chamber saturated with water-vapour at the working temperature. Distilled water of good quality should be used in the thermostat, which should be cleaned regularly and receive fresh water; an appreciable bubbling-time is required to equilibrate the fresh water with an incoming O_2–CO_2 mixture. The thermostat bath is heated by a coil of nickel-copper wire; a thermistor detects the difference between the bath temperature and the intended working temperature and controls accordingly, and gradually, the current supplied to the heating coil.

Other thermostat arrangements have supported the slice-chamber in a relatively small water bath or jacket supplied from a large thermostat (Li & McIlwain, 1957). This arrangement has the advantage of keeping the electrically-operated thermostat and pumps distant from the recording electrodes and amplifiers, so minimizing electrical interference. A thermostat of 20 l, with its associated pump, was used as source of temperature-controlled water and maintained a few degrees above the temperature required in the slice chamber. In trial runs a thermometer was put in the incubation medium to find the setting needed to keep it at 38°. Connection between the thermostat and the bath of the slice-chamber was by thick-walled rubber tubing about 7 mm in internal diameter. With this arrangement it is not satisfactory to moisten the incoming gas in the water of the jacket; this may also be the case when plastic is used to construct the chamber and thermostat.

The thermostat bath carrying the tissue chamber is mounted rigidly to the upper moving plate of two heavy lathe-slides which provide horizontal motion in two directions at right angles to each other. The lower lathe slide is bolted to a steel plate which also carries vertical pillars and a crossbar, on which are fixed the electrodes and pipettes of which the tips are to enter the slice chamber. Also carried on these pillars or crossbar is a binocular microscope of long working-distance, which is used to view the tissue while the electrodes are being lowered towards it. The micromanipulator carrying the electrodes was of a type resembling the controls of a microscope; it permitted horizontal movement of the electrode over the area occupied by the tissue, and vertical movement of the electrode into the tissue, by both coarse and fine adjustments, over a distance of some 6 cm. The vertical movement was calibrated and gave a measure of the depth of penetration of the electrode into the tissue, and thus of the size of the regions from which potentials were observed (see below).

Oxygen–CO_2 or other gas came from a cylinder through a flow meter (Rotameter; calibrated for flow of 5–150 ml/min) at about 75 ml/min, and then a sintered-glass bubbler, to the slice-chamber.

Local additions of small, measured amounts of material to incubating tissues have been made by micrometer syringes. Volumes of 0·1–2 μl can be consistently applied to the surface of the tissues in this way; the behaviour of KCl and of glutamate solutions so added is described by

Gibson & McIlwain (1965), with methods of calculating the penetration of the substances into the tissue. The calculations took into account the distance between the point of addition and the point at which the effect of the K^+ or glutamate ions were manifested, and their diffusion coefficients. It is to be expected that micro-electrophoresis (see Zachar, 1967; Clarke *et al.*, 1973) would be applicable to the injection of small amounts of suitable materials into a tissue specimen. Volatile compounds, for example anaesthetics, have been applied in the gas phase (Campbell *et al.*, 1967; Richards & Smaje, 1974).

Electrodes for stimulating and extracellular recording

Extensive experience in the preparation, testing and use of such electrodes has been gained in general electrophysiological work (see Bureš *et al.*, 1967; Purpura *et al.*, 1972). The *stimulating electrodes* most frequently used with isolated tissues consist of wires, usually a pair, the tips of which are lowered gently to touch the tissue surface. The electrodes are usually of silver; platinum and stainless steel have also been employed but the comments made in Chapter 8 on these materials should be noted. A typical wire-diameter is 0·1–0·3 mm, and often the tip of the wire is formed into a small ball, some 0·5 mm in diam., by fusion in a flame (Fig. 9.1; Yamamoto & McIlwain, 1966). Stimulation may be unipolar with one such wire, the other connection being to earth through the incubation fluid, but more typically is bipolar when two connections to the stimulator are brought to two similar wires resting 0·1–1 mm apart on the tissue surface. This arrangement is particularly appropriate for stimulating a defined region, as the lateral olfactory tract or optic tract, when the two electrodes are mounted in a single holder with their tips 0·5 mm apart, and the holder (Fig. 9.1) moved by one micromanipulator to bring the tips to the tract. Bipolar electrodes may also be concentric, when an insulated wire is coaxially placed in a narrow steel tube.

For stimulating at defined depths within a neural tissue, electrodes may be of metal or glass and are mounted on a calibrated micromanipulator which allows measurement of the depth of penetration of the electrode into the tissue. The metal electrodes, e.g. of stainless steel or tungsten, are made from wires of which one end is pointed electrolytically or by abrasive, and the end of the wire is subsequently insulated except at the extreme tip (see Bureš *et al.*, 1967). Glass capillary microelectrodes are described more fully below; when used as extracellular stimulating electrodes, capillaries of relatively large tip-diameter (5 μm) are used and are filled with a salt solution into which dips a platinum or silver wire giving connection to the stimulator.

As electrodes for *recording extracellularly* the effects of cell-discharge, silver wires are again simple and convenient. For recording from the tissue-surface they may be of the size and form of those used for stimulating. Such electrodes, ending in 0·5 mm diam. silver balls made by fusion, were used in the recordings of Fig. 9.3. Before use such electrodes are 'chlorided' by brief immersion in 0·9% NaCl while connected to a small positive d.c. potential; as cathode in the saline a larger silver plate is used. Distinct types of electrode with smaller tips are used for recording more localized electrical events in tissue samples. The finer tips necessary have been obtained mechanically or electrolytically (see Purpura *et al.*, 1972). The mechanical strength sometimes needed in using recording-electrodes *in vivo* is not essential in a tissue-chamber, and glass capillary electrodes filled with salt solutions are valuable for recording (see below). When used extracellularly (Table 9.1) they are typically of 1–20 μm in tip diameter and filled with NaCl.

Micropipette electrodes and intracellular recording

General descriptions of the glass microelectrodes of tip diameter about 0·5 μm, for recording changes in electrical potential from cell regions some 3–50 μm in diameter, are given by Bureš *et al.*, (1967) and Purpura *et al.* (1972). The Ling-Gerard type based on glass capillaries have been most frequently used in work with isolated tissues of the brain (Li & McIlwain, 1957; Hillman *et al.*, 1963; Gibson & McIlwain, 1965). These electrodes are prepared from lengths of about 15 cm of borosilicate glass tubing, some 2 mm in diam and 1–1·25 mm bore, which have been specially manufactured and selected for uniformity (Pyrex 'electrode tubes': Jencons, Hemel Hempstead, Herts).

Each length of electrode tubing is drawn to capillaries by heating it to redness and sharply pulling; this was initially done manually but the yield of satisfactory electrodes is increased by using a properly adjusted electrode-puller. These instruments are commercially available (Palmer's microelectrode puller: Baird & Tatlock, Romford, Essex RM1 1HA) and yield electrodes the characteristics of which can be controlled by selecting particular settings of the controls of the puller. The adjustments are: (a) of the temperature reached by a platinum wire or strip which forms a loop around the centre of the tube to be drawn and is heated electrically; (b) of the force applied by solenoid to draw the tubing and (c) of the time-relations between the heating, the pulling and switching off the heating. Settings of these controls are chosen to yield an electrode tip of less than 0·5 μm, approached by a tapering section relatively short in length (Fig. 9.1). The shorter tips give better-conducting electrodes for a given tip diameter; a typical resistance is 20 MΩ, most of which is met in the terminal 10 μm.

In preparing for experiments with the microelectrodes, batches of electrodes may be prepared weekly; cleanliness, care and freedom from dust are essential. About 20 electrodes are drawn in a batch and examined microscopically; those conforming to the sizes quoted are fixed to a holder, tip downward, below 2·7 M-KCl in a 1 l flask which is heated and then evacuated under a bell jar. The flask is taken to about 20 mmHg so that the KCl boils for about 2 min, and the process repeated 3 times. Immediately before use, the electrode tips are re-examined with a water-immersion objective at a magnification of about 1200, and their tips confirmed as being <1 μm; their d.c. resistance is measured and values between 15 and 50 MΩ taken as satisfactory.

Electrical connections from the electrode and chamber are outlined in Fig. 9.3. The microelectrode shank was pushed into flexible tubing of the electrode carrier, where its 2·7 M-KCl made contact with 2·7 M-KCl in agar in a tube which also held a chlorided silver wire, constituting a half-cell. The silver wire led immediately to a cathode follower in the same shielding as the half cell and only a few centimetres from the electrode. The cathode follower led to one terminal of the d.c. amplifier and was matched with a similar connection between the amplifier and earth.

Connection to earth from the fluid of the slice-chamber was made through a chlorided silver half-cell similar to that just described, and through a calibrator. The calibrator enabled a pulse of defined potential to be applied through the microelectrode and recorded on the oscilloscope. This was done at frequent intervals during the experiments and allowed the oscilloscope readings to be converted to values for voltage at the electrode tip (Fig. 9.3). Used in the fashion described by Frank & Fuortes (1955) the calibrating signal gave also an indication of change in impedance of the electrode while it was in use. Resistance of the electrode was measured by short-circuiting it to earth through a known resistance.

For observing resting potentials, a d.c. amplifier has been employed; of two stages, push-pull, and with negative feed-back; it led to the amplification stages of an oscilloscope. The oscilloscope was used either visually or with a camera; in either case a monitor screen is desirable, conveniently visible by the observer who is handling the micromanipulator. The main screen is then available for reading or photographing the magnitude and time-course of the deflections produced. For further details relevant to such experiments, see Prince (1972), Zachar (1967) and Massopust et al. (1973).

PROCEDURE FOR MEASURING INTRACELLULAR POTENTIALS

Experiments are most satisfactorily carried out by two workers; before their commencement a batch of microelectrodes is prepared, the thermostat and amplifiers are switched on and adjusted, the incubation medium is made up and an excess, reaching above the grid, introduced to the slice-chamber. The chamber is closed and the gas mixture, normally O_2 or O_2–CO_2 is passed at about 75 ml/min. Tissue is then taken: for example, the cerebral cortex of a guinea pig or rat as described above, and 1–3 slices of 0·35 mm in thickness and totalling about 40–80 mg or 1·5 sq cm, are cut. These are washed from the cutting blade or guide directly into the incubation medium of the chamber and the cover of the chamber replaced, maintaining the gas flow during this. Some 20 min at 37° has been found a minimal time for the appearance of stable resting potentials and preincubation of 30–40 min is normally allowed.

Electrical measurements may then be commenced and Fig. 9.3 illustrates some of the events observed. An electrode is first advanced by its micromanipulator until it touches the solution near the slice. The amplifier is then balanced so that the oscilloscope trace is visible; inability to balance may be due to large tip potentials of the microelectrode used, which should then be replaced. The electrode is then moved to a position above the slice and advanced into it with the fine adjustment of the micromanipulator, which moves it about 0·3 μm/degree of rotation. In normal tissue, entry of the electrode by 10–50 μm

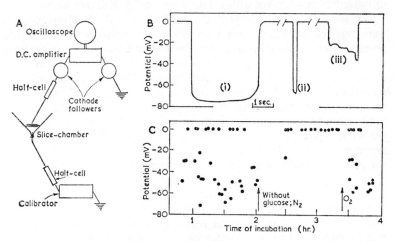

Fig. 9.3 *A*: Electrical connections and components for measurement of resting membrane potentials in isolated cerebral cortex.
B: Voltage-time records made on penetrating slices of guinea-pig cerebral cortex (Hillman & McIlwain, 1961; see Gibson & McIlwain, 1965). (i) Negative potential of 72 mV observed about 50 μm from the outer surface of the cortex, and stable to small changes (1 to 5 μm) in position of the electrodes; with further movement, the potential returned to zero. (ii) A transitory potential of —65 mV, returning to zero without movement of the electrode. (iii) Negative potentials registered during movement of about 30 μm by the electrode, and accompanied by increased electrode resistance; for appraisal, see text.
C: The course of an experiment of 4 h with a single specimen of cerebral cortex, illustrating the loss of potentials in absence of glucose and oxygen. Each point is the maximum stable potential observed during a penetration of the slice to a depth of 200 μm. After zero potentials, electrodes were examined and replaced as described in the text.

o

results in a negative deflection of the oscilloscope trace (Fig. 9.3). This may be quite brief or may last for some minutes. On further movement the electrode normally returns to its original potential and negative potentials are again observed when the electrode is again advanced. In such experiments all potentials may be photographed, or potentials of chosen characteristics (e.g. lasting more than 0·2 s) may be read from an oscilloscope and recorded manually. Concomitant records may be made of time and of the movements made by the micromanipulator to indicate the time and distance for which the electrode remains within a region of negative potential. After penetration of the slice from one position, the electrode was withdrawn, moved elsewhere, and again advanced into the tissue. In typical experiments 5–25 penetrations through most of the thickness of the slice have been made, recording either all potentials, or only the maximal potential attained in each penetration. The solution bathing the slice may then be changed or additions made to it, and further series of readings taken.

Recording with microelectrodes is susceptible to various artefacts, discussed in connection with cerebral tissues by Hillman & McIlwain (1961; see also Gibson & McIlwain, 1965). In this study, measurements were rejected (a) when made with electrodes which showed large tip potentials or increased resistance; these were usually due to blocking with tissue; (b) when the potential was transitory only, lasting less than 0·1 s, when it was doubtful whether the maximum potential had been recorded. (c) Also, after three successive penetrations of a tissue without recording a potential above 20 mV, the electrode was replaced and the used electrode examined microscopically. In normal tissue, such electrodes were usually visibly damaged or blocked; the previous low readings were then discarded. If however the electrode appeared normal, the previous values were regarded as a property of the tissue.

Comment; associated chemical measurements

The resting membrane potentials which have been observed in sections of mammalian cerebral cortex under these conditions, ranged from −40 to −90 mV with a mean value of about −60 mV, and compare favourably with those observed in the cortex *in situ*. Membrane potentials observed in the tissue chamber do not change in mean value after two or three changes of incubating medium or during a period of up to 3 h. Examination of media constituents showed the necessity for glucose and oxygen in maintaining membrane potentials (Fig. 9.3); with progressive increase in the concentration of potassium salts in the medium above the normal 6 mM, potentials fell and the tissue was depolarized at 20–50 mM-K^+ (see below). The tissue could be variously pretreated before making the measurements of potentials, and after some instances of depolarisation in adverse conditions, repolarization could be observed. The methods are applicable to the study of centrally-acting drugs.

The conditions of incubation which have been described were adopted as a result of both electrical and chemical observations. Chemical measurements to indicate the metabolic status of the tissue are an essential part of such experiments, and when conditions of incubation are altered it is advisable to measure the potassium, adenosine triphosphate or phosphocreatine content of the tissue. These constituents are dimished while tissue is being prepared for the experiments, but largely regained under satisfactory metabolic conditions. They approached normal levels (see Chapters 2 and 3) in experiments carried out in the fashion just described. Metabolic inadequacies such as poor oxygenation are quickly reflected in lower phosphocreatine levels. For analysis, tissues are lifted from the slice-chamber with a mounted, bent wire, rapidly rinsed if necessary in a cold solution free from the substance to be estimated, and transferred to fixing agent. Observations in which resting potentials were correlated with the potassium levels of media and tissues, are reported by Hillman & McIlwain (1961) and transmission in piriform cortical preparations was found sensitive to conditions which diminished their content of adenosine triphosphate (Yamamoto & Kurokawa, 1968).

Glutamic acid salts, added to neocortical tissues maintained in glucose-salines, promptly depolarized constituent cells of the tissue (Bradford & McIlwain, 1966). The depolarization was correlated with the excitatory properties shown by a series of acids related to glutamic, when they were administered *in vivo* to experimental animals (Curtis & Watkins, 1965). Concomitant measurement of the Na content of the incubated tissues suggested their depolarization to be due to entry of Na^+ to the tissues. When glutamate salts were added quickly in 20 μl of fluid locally to the surface of tissues in the apparatus of Fig. 9.1, the speed of depolarization observed by micropipette electrodes was correlated with the depth below the tissue surface at which recording was taking place. Results were consistent with a diffusion of glutamate extracellularly to the tissue elements which were depolarized (Bradford & McIlwain, 1966).

Other findings which illustrate the use of the technique described are summarized in Table 9.1.

Cell-firing observed intracellularly

Occasionally on penetrating regions of negative potential in incubated sections of guinea pig and rat neocortex, spike discharges are recorded (Li & McIlwain, 1957; Hillman *et al.*, 1963; Gibson & McIlwain, 1965). The discharges were initiated at levels of resting membrane potential of -32 to -60 mv, and the spike amplitude was 27 to 70 mv above the immediately preceding resting potential. In some cases a positive overshoot was observed, the region transitorily becoming positively charged in relation to surrounding fluid. The rising phase of the spike was occasionally interrupted by a notch of 5–10 mv. The cell-firing usually occurred in bursts of 2–15 spikes at inter-spike intervals

Table 9.1 Electrical measurements in cerebral tissues; illustrative findings

Measurement	Values measured found to be modified by or correlated with:
1. Intracellular micropipette* Voltage and impedance change on penetration; incidence, persistence and magnitude of voltage recorded; time and extent of recovery of potential after displacement	Oxygen tension; glucose, K, Na, Mg, Ca concentrations; electrical stimulation, cocaine, phenobarbitone, chlorpromazine, glutamate and other excitatory amino acids, basic proteins; tissue and species; *in vivo* measurements
Tissue location of regions of negative potential	Histological structure
Size of regions of negative potential	Cell size
Cell discharge: amplitude, frequency, time-voltage relationships; after discharges	Magnitude of resting potential; electrode movement; tissue preparation; tissue-type; extracellular fluid composition; *in vivo* measurements
Excitatory and inhibitory postsynaptic potentials	Frequency of stimulation; Mg; tissue structure
2. Extracellular micropipette* Unit discharge	Ca, Mg, barbiturates in bathing fluids; halothane. Dopamine, methamphetamine, phenoxybenzamine, prostaglandin E, cyclic AMP (in caudate nucleus)
Latency of discharge	Coupling-type: electrical or chemical
Field potentials	Conditioning stimuli; Mg, Ca, barbiturates; serotonin, lysergic acid diethylamide, mescaline, bufotenine (in superior colliculus)
3. Extracellular, surface electrodes* Conduction velocity	Tissue structure; diameter, myelination and nature of fibre
Amplitude and complexity of postsynaptic response	Adenosine triphosphate of tissue; composition of bathing fluid: Ca, Mg, γ-aminobutyrate; barbiturates, ether, halothane, chlorpromazine
	Part of the brain (Table 12.2); unit discharges in depth of tissue; *in vivo* measurements
Frequency of cell discharge; speed of propagation	Extracellular Cl^-; simulated seizure discharges (Fig. 9.5); modified by phenobarbitone

*See references, p.203.

References: *1*, Li & McIlwain, 1957; Hillman & McIlwain, 1961; Hillman et al., 1963; Gibson & McIlwain, 1965; Bradford & McIlwain, 1966; Yamamoto, 1972. *2*, Yamamoto & Kawai, 1967; Kawai & Yamamoto, 1969; Richards & Sercombe, 1970; Yamamoto, 1972, 1973; Richards, 1973. *3*, Yamamoto & McIlwain, 1966; Campbell, McIlwain, Richards & Somerville, 1967; Yamamoto & Kawai, 1967, 1968; Richards & McIlwain, 1967; Richards & Sercombe, 1970; Yamamoto, 1972.

of a few ms, corresponding to a frequency of cell-firing of 40–170/s. After the burst of firing, the negative membrane potential usually persisted.

In hippocampal sections, intracellular electrodes showed spike discharge of some 50 mV to be evoked by near-threshold stimulation in normal media (Yamamoto, 1972). In chloride-deficient media, stimulation some 30% above threshold produced a series of discharges which were investigated more extensively by extracellular recording.

PROCEDURE FOR EXTRACELLULAR OBSERVATION OF CELL-DISCHARGE

The glucose-bicarbonate saline or other incubation fluid is prepared, oxygenated and the tissue-chamber brought to the working temperature. Electrodes are positioned above the chamber so that they can readily be lowered to the tissue-holding grid; incubation fluid is added and gas flow to the chamber commenced. The chosen tissue (Chapter 12) is prepared by slicing or dissection and transferred immediately from the cutting apparatus to the tissue-chamber, using if necessary a Pasteur pipette and the incubation fluid to aid the transfer. In the chamber the tissue is preincubated for 20–40 min, and at this point isotopically-labelled metabolites or drugs may be added. During preincubation the tissue is usually floating freely, and subsequently it is manoeuvered into position on the grid *g* of Fig. 9.1 and 9.2. In alternative apparatus (Kawai & Yamamoto, 1969; Heller & McIlwain, 1973), the tissue is superfused during this preliminary incubation. Also during the incubation, electrical connections are made from electrodes to the stimulating, amplifying and recording apparatus which is being used.

After preincubation, incubating fluid is lowered to the level of the grid on which the tissue lies. (In experiments with isotopically labelled compounds, the fluid may be removed altogether and the tissue washed with isotope-free fluid at the working temperature; superfusion systems, q.v., are convenient for this purpose). Recording electrodes are then lowered to meet a chosen part of the upper surface of the tissue, which should carry very little adhering fluid. Spontaneously occurring cell discharge may occasionally be seen at this stage; this is rare from neocortex but may be found in cerebellar preparations (Richards & McIlwain, 1967; Okamoto & Quastel, 1973), and may then be recorded and its modification sought by added agents (see below).

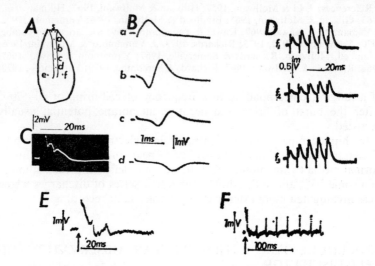

Fig. 9.4 Extracellular recording of response to stimulation of the lateral olfactory tract, the piriform cortex and the neocortex of guinea pigs and rats (Yamamoto & McIlwain, 1966; Campbell et al. 1967; Richards & McIlwain, 1967; McIlwain, 1972).
A: A tissue preparation of the piriform lobe, indicating placements of stimulating electrodes (s) and recording electrodes (a–f); all electrodes were fine silver wires, ball-tipped and placed at the surface of the tissue. B: Records from sites a–d after stimulation at s, showing conduction along the lateral olfactory tract. C: A surface response from the piriform cortex at e (a surface-negative potential giving upward deflection as illustrated), showing the large negative wave with superimposed positive notches. D: Records from site f following stimulation at s and showing at f_1, recruitment of response to successive stimuli delivered at 5/s; at f_2 a diminished response to the same stimuli after 10 min exposure to 1 mM phenobarbital, and at f_3 the original response restored after 15 min in saline in absence of the drug. E: Unit discharges from a neocortical preparation stimulated with surface electrodes and observed with a micropipette electrode, tip diameter 2 to 4 μm about 2 mm distant and 200 μm below the cortical surface. F: Unit discharges observed as in E, from tissue in media in which sodium ethyl sulphonate partly replaced the normal chloride content. A single stimulus at the arrow (the 40th in a series applied at 1/s) evoked a train of responses.
(Reproduced by courtesy of Raven Press; see McIlwain, 1972.)

Stimulating electrodes are next placed in position, for example as shown in Figs. 9.4 and 9.5, and stimulation commenced at potentials rising from 0·5 to 20 V, from 10–200 μs in duration and initially as single, sparse pulses. The stimulation may generate a single spike potential in response to each stimulus, or generate a burst response, or modify the frequency of pre-existing spontaneous activity.

When a discrete response is obtained, its character may be investigated further in a number of ways which include the following. (1) By reversing

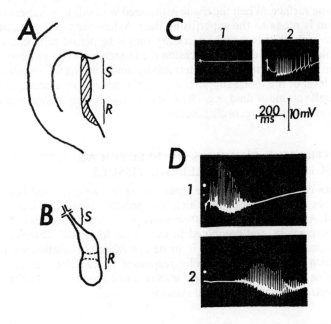

Fig. 9.5 Electrical responses to stimulation observed in systems isolated from the guinea pig brain, showing the positions of stimulating (S) and recording (R) electrodes (Yamamoto & Kawai, 1967, 1968).
A: A section of the hippocampus which includes part of the dentate gyrus, alveus, stria oriens, and pyramidal cell layer. B: A preparation of the optic tract with contralateral brachium and superior colliculus. C: Response of the hippocampal preparation to stimulation in normal medium (1) and in medium containing propionate in place of chloride ions (2). D: Propagation of the after-discharge generated as in C. Distance of recording electrode from stimulating electrode: (1) 1 mm; (2) 5 mm.
(Reproduced by courtesy of Raven Press; see McIlwain, 1972.)

the pulse polarity at the stimulating electrodes; if the essential components of a response persist, freedom from stimulation artefacts is likely. (2) By applying paired stimuli. The first of these is a conditioning stimulus, and when followed within a few ms by a second similar stimulus, the response given to this second stimulus is modified. A third, and subsequent stimuli, can take the modification further; Fig. 9.4D gives an example. (3) By altering the position of the recording electrodes. This can give values for the velocity of transmission of a particular response or component of the response. It may also enable components of a complex response to be differentiated. The surface of a region may be explored to indicate the direction taken by impulse-conducting tracts. Using microelectrodes (glass or sharpened metal), depth of insertion may be modified to give information on the position of the discharging tissue-elements. (4) By addition of reagents to the incubating fluid or to the tissue surface. When the tissue is supplied with saline by superfusion, addition is made to the superfusing fluid. When the tissue rests at the surface of fluid, it is usually necessary that it be periodically immersed in the fluid, and on one such occasion the immersion fluid is arranged to contain the added substance. Alternatively, and with the fluid withdrawn to the level of the tissue-supporting grid, the added substance is applied in a small volume of fluid, e.g. 20 μl, to the tissue surface between or near the stimulating and recording electrodes.

ILLUSTRATIVE FINDINGS BY ELECTRICAL MEASUREMENTS IN CEREBRAL TISSUES

Notes on several studies are summarized in Table 9.1 and illustrate the types of experiment in which the present techniques have been applied. As can be seen, some 20 or more types of electrical measurement have been made, and combined in numerous ways with chemical and histological studies. Table 9.1 is to be considered in relation to Table 8.4 which summarizes metabolic responses to electrical stimulation, and Table 12.2 which is concerned with regional aspects of both chemical and electrical events in cerebral systems.

REFERENCES

Bradford, H. F. & McIlwain, H. (1966) *J. Neurochem.* **13,** 1163.
Bureš, J., Petraň, M. & Zachar, J. (1967) *Electrophysiological Methods in Biological Research.* Prague: Czechoslovak Academy of Sciences.
Campbell, W. J., McIlwain, H., Richards, C. D. & Somerville, A. R. (1967) *J. Neurochem.* **14,** 937.
Clarke, G., Hill, R. G. & Simmonds, M.A. (1973) *Br. J. Pharmacol.* **48,** 156.
Curtis, D. R. & Watkins, J. C. (1965) *Pharm. Rev.* **17,** 347.
Doré, C. F. & Richards, C. D. (1974) *J. Physiol.* **239,** 83P.
Frank, K. & Fuortes, M. G. F. (1955) *J. Physiol.* **130,** 625.
Gibson, I. M. & McIlwain, H. (1965) *J. Physiol.* **176,** 261.
Heller, I. H. & McIlwain, H. (1973) *Brain Res.* **53,** 105.
Hillman, H. H. & McIlwain, H. (1961) *J. Physiol.* **157,** 263.
Hillman, H. H., Campbell, W. J. & McIlwain, H. (1963) *J. Neurochem.* **10,** 325.
Kawai, N. & Yamamoto, C. (1969) *Int. J. Neuropharmacol.* **8,** 437.
Li, C.-L. & McIlwain, H. (1957) *J. Physiol.* **139,** 178.
McIlwain, H. (1972) In Purpura *et al.* 1972, p. 269.
Massopust, L. C., Wolin, L. R., Kadoya, S., White, R. J. & Taslitz, N. (1973) *Meth. Neurochem.* **4,** 288-303.
Okamoto, K. & Quastel, J. H. (1973) *Proc. R. Soc. B.* **184,** 83.
Prince, D. A. (1972) In Purpura *et al.* 1972, pp. 71-72.
Purpura, D. P., Penry, J. K., Tower, D. B., Woodbury, D. M. & Walter, R. D. (Ed.) (1972) *Experimental Models of Epilepsy.* New York: Raven Press.
Richards, C. D. (1973) *J. Physiol.* **233,** 439.
Richards, C. D. & McIlwain, H. (1967) *Nature,* Lond. **215,** 704.
Richards, C. D. & Sercombe, R. (1970) *J. Physiol.* **211,** 571.
Richards, C. D. & Smaje, J. C. (1974) *J. Physiol.* **239,** 103P.
Yamamoto, C. (1972) *Exp. Brain Res.* **14,** 423; *Exp. Neurol.* **35,** 154.

Yamamoto, C. (1973) *Adv. Neurol. Sci.* **17,** 64.
Yamamoto, C. & McIlwain, H. (1966) *J. Neurochem.* **13,** 1333.
Yamamoto, C. & Kawai, N. (1967) *Experientia* **23,** 821.
Yamamoto, C. & Kawai, N. (1968) *Jap. J. Physiol.* **18,** 620.
Yamamoto, C. & Kurokawa, M. (1968) *Proc. Japan Acad.* **44,** 1084.
Zachar, J. (1967) In Bureš *et al.*, 1967, p. 209.

10. Cell-Free Preparations and Subcellular Particles from Neural Tissues

R. M. MARCHBANKS

Disrupting cell structure by grinding or homogenizing	210
Homogenizer, pestle and motor	211
Suspending medium	212
Procedure	213
Separation by centrifuging	214
Apparatus	217
Differential centrifugation	217
Density-gradient separation	218
Apparatus	221
Procedure	222
Isolation and properties of subcellular organelles	223
Nuclei	226
Myelin	226
Synaptosomes	227
Mitochondria	228
Lysosomes	228
Postmitochondrial fraction	229
Vesicles	230
Metabolic properties of subcellular fractions	232
Transport experiments with synaptosomes	233
Release experiments with synaptosomes	236
Respiration of synaptosomes and mitochondria	237
Notes on associated microscopic and analytical methods	238
References	240

It is a reflection of the cell as a fundamental entity in biology, that many aspects of the chemical activities of neural systems are reproduced in very small, cell-containing fragments from it but are lost or transformed when cell structure is broken. On the other hand, the cell-free systems resulting may show activities not previously apparent. Many problems of chemical mechanism and metabolic route can be posed only from data obtained in cell-containing systems and solved only by study of cell-free systems.

The isolation of many subcellular organelles from neural tissues in fractions of reasonable purity has been achieved within the last decade. Most data concern rodents, but other species have given similar results. The first systematic investigations were carried out independently by Whittaker (review; 1965) and De Robertis (review; 1967) and since then many special methods have been developed. From the point of view of subcellular fractionation, neural tissues, especially those of the brain differ from other tissues in the facility with which the membranes reseal after cell rupture and so form discrete particles. These particles can then by isolated and indeed considerable attention has been given to the isolation of an artificial particle formed in this way from the nerve terminal; the synaptosome (Whittaker, Michaelson & Kirkland, 1964).

Most fractionation schemes for parts of the brain are based on those originally developed for liver. The particles resulting from homogenization are separated as a result of the differing speeds (due to density, size and shape) with which they sediment in a gravitational field. Since neural tissue is more complex than liver the primary fractions separated by differential centrifugation are more heterogeneous. Thus the nuclear spin (10^4 g min) precipitates nuclei, cell debris and large fragments of myelin. The mitochondrial spin (2–5×10^5 g min) sediments small fragments of myelin, synaptosomes, mitochondria, large membrane fragments and lysosomes, while the post-mitochondrial spin (3–6×10^6 g min) precipitates ribosomes, fragments of endoplasmic reticulum (microsomes), vesicles and small fragments of plasma membrane.

The preparation of pure fractions demands, therefore, a further separation stage and this is usually achieved by density gradient centrifugation. As its name implies this process separates the particles on the

Table 10.1 Buoyant density in sucrose, and isolation methods for subcellular organelles

Organelle	Molarity of sucrose having approximately the same density	Reference to isolation method
Nuclei (neuronal)	2·0–2·2	Thompson, 1973
Nuclei (glial)	2·3–2·6	Thompson, 1973
Myelin	0·6–0·7	Autilio, Norton & Terry, 1964
Synaptosomes	1·0–1·2	Gray & Whittaker, 1962
Mitochondria	1·4	Stahl et al., 1963
Lysosomes	1·7	Koenig et al., 1964
Membrane fragments	0·7–1·1	Arnaiz, Alberici & De Robertis, 1967
Ribosomal particles	1·6	Zomzely, Roberts & Rapoport, 1964
Microtubules	1·4	Kirkpatrick et al., 1970
Synaptic vesicles from:		
Cerebral cortex	0·4	Whittaker, Michaelson & Kirkland, 1964
Superior cervical ganglion	0·4	Wilson, Schulz & Cooper, 1973
Electric Organ	0·4–0·5	Israël, Gautron & Lesbats, 1970
Spleen	0·6 & 1·1	Bisby & Fillenz, 1971
Neurosecretory granules	1·7	Dean & Hope, 1968

basis of their equilibrium density, the supporting medium being solutions (usually of sucrose) of differing density. A list of the approximate buoyant densities in sucrose of various subcellular components is given in Table (10.1). The gravitational fields necessary to bring most particles in this range to density equilibrium are between 10^7 and 10^8 g min Within this general format considerable experimental variation has been introduced. For example if the reduction of microsomal contamination is important (as in the isolation of synaptic vesicles from cerebral cortex) the mitochondrial spin can be curtailed to $1-2 \times 10^5$ g min thus resulting in a lower yield of synaptosomes but almost completely eliminating microsomal contamination.

In this chapter only some of the more general and comprehensive fractionation schemes will be considered in detail; references to the methods of choice for the isolation of particular organelles are given in Table 10.1 and more extensive reviews of principle and practice are given by Cotman, 1972 and by Marchbanks, 1975. Here we consider (i) methods of rupturing cells to release their cell contents in a form as close as possible to that of the intact tissue; (ii) the isolation of the particles so released and (iii) properties of some of the separated fractions, which are of value in characterizing them. The further extraction of the particles and the separation of enzymes concerns Chapter 11.

DISRUPTING CELL STRUCTURE BY GRINDING OR HOMOGENIZING

Although it is relatively easy to damage the cell structure of neural tissues, the breakdown of the cells in such a way that they freely release their contents in separable form has required specific investigation. In the unusual instance of the giant axon, about 1 mm in diameter, axoplasm can be extruded from the cell wall and associated structures by squeezing it like a tube of paste. The more typical neural cells are however only 20–50 μm across and their long, thin and entangled processes are some 1–10 μm in diameter. Systematic investigation of the optimum conditions of homogenization to release any particular organelle (see for example Whittaker & Dowe, 1965) is always useful. The maintenance of these conditions is more critical for the success of subcellular fractionation in the case of neural tissues than in the case of other tissues.

The most generally used methods of breaking tissues to yield cell-free preparations, have applied shearing forces to the tissue sample. This has been done by pressing the tissue with a suspending fluid into a narrow space between two surfaces, one usually rapidly moving. In general the weaker the shear forces the larger the resultant membrane fragments. Thus the relatively weak shear forces produced by a pestle and mortar homogenizer with a wide clearance rotating at a speed below 1000 rev/min produces synaptosomes and large membrane fragments having diameters about 1 μm. The stronger shear forces produced by piston-press homogenizers or Waring blenders tend to reduce the membrane fragment size to that of microsomes (100–500 nm).

Homogenizer, pestle and motor

The test-tube homogenizer was originally designed as an all-glass apparatus by Ten-Broeck and by Potter & Elvehjem. Made from a tube about 15 × 1·5 cm (internal diameter) and of total capacity about 25 ml, it will take 8–10 ml of liquid and 1–2 g of tissue. The pestle is driven by a motor capable of 2000 rev/min, and with sufficient power (1/4 h.p.) so that the speed of rotation does not diminish when the tissue is broken up. The motor is usually mounted on a stand with its axis vertical and ending in a chuck or a stout rubber tube to take the pestle. A resistance or variable transformer controls the speed of the motor.

The inside of the homogenizer tube must be ground or otherwise made to a true cylindrical form to give the goodness of fit necessary for it to be effective; it should nevertheless be fairly smooth. Tubes fitted with glass pestles are supplied by several manufacturers; usually they must be selected individually according to their performance with the tissue being studied. Each occasion on which a motor-driven all-glass homogenizer is used with the vigour necessary to break cells, causes appreciable wear. This will be seen if the pestle is operated in the tube with water only, for the water becomes opalescent with ground glass.

Plastic pestles do not have this disadvantage, and continue to fit their corresponding tube for a much longer time. Such a homogenizer is described by Aldridge, Emery & Street (1960) and illustrated in Fig. 10.1A. Precision-bore glass tubing (Veridia tubing, Chance Bros. Ltd., Malvern Link, Worcestershire) of diameter 3 cm is fitted with a detachable Perspex base and sealed with a silicone rubber cement (Silastomer, Hopkins & Williams Ltd., Chadwell Heath, Essex). The pestle is machined from Perspex to fit the mortar with the required clearance and mounted on a stainless steel shaft.

For the most effective breaking of cells, the difference in radii between tube and pestle has been quoted by different workers as 0·12–0·2 mm. It has been specifically noted that the clearance for complete disruption of cerebral tissues should be less than for liver (Aldridge & Johnson, 1959). For optimal production of synaptosomes an Aldridge type homogenizer, 3 cm diameter with clearance 0·25 mm should be used. For homogenizers of smaller diameter the same relative clearance should be maintained. These values, however, depend to some extent on the form and handling of the apparatus; synaptosomes can be produced with a homogenizer having a ball-ended plunger (Dounce et al., 1955) with a clearance of only 0·01–0·03 mm which is used entirely by vertical motion of the plunger in the tube. Even when the homogenizer has a rotating pestle its action depends greatly on the vertical passage of tissue suspension between tube and pestle. Thus it can be calculated that if 10 ml of suspension is forced in 1 sec through an 0·15 mm gap round a pestle 2 cm long and 1·5 cm in diameter and running at 1500 rev/min the vertical motion of the liquid through the annulus takes place at about the same speed as the motion of the surface of the pestle. It is thus important that a worker should standardize the speed and number of times that the suspension is forced past the pestle.

Where large amounts of tissue need to be homogenized, as for example in zonal centrifugation, a flow-through homogenizer (Rodnight et al., 1969) should be utilized. In this apparatus the result of a preliminary coarse homogenization is pumped through a cylinder containing a rotating spindle which has regions offering clearances of 0·25 mm with the cylinder wall.

Piston-press homogenizer. This apparatus of Emanuel & Chaikoff (1957) depends purely on the motion of a tissue suspension through a very narrow annular space around a rod about 1 cm in diameter, the space being about 0·02–0·04 mm wide and 1 cm long. Pressures of 2000–10,000 lb/in^2 are required for this and can be obtained with a hydraulic press. The instrument has been applied to guinea pig cerebral tissues, either freshly obtained or after incubation in metabolic experiments, as follows (Wolfe & McIlwain, 1961). The apparatus with its chosen piston-rod is cooled overnight in a refrigerator. Rods of different diameters are supplied and condition the width of the annular space; that giving a width of 0·025 mm is suitable. The tissue (1–30 g) is prepared in suspension at 0° by a 'loose' plastic pestle in a large glass-tube homogenizer, at a concentration of 1 g tissue/10 ml sucrose. The suspension is then poured or sucked into the cylinder of the piston-press homogenizer and the assembled homogenizer fitted to the press. The homogenizer is equipped for brine cooling, but without this, three suspensions of about 100 ml may be prepared in quick succession without serious rise of temperature.

Suspending medium

Almost all preparations of subcellular particles from neural tissues have employed approximately iso-osmotic sucrose, i.e. 0·25–0·32 M. Stronger sucrose solutions are used in density gradient separations, and a few hundred millilitres of 2 M or 1·6 M-sucrose can be prepared and kept ready at 2° for use, as such or after dilution. The concentrated solutions do not encourage microbial growth. The sucrose may be made up with 0·5–1 mM-ethylene–diaminetetraacetate (see below), but additions substantially increasing the ionic strength should be avoided and the pH brought to 6–7 with 0·1N-NaOH.

With regard to the tissue to be homogenized, it is desirable that a defined tissue be taken rather than, for example, the whole brain. Different cellular elements preponderate in different parts of the brain and are likely to differ in the conditions which are optimal for breakdown and subsequent fractionation. It is more satisfactory to work with the cerebral cortex than with cerebral hemispheres which include white matter. The white matter can be readily removed by everting the cerebral cortex on a moistened filter paper and lightly scraping off the exposed myelinated tracts. Subcellular fractions from specific parts of the brain, such as the pituitary may present special problems and opportunities in relation to their content of hormonal or other active constituents (McIlwain & Bachelard, 1971).

Tissues should be homogenized promptly after the death of an animal or the stopping of blood supply to the part concerned. Keeping for an appreciable time, even at $0°$, permits redistribution of material and can alter the properties of subcellular particles from the point of view of their aggregation; this results in major differences in sedimentation and fractionation. When an interval between death and homogenizing is inevitable, as with material brought from a slaughterhouse and in some cases with surgical specimens, the material is best placed in a small watertight glass or plastic vessel in an ice-water mixture in a vacuum can.

Procedure

The apparatus and supplies needed should be ready to hand; they include the following items. (a) Homogenizer tubes of sizes appropriate to the specimens expected, and fitted with pestles of known performance. (b) An excess of the chosen suspending fluid cooled to $0°$, in a measuring cylinder or flask with a suitable pipette. (c) Centrifuge tubes or other vessels for subsequent manipulation of the homogenate. (d) A vessel of chipped ice sufficiently large to take all the preceding apparatus. (e) Containers or instruments for collecting or dissecting the chosen tissue are needed. If the dissection is likely to occupy longer than 20 to 30 sec, it is best carried out on a cutting table (Fig. 6.3) which has been cooled in the ice bath and taken from the bath just before use. Otherwise the cutting table at room temperature or the dish carrying the tissue, suffice; a scalpel, dissecting scissors, and a spatula (20 cm long; blades 0.4×1 cm) are likely to be needed for handling the tissue. (f) For weighing up to a gram of tissue, the rider of Fig. 6.5 used with a torsion balance is probably the most rapid; larger amounts may be weighed, accurate to $1-5\%$ on a watch glass with a simple direct-reading balance. (g) The homogenizer motor, with a means of controlling its speed, preferably by a foot pedal so that both hands are free for the tube and pestle.

The tissue is now collected or dissected, the chosen sample weighed and placed in a homogenizer tube which is returned to the ice-bath. The suspending medium at $0°$ is measured and added so that it washes the tissue to the bottom of the tube. Usually a definite quantity of medium is used per unit weight of tissue, typically 10 ml/g tissue. The homogenizer pestle is replaced in its tube, both removed from the ice bath and the pestle fitted to the motor. The motor is operated at a chosen speed, usually about 1000 rev/min and controlled by the numerical setting of the resistance or variable transformer fitted to the motor; this should be reproducible in successive runs of the instrument. The homogenizer tube is held in the hands so that it does not rotate, but is moved upwards so that the pestle eventually meets the bottom of the homogenizer tube. All the tissue and fluid in the tube should pass the pestle in about 5–10 sec, while the tube is being forced up; it is then pulled down slowly and halted momentarily when all the suspension has again passed the pestle. Creating a vacuum by pulling the tube down too fast is highly deleterious to most subcellular organelles and should be avoided. The foaming caused by allowing the pestle to rotate above the suspension should also be avoided. The suspension should in this way be made to traverse the rotating pestle about 10 times in each direction in the course of 2 min.

The motor is then stopped and the homogenizer is returned to the ice bath. The suspension is used immediately in subsequent manipulations, a small sample being kept for microscopical or other examination.

In a few instances, homogenizing for longer periods has been recommended, and arrangements made for cooling the tube during the process. When more than one specimen is to be homogenized, time is saved by preparing the corresponding number of homogenizers. There can however be an advantage, in comparative studies, in using the same homogenizer for preparing successive specimens.

SEPARATION BY CENTRIFUGING

The experimental advantages offered specifically by cell-free preparations such as those made by homogenizing are two fold. (1) There is the opportunity of studying as reactants substances whose access to the cell is normally restricted. (2) There is further, the opportunity of separating entities which in the intact cell are almost always in association but which prove to be very distinct chemically and metabolically. The present section concerns one fashion of separating distinct subcellular entities: centrifuging to obtain the particulate components of the cell. This has been almost the only process employed for such separations in neural tissues, although the potentialities of other processes, such as migration in electrical fields or distribution in aqueous two-phase systems, should be remembered.

Biochemical and microscopical examination of the fractions is always advisable and is mandatory when the procedures are adapted for use on tissues other than those described. There are two biochemical measures of use in this context from which can be derived two different kinds of conclusions. If it is required to establish whether a particular substance is present in a given organelle it is necessary to determine the percentage distribution of the compound and show that the fraction supposed to contain the organelle contains the highest percentage (of total in the tissue) of the substance in question. If the fraction can then be shown by microscopical examination or the presence of other biochemical markers to be enriched in the given organelle then one may conclude that the compound in question is concentrated within the organelle although of course it may be present also in other parts of the cell, and may not be the sole constituent of the organelle.

If however, it is required to show that a particular organelle is concentrated in a particular fraction the relative specific activity is the appropriate measure. The relative specific activity is the ratio of marker substance (as percentage total recovered) to the amount of protein (as percentage total recovered). Clearly this ratio should be greater than 1 if purification of a particular fraction has been achieved, and less than 1 for markers associated with other fractions.

In practice both measurements are usually made but the inferences that can be drawn from them should not be confused. The quotation of activities of fractions on a protein basis is barely meaningful unless supported by other data. A list of biochemical markers for various subcellular organelles is given in Table 10.2 along with references to suitable assay methods. It is always useful to evaluate the percentage recovery of a marker compound throughout the fractionation procedure, i.e., the ratio of the total recovered in the fractions to the amount originally present in the tissue. If this falls below about 80% it suggests that the compound is being destroyed during the fractionation procedure or that the assay method is being inhibited by, for instance, the high sucrose concentrations in the subfractions. This situation can then be fairly easily rectified, or taken into consideration in the interpretation of the results.

The small size of subcellular organelles requires the use of the electron microscope for their adequate examination. For rapid morphological evaluation of synaptic vesicle and ribosomal fractions, negative staining (Horne & Whittaker, 1962) is recommended but the other fractions require fixation, embedding and the cutting of their sections. (See Notes at the end of this chapter.)

The results of a differential centrifugation of a cerebral cortex homogenate are summarized in Table 10.3 together with the means by which they are obtained and characterized. The initial centrifuging is to be regarded as a primary fractionation only, yielding mixtures to be further separated or purified for an at all detailed study. Also, the fractions as first obtained inevitably overlap; contributing to this are associations between the particles, and the characteristics of the centrifuges ordinarily used. Thus centrifugal force increases with distance from the axis of spin, and because of this a lighter particle initially at the bottom of a tube can be acted on by a sedimenting force greater than that on a heavier particle initially at the top; thus these two can come together at the middle of the tube. Also, various factors can cause fluid movement

Table 10.2 Biochemical markers for subcellular organelles

Organelle	Marker substance	Assay method
Nuclei	DNA	Croft & Lubran, 1965
	RNA nucleotidyltransferase, EC.2.7.7.6	Austoker, Cox & Mathias, 1972
Myelin	Carbonic anhydrase, EC.4.2.1.1.	Roughton & Booth, 1946
	2′,3′-cyclic nucleotide 3′-phosphohydrolase	Olafson, Drummond & Lee, 1969
	Butyryl cholinesterase, EC.3.1.1.8	Whittaker & Barker, 1972
	Cerebroside & sulphatide	Eichberg, Whittaker & Dawson, 1964
Synaptosomes	Bound neurotransmitters	See synaptic vesicles
	Occluded lactate dehydrogenase, EC.1.1.1.27	Marchbanks, 1967
Mitochondria	Cytochrome oxidase, EC.1.9.3.1	Tolani & Talwar, 1963
	Succinic dehydrogenase, EC.1.3.99.1	Porteus & Clark, 1965
Lysosomes	β-glucuronidase, EC.3.2.1.31	Gianetto & De Duve, 1955
	Acid phosphatase, EC.3.1.3.2	Schmidt, 1955
Membranes:		
Plasma membrane	Na^+, K^+-activated ATPase, EC.3.6.1.4	Hosie, 1965
	5′-Nucleotidase, EC.3.1.3.5	Israël & Frachon-Mastour, 1970
	Acetylcholinesterase, EC.3.1.1.7	Ellman et al., 1961
Endoplasmic reticulum	NADPH-cytochrome-c reductase, EC.1.6.2.3	Sottocasa et al., 1967
Ribosomes	RNA	Fleck & Munro, 1962
Microtubules	Colchicine-binding activity	Wilson, 1970
Synaptic vesicles (small)	Bound transmitter:	
	acetylcholine	Szerb, 1961
	noradrenaline	Häggendal, 1963
	dopamine	Cuello, Hiley & Iversen, 1973
	5-hydroxytryptamine	Vanable, 1963
	amino-acids	Rassin, 1972
Large granular vesicles	Bound noradrenaline	Häggendal, 1963
	Dopamine-β-hydroxylase, EC.1.14.2.1	Molinoff, Weinshilboum & Axelrod, 1971
Neurosecretory granules	Vasopressin and oxytocin	Dean & Hope, 1968
Cytoplasm	Free lactate dehydrogenase, EC.1.1.1.27	Kornberg, 1955
	Potassium ions	Flame photometry

within the centrifuge tube. For these reasons the fractions first obtained can almost always be further purified by resuspending them and again centrifuging under the conditions of their initial sedimentation. Resuspension and recentrifugation is known as 'washing' and is almost always carried out on the nuclear fraction and frequently also the mitochondrial fraction. The pellets and supernatant fluids from the two centrifugations can then be combined and the fractionation proceeded with.

Apparatus

The centrifuge must be refrigerated for almost all work with subcellular particles. Expedients such as placing an ordinary centrifuge in a cold room, or placing centrifuge tubes in large buckets packed with ice, are only in part effective. The centrifugal forces needed for all except the larger particles require rotors to operate at speeds at which air friction is considerable; the rotors are thus best run *in vacuo* and an apparatus of the status of an ultracentrifuge is needed. Most experience has been obtained, and all the following procedures can be carried out, with the preparative ultracentrifuges manufactured by Beckman Instruments Ltd, California or M.S.E. Ltd, Crawley, Sussex. The instrument handbooks must be consulted for details specific to the machine employed, its maintenance, and for the speeds which, with the particular rotors concerned correspond to the centrifugal forces required. In the following description the centrifugal forces quoted are mean values established at the midpoint of the tube. A variety of centrifuge tubes are available, the most satisfactory for differential centrifugation are those made from polycarbonate since they are transparent and sufficiently strong to sustain medium centrifugal forces (5×10^4 g). For high centrifugal forces the tubes should be sealed with the cap assemblies supplied by the makers.

Choice of centrifuge heads is conditioned by the type of separation desired. An *angle* head is usually selected for separations which depend solely on the centrifugal forces applied, as in the differential centrifuging described first below. It allows a given volume of solution to be accommodated within a smaller range of radii, from the axis of spin than is possible with swing-out heads; consequently the centrifugal field is more uniform.

Procedure yielding primary fractions by differential centrifuging

The rotor to be used should already be at 0°, and the refrigerator of the centrifuge is started before use. The ground tissue suspension immediately after preparation and still at 0° is filtered quickly through gauze in a filter funnel in order to remove coarse tissue debris, such as vascular structures. The filtrate is added to three-quarters fill the centrifuge tubes and these are balanced and placed symmetrically in the rotor, and the rotor in the centrifuge.

Nuclear fraction (Table 10.3). The rotor is taken to a speed corresponding to 1000 g and run for 10 min without vacuum. The tubes are removed and opened. A firm pellet is found at the bottom of each tube, usually with a red layer of erythrocytes that is visible although the cells represent only a small proportion of the sediment. The supernatants are still heavy suspensions, and are removed by pipette or by decanting without delay to a further set of tubes. The pellets are resuspended by pipetting with a small volume of the homogenizing medium (2ml/per g original tissue). Both the resuspended pellets and the original supernatant fluids are recentrifuged as before. The supernatant fluids after recentrifugation are combined for precipitation of the mitochondrial fraction, and the pellets representing the washed nuclear fraction may be resuspended and combined for further study.

Mitochondrial fraction. The centrifuge tubes containing the supernatant from the nuclear fraction are balanced and fitted symmetrically to their rotor which is sealed and placed in the centrifuge chamber. This is now evacuated and the rotor taken to a speed corresponding to 12,000 g for 30 min (but see above for comment on the centrifugal forces to be used at this stage). After removing the tubes from the apparatus the tubes should show cream-coloured opaque pellets with opalescent supernatants. The supernatants are removed by pipette or decantation to a further set of centrifuge tubes for preparing the microsomal fraction. The mitochondrial fraction is washed by resuspending it in the manner described in the previous paragraph, and again sedimenting at 12,000 g for 30 min.

Post mitochondrial fraction. The tubes containing the supernatant from the mitochondrial fraction are sealed and fitted to their rotor, which is run *in vacuo* at 100,000 g for 30 min. This yields a semitransparent, gelatinous, cream-coloured deposit. The *supernatant* fraction now appears quite clear and nearly colourless, and it can be removed by decantation.

DENSITY-GRADIENT SEPARATION

Subcellular particles differ markedly in density (see Table 10.1). Higher density is only one factor causing more rapid deposition of a given type of particle in ordinary centrifuging. It can be made the major factor in separating two types of particles when the centrifuging is carried out in a fluid whose density is between that of the two particle-types. One will then sink and the other float. The speed of such separation is conditioned by the shape and size of the different particles, but as these factors do not condition whether the particles sink or float, the resulting separation differs from that available by ordinary centrifuging. In practice, three to eight successive layers of liquid of different densities

Table 10.3 Primary fractions obtained from cerebral homogenates

Sedimenting conditions (g/min)	Morphology	Approx. % of tissue content					Relative specific activity			
		Protein	DNA	Succinate dehyrdogenase	NADP-cyt-c reductase	Lactate dehydrogenase	DNA	Succinate dehydrogenase	NADP-cyt-c reductase	Lactate dehydrogenase
10^4	Nuclei, small nerve cell bodies, red blood cells, large myelin fragments and cell debris	16	77	8	10	11	4·8	0·5	0·6	0·7
$2–5 \times 10^5$	Synaptosomes, small myelin fragments, large membrane fragments and mitochondria	41	15	84	21	20	0·4	2·0	0·5	0·5
$3–6 \times 10^6$	Small membrane and endoplasmic reticulum fragments, ribosomes	16	5	6	47	8	0·3	0·4	2·9	0·5
Not sedimenting	No structure, soluble cytoplasm from disrupted cells	26	3	2	22	61	0·1	0·1	0·8	2·3

Fractions are obtained by centrifuging under the conditions described in the text; results refer to laboratory rodents.

are employed in an ordinary plastic centrifuge tube. Powerful centrifugal fields giving some 5×10^6 g min are required, for separation is most effective when the difference in density between successive layers of liquid is as small as practicable, and centrifugal force has minimal effects on a particle in a liquid of nearly its own density.

The use of this technique with neural tissues (Hebb & Whittaker, 1958 and many subsequent workers; see review, Marchbanks, 1975) has largely adopted procedures developed by Kuff & Schneider (1954) employing sucrose gradients. The solutions used are between 0·8 M of density 1·11, and 2 M of density 1·33. They are therefore markedly hypertonic. The polydextran Ficoll (Pharmacia, Uppsala, Sweden) has been used (see for example, Abdel-Latif, 1966) to make the density gradients, thus avoiding the imposition of hypertonic conditions on particles which travel through the gradient. The Ficoll needs to be purified by precipitation with ethanol to remove salt. Ficoll solutions are considerably more viscous than sucrose solutions of equivalent density so the approach to sedimentation equilibrium is slower. In fact the deleterious effects of hypertonic sucrose appears not to be extensive (Marchbanks, 1974; Pull & McIlwain, 1974) and for most purposes the sucrose gradients originally devised are adequate. Other compounds have occasionally been used to form the density gradients, for example sodium diatrizoate (Tamir et al., 1974).

Although the 'step' or discontinuous gradients described here are simpler to construct and adequate for preparative purposes continuous gradients of linear or exponential form have been employed in many studies. They are obviously applicable for analytic purposes, i.e. when it is desired to establish the density of a particular subcellular organelle. A simple device for making linear continuous gradients is described by Salo & Kouns (1965).

Recently the technique of zonal centrifugation has been applied to neural tissue (Rodnight, Weller & Goldfarb, 1969; McBride et al., 1970). The zonal rotor is a hollow bowl shaped vessel with facilities for filling and removing its contents through a rotating seal while it is still spinning. The bowl has a central core (its axis of rotation) and is divided into compartments by septa. The gradients which may either be continuous or discontinuous are pumped in through the hollow seal while the rotor is spinning (2000–5000 rev/min), followed by the sample. Under the gravitational field (30,000 g) the particles separate according to their density in concentric layers and the heaviest being nearest the periphery. The separated fractions are recovered by displacement with 2·5 M sucrose, pumped in so that it enters at the periphery while the rotor is still spinning. The technology of zonal centrifugation is more complex than bucket centrifugation and the reader is referred to the article by Anderson (1968) for further details.

The density gradient separations to be described here have been widely used to separate isolated nerve terminals (synaptosomes) (Gray

& Whittaker, 1962) and the intraterminal constituents (Whittaker, Michaelson & Kirland, 1964).

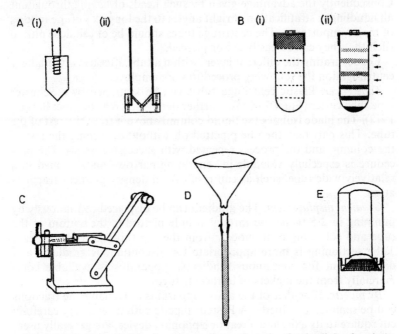

Fig. 10.1 Apparatus used in the preparation and study of subcellular fractions. *A:* (i) Pestle and (ii) mortar of the tissue homogenizer described in detail by Aldridge, Emery & Street (1960). The tube is Chance Veridia glass 3 cm in diameter, and the plug, base plate and pestle tip are Perspex. *B:* Gradient centrifugation tubes before (i) and after (ii) the separation of synaptic vesicles, etc., from osmotically disrupted synaptosomes. The molarities of sucrose (i) are as described in the text, and the separated fractions (ii, shown as stipple) correspond to these described from top to bottom of Table 10.5. The cuts to be made with the tube slicer are indicated by arrows. *C:* Schuster-type tube cutter in diagrammatic form to illustrate operation. Full constructional details are given by Whittaker & Barker (1972). *D:* Small columns containing 5·0 ml G–50 (coarse) Sephadex slurry used to manipulate synaptosome suspensions as described in text. Columns 11 cm by 0·8 cm internal diameter and 15 cm long can be fabricated with a B10 'Quickfit' joint so that a 75 ml funnel similarly jointed can be fitted on top to contain the solution used to rewash the column; 2–3 mm diam. glass beads are used at the bottom of the column to prevent escape of the Sephadex. *E:* Perspex cell 3 cm in diameter used to prepare synaptosome beds for superfusion studies as described in the text. The bottom and top half of the cell are joined by a push-fit connection. Details of the construction and use of the cell are given by De Belleroche & Bradford (1972).

Apparatus

Swing-out heads are used for density-gradient separations, for here the sedimentation is taken closer to equilibrium conditions; particle density is the major differentiating factor and lack of uniformity in the centrifugal field is less important than in differential centrifugation. Consequently the advantage given by such heads of having throughout all handling, a stratification at right angles to the long axis of the tube, is of major importance. The centrifuge tubes should be of cellulose nitrate such that they can be easily cut or pierced.

For separating the different layers within a tube after density-gradient centrifugation the following procedures are employed.

Tube slicer. Plastic centrifuge tubes can be sectioned by a Schuster type tube cutter (Fig. 10.1C; for further details see Whittaker & Barker, 1972). The blade isolates the liquid column above it from the rest of the tube. This part may then be pipetted off without disturbing the rest of the column, and the process repeated with successive layers. The procedure is especially valuable in separating narrow bands formed in a relatively wide tube, such as can result from density gradient fractionation.

Gradient displacement. The gradient can be displaced continuously by pumping in 2·5 M-sucrose through a hole pierced in the bottom of the tube and collecting continuously from the top through a pierced cap. Tube sectioning is more appropriate for discontinuous gradients and displacement for continuous gradients. Apparatus is available commercially from the makers of ultracentrifuges.

By pipette. If neither of the above apparatus is available the fractions can be sampled by pipette. A Pasteur pipette with its end very carefully cut square to its axis, and a teat or siphoning device, are generally used. The end of the pipette may be bent to face vertically, to minimize disturbance of the precipitate. In difficult separations it can be advantageous to clamp the tube vertically to a stand which also carries the pipette on a rack and pinion.

PROCEDURE IN DENSITY-GRADIENT CENTRIFUGATION

In a basin of crushed ice are placed the centrifuge tubes which are to receive the specimens, a flask of distilled water, and tubes in which dilutions of sucrose will be prepared. The stock solution of sucrose, of 1·6 or 2 M, is already kept cool in a refrigerator. Excess of the chosen dilutions of sucrose are prepared; those used for the isolation of synaptosomes need to be 1·2 M and 0·8 M, and those for the separation and isolation of the synaptosome constituents need to be 1·2 M, 1·0 M, 0·8 M, 0·6 M and 0·4 M. The volume of each density layer should be the same and such that with the addition of an equal volume of suspension the total nearly fills the centrifuge tube. The densest solution is pipetted first into the centrifuge tube following which the tubes are removed one by one from the ice, held at eye level and an equal volume of the next densest solution slowly and carefully run from a pipette down the side of the tube to form a layer above the first. The tubes are each returned to the ice after pipetting, and the next densest sucrose solutions are similarly added. The position of the interfaces are marked on the side of the tube and they should then be left for at least 1 h so that the sharp interface disappears.

For the isolation of synaptosomes a mitochondrial fraction is prepared, washed once and resuspended in 2ml/g original tissue of 0·32 M-sucrose. This suspension is layered onto the 1·2–0·8 M step gradient. When it is required to isolate synaptic vesicles and other intra-terminal constituents the procedure is modified to include a period of osmotic rupture. A 'low cut' mitochondrial fraction ($10,000\,g \times 20$ min) is washed once and then resuspended in 2ml of water at 2°C./g original tissue. The resuspension is carried out by sucking up and down with a wide-bore pipette. This suspension is layered on the 1·2 M, 1·0 M, 0·8 M, 0·6 M and 0·4 M gradient.

The tubes are capped, placed symmetrically and with care not to disturb the gradient in the swing-out buckets of the rotor and the rotor placed in the centrifuge with its refrigerator running. The rotor is taken slowly in a vacuum to a speed sufficient to give $5 \times 10^6\,g$ min in 1–3 h. No braking is used during deceleration and after removing the tubes from the centrifuge a successful separation has the appearance of Fig. 10.1*B* (see also Tables 10.4 and 10.5). The tubes are uncapped and returned to the ice.

Separation of the layers is done preferably by slicing the tubes. The layers to be separated are marked on the side of the tube and the tube is placed in the cutter, the vertical rack is adjusted so that the blade is aligned with position of the first cut at the top of the tube. The tube is steadied by holding it *at the top*, the blade is moved so that it just pierces the tube and pressure is then exerted so that it completely cuts the tube separating the top layer. The top layer is removed by pipette to a collection tube, the surface of the blade is wiped clean and it is then withdrawn. The first few millilitres of the next layer are carefully withdrawn by pipette to another collection tube. The centrifuge tube is then elevated to the next cutting mark and the process repeated until all the layers have been collected. The alternative process of collection by pipette is carried out by inserting the bent end pipette to the level of the layer to be collected and sucking the suspension up until the less turbid zones above and below have joined. It is less satisfactory than cutting because it is difficult to avoid some turbulence and mixing. The isolated fractions can be submitted to various forms of biochemical and morphological analysis, typical results of which are shown in Tables 10.4 and 10.5

ISOLATION AND PROPERTIES OF SUBCELLULAR ORGANELLES

The schemes of differential and density gradient fractionation described in the preceding sections are derived from the methods developed and

Table 10.4 Subfractions obtained by density gradient centrifugation of a crude mitochondrial fraction

Molarity of sucrose at which equilibration occurs after 120 min at 57,000 g	Morphology	Approx. % of content of parent fraction				Relative specific activity			
		2',3'-Cyclic nucleotide 3'-phosphohydrolase	Occluded lactate dehydrogenase	Acetylcholine	Succinate dehydrogenase	2',3'-Cyclic nucleotide 3'-phosphohydrolase	Occluded lactate dehydrogenase	Acetylcholine	Succinate dehydrogenase
<0.8	Myelin fragments	79	20	18	4	2.6	0.7	0.6	0.1
>0.8 <1.2	Synaptosomes and membrane fragments	21	71	74	35	0.4	1.3	1.4	0.7
>1.2	Mitochondria	0	9	8	61	0	0.6	0.5	3.8

Results refer to laboratory rodents, and specific activities are relative to the activities of the same enzymes in the crude mitochondrial fraction.

Table 10.5 Subfractions obtained by density-gradient centrifugation of an osmotically ruptured crude mitochondrial fraction

Molarity of sucrose at which equilibration occurs after 120 min at 57,000 g	Morphology	Protein	Approx. % of content of parent fraction				Relative specific activity			
			Lactate dehydrogenase	Acetylcholine	Na$^+$, K$^+$-ATPase	Succinate dehydrogenase	Lactate dehydrogenase	Acetylcholine	Na$^+$, K$^+$-ATPase	Succinate dehydrogenase
<0.3	No structure, soluble intraterminal cytoplasm	26	65	2	0	1	2.5	0.1	0	0.1
>0.3, <0.4	Synaptic vesicles	10	10	33	0	1	1.0	3.3	0	0.1
>0.4, <0.6	Synaptic vesicles and small membrane fragments	7	3	6	3	2	0.4	0.9	0.4	0.3
>0.6, <0.8	Larger membrane fragments and occasional myelin	12	1	4	33	2	0.1	0.3	2.7	0.2
>0.8, <1.0	Empty synaptosome profiles and large membrane fragments	13	2	11	35	4	0.2	0.9	2.7	0.3
>1.0, <1.2	Undisrupted synaptosomes	20	16	35	25	29	0.8	1.7	1.3	1.4
>1.2	Intraterminal mitochondria	10	3	9	4	61	0.3	0.9	0.4	6.1

Results refer to laboratory rodents, and specific activities are relative to the activities of the same enzymes in the crude mitochondrial fraction.

reviewed by Whittaker (1965) and De Robertis (1967). These procedures were developed to produce pure preparations of nerve terminals and synaptic vesicles. They are also fairly comprehensive in that most of the other organelles of neural tissues are isolated in samples of varying purity. These schemes are therefore specially advantageous when it is desired to establish the intracellular localization of a substance in neural tissues because the percentage distribution (cf. p. 214) in each fraction can be compared. However an investigation may demand the preparation of a particular organelle in a pure sample so that detailed studies can be made of its properties without any particular reference to other organelles. Specialized procedures have been developed over the last decade to isolate various subcellular organelles of neural tissue. In this section we consider some of the specialized methods and fractions resulting from them.

Nuclei

Neuronal and various types of glial nuclei differ from each other in density and appearance. Neuronal nuclei have a single well-defined nucleolus and under light microscopy a large pale nucleoplasm. Astrocytic glial nuclei are similar but with more than one nucleolus whereas oligodendrocytic glial nuclei are smaller, more dense and have a darker nucleoplasm with no discrete nucleolus. Separation from cell debris and other materials and isolation of glial and neuronal nuclei is best carried out by homogenizing in 2 M-sucrose containing 1mM-$MgCl_2$ (Thompson, 1973). This molarity of sucrose prevents sedimentation of the cell debris and myelin but the nuclei can be precipitated by centrifuging at 64,000 g for 30 min. The pellet can then be resuspended in 2·4 M-sucrose (containing 1 mM-$MgCl_2$) and 1·8 M-sucrose (also containing 1mM-$MgCl_2$) is layered over it. A density gradient spin at 10^5 g for 45 min leaves the neuronal nuclei in the 2·4 M layer while the glial cell nuclei sediment.

Myelin

Myelin fragments although light are rather large and sediment with the nuclear and crude mitochondrial primary fractions. They may then be purified further by the density gradient centrifugation described previously. When pure myelin only is required a scheme of repeated centrifugation over 0·656 M-sucrose followed by osmotic shock and density gradient centrifugation can be used. The homogeneity of the myelin fractions obtained by density gradient centrifugation is uncertain.

In the procedure described by Autilio, Norton & Terry (1964) the homogenate in 10 volumes 0·32 M-sucrose is layered over 0·656 M-sucrose and centrifuged at 40,000 g for 30 min. The interfacial layer is removed and resuspended in 10 vol of sucrose solution so that final concentration of sucrose is 0·32 M. This is then layered over 0·656 M-sucrose and the centrifugation repeated. The interfacial layer is again resuspended in 16 vol of 0·25 M-sucrose and centrifuged at 1000 g for 10 min. The pellet is resuspended in distilled water and centrifuged at 40,000 g for 10 min. Again the pellet is resuspended in 2 vol of 0·32 M-sucrose and layered over a continuous gradient of sucrose from 0·32 to 0·8M. After centrifuging at 53,000 g for 60 min, the 'light myelin' fraction can be recovered from the middle of the continuous gradient and the 'heavy myelin' from the bottom.

Synaptosomes

The isolation of nerve terminals (synaptosomes) as described in the previous section has obvious importance for studies of the biochemical mechanism of transmission, and the method has been much investigated. Synaptosome formation occurs in a wide variety of central nervous tissues and many different species have been employed. Modification to the original procedure to enable isolation from other parts of the central nervous system have been reported, e.g. from cerebellum (Israël & Whittaker, 1965), from spinal cord (Ross, Andreoli & Marchbanks, 1971) and from posterior pituitary gland (Labella & Sanwal, 1965). The yield of synaptosomes from peripheral tissue is much lower and variable, probably because of unfavourable mechanical conditions. However intact varicosities have been isolated from vas deferens (Bisby & Fillenz, 1971) and by prior treatment of the tissue with collagenase they may be isolated from superior cervical ganglion (Wilson & Cooper, 1972).

In general synaptosomes sediment with the crude mitochondrial fraction; an exception to this is the very large mossy fibre endings which sediment with the nuclear fraction (Israël & Whittaker, 1965).

The principle contaminants of the synaptosome fraction as isolated by a subsequent density gradient are myelin fragments carried down from the gradient layer above and empty membrane sacs (0·2–1·0 μm diameter) of unknown origin. Myelin contamination is best avoided at source by removing as much myelinated tissue as possible by blunt dissection before homogenization. The membrane sacs are difficult to remove and no effective separation has been developed. Some of the membranous sacs are imperfectly re-sealed nerve endings since serial sectioning (C.D. Voorhoost quoted by Whittaker, 1968) revealed remnants of the internal constituents of terminals. Neuronal and glial (Cotman, Herschman & Taylor, 1970) plasma membranes partially vesiculated probably account for some of the sacs, and possibly astrocytic end feet are represented.

Because they are complex particles a variety of biochemical markers can be used to characterize synaptosomes. Almost by definition the preparation should contain bound transmitter substances; these, however, are often tedious and difficult to assay. The presence of occluded lactate dehydrogenase (Marchbanks, 1967) is easy to establish but does not distinguish between synaptosomes and other bodies capable of occluding an enzyme. Membrane markers such as acetylcholinesterase and the Na^+, K^+-ATPase suffer from the same disadvantage. Probably the best strategy is to use a combination of markers (see Table 10.2).

A point of interpretation needs to be mentioned. The synaptosome is a membrane bounded structure which contains among other things a sample of the soluble neuronal preterminal cytoplasm. Therefore cytoplasmic constituents (and this includes materials taken up nonspecifically into the cell) will appear to be concentrated in the synaptosome fraction of the density gradient. Such concentrations should not therefore be interpreted as having any special relevance to synaptic transmission or other events at the terminal.

Mitochondria

Although mitochondria can be purified from the crude mitochondrial fraction by density gradient centrifugation as described previously they will have been exposed to hyperosmotic conditions and the preparation is rarely uncontaminated by synaptosomes. The most satisfactory procedure for producing brain mitochondria with respiratory control ratios in the range 3–8 and normal morphology is repeated differential centrifugation with careful decantation of the white fluffy material (see for example Stahl, Smith, Napolitano & Basford, 1963).

For this preparation it is advantageous to use a homogenizer with a small clearance (0·1mm), and from the suspension a 'low cut' mitochondrial fraction (9000 g × 15 min) is precipitated after the nuclear spin and wash. This pellet is resuspended in homogenizing medium to which 8% (w/v) Ficoll has been added, and centrifuged for 9000 g × 30 min. Decant as much of the white fluffy material off as possible and resuspend in the original homogenizing medium and repeat the previous centrifugation. The pellet should contain the activities of characteristic mitochondrial markers (Table 10.2) with little occluded lactate dehydrogenase.

Lysosomes

Components of the crude mitochondrial fraction show acid phosphatase and β-glucuronidase activity. In other tissues these have been associated with a distinct lysosome particle, and in cerebral tissues the two activities are separable by sucrose gradients (Whittaker, 1959). Lysosomes have a buoyant density rather greater than that of mitochondria and this fact has been utilized by Koenig et al. (1964) for an isolation procedure from the crude mitochondrial fraction, by density gradient centrifugation. The crude mitochondrial fraction is washed, resuspended in 0·32 M-sucrose and then layered on a density gradient of 1·4 M, 1·2 M, 1 M and 0·8 M-sucrose and centrifuged at 60,000 g for 2h. The lysosomes sediment completely through the gradient and are recoverable (with some mitochondrial contamination) from the pellet.

Postmitochondrial fraction

The postmitochondrial or microsomal fraction from cerebral tissues has usually been obtained as indicated in Table 10.3, by centrifuging at speeds approaching the maximal available. As from liver it contains ribosomes, fragments of endoplasmic reticulum and in addition small pieces of neuronal and glial plasma membrane and often vesicles. The *membrane fragments* produced by homogenization sediment on differential centrifugation over the range 10^5–$10^7 g$ min; larger fragments sediment first with the crude mitochondrial fraction and smaller fragments later in the postmitochondrial fraction. Size of fragment seems to be the only determinant of behaviour on the initial differential centrifugation, and so the separation methods depend, in effect, on the way the membranes fragment when the tissue is homogenized. On the density gradient the fragments sediment to sucrose molarities in the range 0·7–1·1 M. Probably the best strategy for producing plasma membrane relatively uncontaminated by endoplasmic reticulum is to utilize the synaptosome or crude mitochondrial fraction as a starting material and then purify by density gradient centrifugation after hypoosmotic shock as described previously. These membranes are probably more representative of the terminal region than other parts of the nerve cell because of the predominance of synaptosomes in the starting material.

The partial purification of *endoplasmic reticulum* can be accomplished by density gradient centrifugation (125,000 g × 180 min) of the postmitochondrial fraction over 1·5 M, 1·2 M and 1 M-sucrose containing 1mM-$MgCl_2$. (Hanzon & Toschi, 1960). Microsomal membranes containing endoplasmic reticulum can be recovered from the 1 and 1·2M sucrose layers while ribosomes sediment right through the gradient. If the preparation of ribosomes is the sole concern the membranes can be dispersed by solubilization with desoxycholate. The postmitochondrial fraction, prepared in sucrose containing 4 mM-$MgCl_2$ is sedimented at 105,000 g for 120 min and then resuspended in $MgCl_2$ and sucrose containing a final concentration of 0·25% desoxycholate (Zomzely, Roberts & Rapoport, 1964). Ribosomes can then be sedimented at 105,000 g × 120 min and washed, preparatory to experiments on protein synthesis.

The *postsynaptic membrane* containing transmitter receptors is of fundamental interest. Various attempts to isolate it from cerebral cortex have been made using detergents to solubilize other contaminants (Fiszer & De Robertis, 1967; De Robertis, Azcurra & Fiszer, 1967).

Cerebral cortex is not a very satisfactory starting material because it is heterogeneous with respect to transmitters and an isolation method from the purely cholinergic electric organ of *Electrophorus* is reported by Kasai & Changeux (1971).

A method for isolating *microtubules* from brain homogenates has been described that depends on their stabilization at low temperature (1°C) in the presence of 1 M-hexylene glycol buffered at pH 6·4 (Kirkpatrick *et al.*, 1970). The microtubules in a $1·5 \times 10^6 g$ min supernatant fluid from brain homogenates in the above medium can be further purified by density-gradient centrifugation.

Vesicles

Synaptic vesicles are the small organelles approximately 50–100 nm in diameter seen characteristically aggregated in the preterminal ending at most synapses. It was proposed, originally by Del Castillo & Katz (1955) that the vesicles store the quanta of transmitter prior to their release. It has also been proposed that the vesicles are involved in the actual process of transmitter release, exocytosis of the vesicle content being the most favoured mechanism (for review see Hubbard, 1970). Both hypotheses are accessible to biochemical investigation provided that there are methods available for the isolation of vesicles.

There are several different types of vesicles to be seen in neural tissue, not all of which are uniquely associated with the synaptic region. At cholinergic synapses the vesicles have a diameter between 30 and 60 nm and are electron lucid. At noradrenergic terminals (varicosities) fixed with glutaraldehyde and postfixed with osmium or permanganate two kinds of 30–60 nm vesicles are seen; electron-lucid throughout, or having an electron-dense core (granular vesicles). Noradrenergic tissue also contains larger (60–150 nm diameter) granular vesicles. These however are not specifically aggregated at the synaptic region but are also seen in the axon, indeed in comparison to the small granular vesicles they are infrequent in the synaptic region.

Even larger granules (100–200 nm in diameter) similar to those seen in the supraoptic nucleus and containing neurophysin and polypeptide hormones can be isolated from homogenates of pituitary gland (Dean & Hope, 1968). At inhibitory synapses in the cerebellum (Uchizono, 1965) and in other regions of the CNS the vesicles have a flattened or ovoid profile; methods to isolate these vesicles have not been developed. Vesicles having a surround of radiating spikes (basket, coated or complex vesicles) have also been described and isolated from cerebral cortex by differential and density gradient centrifugation (Kanaseki & Kadota, 1969).

Only the 30–60 nm diameter vesicles are uniquely associated with the synaptic region and these can be isolated from cerebral cortex by the methods described in the previous section on density gradient centrifugation. A procedure similar in principle has been used for the isolation of cholinergic vesicles from superior cervical ganglion (Wilson, Schultz & Cooper, 1973).

No unambiguous biochemical marker exists for the vesicles except for the presence of bound transmitter in the fraction. The vesicle fraction from cerebral cortex probably contains vesicles of types differing with respect to their transmitter content. In most fractionation procedures the vesicle fraction is isolated adjacent or close to the top layer of the gradient which contains the soluble cytoplasmic constituents. Diffusion of freely soluble substances into the vesicle fraction can therefore readily occur. There arises then the problem of distinguishing free diffused transmitter from bound transmitter in the vesicle fraction. In the case of acetylcholine this is not difficult since unless anticholinesterases are present free acetylcholine will be hydrolysed by residual esterases. However if anticholinesterases are present other criteria of binding must be used if spurious results are to be avoided. No entirely satisfactory method exists; however, iso-osmotic exclusion chromatography can be used to determine bound transmitter (Marchbanks, 1968a; see also below). Alternatively a layer of intermediate density can be interposed between the soluble layer and the density at which the vesicles equilibrate so that there is less interference from diffused soluble transmitter. Estimates of the amount of transmitter lost from the vesicle fraction by leakage can only be made indirectly by estimating the rate of leakage *in vitro* and extrapolating back to include the preparation period.

The major morphologically indentifiable contaminants of vesicle fractions are myelin, which is best removed at source by blunt dissection before homogenization, and artificially-formed membranous saccules. The latter are particularly troublesome because they may occlude soluble transmitter and give the appearance of binding. Investigators should be alert to this possible explanation of bound transmitter in the vesicle fraction. Vesicle fractions can readily be morphologically monitored by using the negative staining technique described by Whittaker & Sheridan (1965) and also discussed at the end of this chapter.

The isolation of synaptic vesicles from cerebral cortex entails the preliminary isolation of a synaptosomal fraction. There is therefore some assurance that the vesicles have originated from the terminal region rather than some other part of the nerve cell. Vesicles have been isolated from a variety of peripheral neural tissues, but here synaptosomes are not formed on homogenization and so this assurance is lacking.

An interesting intermediate case is the isolation of neurosecretory granules containing oxytocins and neurophysins from the posterior pituitary gland. If this tissue is homogenized with pestle clearances 0·279–0·635 mm the nerve endings reseal and may be isolated rather similarly to synaptosomes (Bindler, Labella & Sanwal, 1967). Osmotic rupture of the terminals releases the neurosecretory granules which can then be isolated on a density gradient. However, vigorous homgenization conditions (pestle clearance 0·09 mm, 2000 rev/min) totally disrupt the nerve endings and the granules can be harvested in a fraction sedimenting between 15,000 g/min and 550,000 g/min (Dean & Hope, 1968). They can then be purified by sedimenting them through a density gradient (1·3, 1·35, 1·4, 1·45, 1·5 M-sucrose, 145,000 g for 5 h). The isolation of vesicles from peripheral nervous tissues generally follows the second format, i.e., the vesicles are released directly and remain with the supernatant fluid during the mitochondrial spin. The postmitochondrial supernatant fluid is then submitted to density-gradient centrifugation to purify the vesicles. Examples to be found are the isolation and separation of small and large noradrenaline-containing vesicles from spleen (Bisby & Fillenz, 1971), the isolation of small noradrenaline-containing vesicles from heart muscle (Michaelson et al., 1964) and the isolation of acetylcholine containing vesicles from the electric organ of *Torpedo marmorata* (Israël, Lesbats & Gautron, 1970). The vesicle populations from these peripheral tissues are homogeneous with respect to chemical transmitter.

METABOLIC PROPERTIES OF SUBCELLULAR FRACTIONS

The subcellular particles form a distinct level of the organization of the structure and activities of neural tissues. After isolation, they have many properties which link them with entities and activities of the whole tissue, as is evident from the preceding account. However, the particles have properties not always apparent in the same form in the intact tissue, and not necessarily exhibited in the same way by individual enzymes extracted from them. In biochemical work with the particles, there may be sought either the retention of this organization or its destruction in order to study individual enzyme reactions.

In this section we consider experiments in which the subcellular organization is retained and attempts are made to recreate some measure of functional activity. Neural tissue contains all the major subcellular organelles common to other tissue such as nuclei, mitochondria, endoplasmic reticulum and ribosomes. Studies of these organelles have not revealed substantial differences in those from neural tissue compared with those from other tissues, and so they will not be extensively discussed. The reader is referred to the following articles for examples of appropriate methods and their adaptation to organelles from neural tissue; nuclei, Austoker, Cox & Mathias (1972); mitochondria, Ozawa et al. (1967) and ribosomes, Zomely, Roberts & Rapoport (1964). Attention will be focused here on the investigation of subcellular organelles peculiar to the nervous system, in particular synaptosomes and synaptic vesicles.

Media. The composition of the normal fluid environment of subcellular constituents is to be taken into account in any work with them. Probably there is more than one category of intracellular fluid. Channels appear to lead from outside cells, including neurons, to their interior, surrounding various structures on their way so that even the nucleus may be in relatively close contact with fluids normally regarded as extracellular, as well as with intracellular ones. The composition of a medium based on the intracellular fluid is shown in Table 10.6. Studies with organelles which are still bounded by the plasma membrane, such as synaptosomes, should be conducted in a medium based on the extracellular fluid, also shown in Table 10.6. Note the differences in the concentration of potassium and sodium salt and pH but the similarity of osmotic strength between the two media. Investigators should be prepared to establish the optimum conditions for the phenomena under study and vary the medium accordingly, as part of a general appraisal of the metabolic status of their preparations.

Table 10.6 Incubation media corresponding to the extracellular and intracellular environments

Constituents	Extracellular (mM)	Intracellular (mM)
Na^+	148	15
K^+	4	130
Ca^{2+}	1	0
Mg^{2+}	2	9
pH	7·2	6·8
Phosphate	10	15
Chloride	148	148
Glucose	10	0
ATP	0	3

Transport experiments with synaptosomes

Synaptosomes show a surprising degree of functional metabolic activity, and they can be regarded as miniature anucleate cells derived from the terminal region of the nerve cell. The functional integrity of the membrane is retained sufficiently for the carrier mediated transport of a number of neurobiologically interesting substances to be demonstrated (Bogdanski, Tissari, & Brodie, 1968; Marchbanks, 1968b; Grahame Smith & Parfitt, 1970; Martin & Smith, 1972). A representative experiment on the uptake of [^{14}C]choline (Marchbanks, 1968b) is described below.

The synaptosome suspension is isolated in sucrose solutions having an osmolarity of 0·8–1·2. Rapid dilution to the lower osmolarity is very deleterious to synaptosome metabolic functioning and it has been found advantageous to dilute into the medium of lower osmolarity in two stages. This can be conveniently done by diluting the isolated synaptosomes slowly 1:1 with the final incubation medium giving an osmolarity of 0·6–0·8. They may then be centrifuged out of their medium (20,000 g for 20 min) and the pellet broken up and resuspended in the final incubation medium. Experiments can be started after about a 15 min recovery period at 4°C followed by about 15 min preincubation at 30°. [^{14}C] Choline (1μCi/ml final concentration) is then added and the uptake followed as a function of time.

Judging from their content of occluded lactate dehydrogenase, synaptosomes cannot sustain incubation at 37° for long periods without breaking up (Marchbanks, 1967). It is preferable therefore for long term experiments (30–90 min) to use incubation temperature of 25–30°C though for shorter periods the higher temperature may be used. Criteria of synaptosomal integrity to be used as controls in metabolic experiments have not been extensively investigated. The lactate dehydrogenase inside the synaptosomes is not accessible to its substrates when these are added in iso-osmotic media. Estimates of occluded lactate dehydrogenase (see Notes at the end of this chapter; and Marchbanks, 1967) can be made rapidly on a recording spectrophotometer and are useful as a rough guide to the integrity of the membrane.

Small Sephadex columns can be used routinely to separate synaptosomes from their suspension medium. The columns (Fig. 10.1D) are filled with 5·0 ml wet volume of Sephadex (G-50, bead form, Pharmacis, Uppsala, Sweden) which has been equilibrated and washed with a sucrose solution iso-osmotic to the synaptosome incubation medium. The column is allowed to drain and 0·2 ml of the synaptosome suspension is applied to the top of the column and elution continued immediately with the equilibration medium. The performance of representative columns can be checked beforehand using suitable markers, e.g. blue dextran and a low molecular weight dye. After 1·3 ml of eluting solution has been added (measured from the addition of the sample) synaptosomes appear and are 85% recovered after collection of a further 1ml. Low molecular weight compounds appear after 2·5 ml of elution and up to 6·0 ml. The contamination by low molecular weight compounds in the synaptosome cut is about 1%. Various modifications to this procedure can be made to suit particular conditions and after calibration the small columns are sufficiently reproducible for batch elution to be used, thus avoiding serial collection of small amounts of effluent. Very concentrated suspensions tend to clump and therefore they do not pass through the column easily. Some amelioration of the clumping can be gained by vigorous shaking of the suspension but it is best avoided by not using suspensions with more than 1–2 mg/ml of protein. Small variation in the recovery of synaptosomes in the effluent can be estimated for, and corected by measuring the protein in the sample.

The method is rapid (2 min) and the columns can be washed for reuse quite quickly. Leakage of low molecular weight compounds out of the synaptosomes while they are passing through the column is small and further manipulation such as density gradient centrifugation can be carried out on the suspension. By eluting the column with a hypoosmotic medium the synaptosomes are ruptured and lose their content of

Table 10.7 Uptake and metabolism of [^{14}C]choline in synaptosomes

Condition	Uptake, d.p.m. $\times 10^{-3}$/mg synaptosomal protein after 30 min at 30°C	Conversion (%) to [^{14}C]acetylcholine in 30 min (d.p.m. acetylcholine/d.p.m. choline, within synaptosomes)
Control, medium as in Table 10.6	14·0	22
Na$^+$ replaced with sucrose (to maintain osmolarity)	5·5	11
Na$^+$ replaced with K$^+$	6·5	12
Added K$^+$, 35 mM	9·0	31
Added choline, 2 mM	4·0	9
Added hemicholinium -3, 50 μM	3·0	8
Added cyanide, 2·5 mM, plus iodoacetate, 2·5 mM	4·5	8

The table quotes typical results of an experiment in which synaptosomes are incubated with 1 μCi [^{14}C]choline/ml as described in the text, and the uptake and conversion to acetylcholine is studied by gel filtration.

low molecular weight compounds. The small amounts then found in association with the synaptosomal protein are non-specifically bound and can be corrected for. The method is equally applicable to studies on the binding of transmitter by synaptic vesicles (Marchbanks, 1968a) but recently synthesized acetylcholine appears to be preferentially lost from the vesicles during passage through the column (Marchbanks & Israël, 1972).

The amount of radioactivity per mg protein of the column effluent can be determined at various times and also when the final steady state level is reached after 30–45 min. This indicates the amount of [^{14}C] choline that has entered the synaptosomes. Typical results for various conditions and inhibitors are shown in Table 10.7. Two to three column effluents can be pooled and an analysis of the metabolic conversion of the [^{14}C] choline made. A suitable procedure for the extraction of choline bases and thin layer chromatographic separation of choline and acetylcholine is given by Marchbanks & Israël (1971). The conversion of [^{14}C] choline to [^{14}C] acetylcholine is characteristically 20–30% after a 30 min incubation of cerebral cortex synaptosomes.

Release experiments with synaptosomes

A major advance in the display of functional activity in synaptosomes has been the demonstration that electrical stimulation of synaptosome suspensions preferentially releases those amino acids that are putative transmitters (Bradford, 1970). Although such stimulation can be applied to the suspension through concentric ring electrodes the development of wafer aggregates of synaptosomes (synaptosome 'beds') facilitates it and the collection of released substances (De Belleroche & Bradford, 1972). The synaptosome preparation is resuspended in Krebs–phosphate medium, and 5 ml containing approximately 2 mg/ml of protein is introduced into a small circular Perspex cell with a detachable base (see Fig. 10.1E). At the bottom of the cell a piece of nylon gauze (110 mesh) is positioned and held between the body of the cell and its base. The cell can then be centrifuged at 550g for 10 min which deposits the synaptosomes as a bed in the nylon gauze. The supernatant fluid is poured off, the cell dismantled and the top surface of the wafer covered with another piece of gauze. The bed can then be positioned within the jaws of a quick-transfer apparatus (Fig. 7.2) which enables rapid changes of incubation medium, and electrical stimulation, to be carried out. In addition, beds of synaptosomes and of other particulate preparations can be superfused by using the system of Fig. 7.3; this has been applied in following the output of [^{14}C] adenine derivatives on applying electrical pulses (Kuroda & McIlwain, 1974; Pull & McIlwain, 1974).

Stimulation conditions found sufficient to release amino acids from synaptosomes either in suspension or as beds are: square wave pulses of 10 V, 0·4 ms duration with alternating polarity applied at frequencies of

100/s for about 30 min. For stimulation of synaptosome suspensions concentric gold ring electrodes are used, while in the quick-transfer arrangement the silver electrode wires are wound around the jaws of the apparatus. The average continuous current is 10–15 mA in the suspension and 30–40 mA in the wafer arrangement. The bed arrangement can be used with a lower fluid: tissue ratio, which facilitates the detection of released substances and it also approximates more closely to the *in vivo* situation. However, it reintroduces some of the problems of the cerebral slice preparation in that there is a diffusion gradient across the bed.

Respiration of synaptosomes and mitochondria

The respiration of brain mitochondria (Nicklas, Clark & Williamson, 1971) and synaptosomes (Verity, 1971) may be studied polarographically. The Clarke-type oxygen electrode assemblies are obtainable from Rank Bros., Bottisham, Cambridge, or the Yellow Springs Instrument Co., Yellow Springs, Ohio, U.S.A. The electrode consists essentially of an Ag/AgCl reference half cell joined to an irreversible Pt/O_2 electrode by a saturated KCl bridge. The rate of transfer of molecular oxygen to the platinum cathode is proportional to the concentration of oxygen in the reaction vessel at polarizing voltages of about 0·6 V and the whole electrode assembly is protected from the contents of the reaction vessel by a Teflon or polyethylene membrane readily permeable only to oxygen. The current flow through the cell can be monitored as a function of time on a suitable recorder. The reaction vessel of volume about 2·5 ml is capped with an airtight lid whose level is adjustable so that air can be eliminated by screwing it down. There is an injection port 1mm in diameter in the top of the lid, and the whole apparatus is placed over a magnetic stirrer.

Having first establishing that there is a plateau in the current/voltage relationship in the region of 0·5–0·8 V under anaerobic conditions and the polarizing voltage being adjusted to this level, the cell may be calibrated. This is done by measuring the current flow across the cell when it contains medium fully aerated, i.e. stirred with the cap off. A crystal of sodium dithionite is now added and the cell capped rapidly, stirring being continued at the same rate. As the medium becomes anaerobic the current will drop to the baseline level. The difference in current flow is then proportional to the oxygen content of the medium when it is saturated in air. This can readily be calculated from the Bunsen coefficient for oxygen in water at the appropriate temperature (see Dawson *et al.*, 1969, p. 609) and is relatively unaffected by salts in the medium. The behaviour of the electrode is very temperature-sensitive so all the solutions should be brought to the working temperature and water at this temperature circulated through the jacket of the reaction vessel.

About 2 ml of the preparation in the appropriate medium is put into the cell and the lid screwed down with the stirrer on. Care is taken to eliminate all the air bubbles through the injection port. After a period (1 min) during which the baseline respiration is monitored a substrate may be introducted through the injection port by means of a Hamilton syringe and the change in respiration rates monitored. Several additions can be made in this way before the oxygen in the cell is exhausted, and so a variety of conditions can be investigated in a single run.

The respiratory quotient of cerebral mitochondria can be investigated by observing the rate of oxygen utilization before and after the addition of ADP, and the P/O ratio calculated by relating the amount of oxygen taken up to the ADP added in the presence of a particular substrate (Nicklas, Clark & Williamson, 1971). After the incubation the suspension can be removed from the cell and changes in chosen metabolites measured.

Manometric incubation (see Chapter 7) may be useful in experiments which are comparable but which involve longer periods of incubation or the use of more synaptosomes. A study of synaptosomal respiration by Warburg manometry is described by Bradford (1969).

NOTES ON ASSOCIATED MICROSCOPICAL AND ANALYTICAL METHODS

Microscopy. The ordinary light microscope, at magnifications up to 1000 can be used with appropriate stains to show the presence and relative numbers of intact cell bodies, nuclei, and tissue debris such as the remains of vascular tissue, erythrocytes and blood platelets. Nuclei can differentiated from cell bodies of similar size; and by staining with 0·5% crystal violet (Thompson, 1973) nuclei originating from neuronal and glial cells can be distinguished. Phase contrast microscopy most valuably supplements or replaces the use of stains, and is more rapid. As observations by phase contrast are made in the normal suspending media, aggregation among particles can be recognized with greater certainty than when dyes are added. Aggregation greatly affects the behaviour of particles in centrifuging; for relevant information the suspension must be examined promptly.

Electron microscopy is essential for knowledge of all the smaller particles which by either of the preceding methods are unresolved dots recognized only by their Brownian movement, or are unnoticed. As previously noted the negative staining technique can be used for examination of synaptic vesicle and ribosomal fractions. Collodion carbon-coated specimen grids and a fine watchmaker's forceps are required. All reagents should be at 4°C. To 0·1 ml of fraction add 0·1 ml of a formaldehyde fixation solution (10% w/v in 0·32 M-sucrose brought to pH 7·4 with NaOH). Wait 5 min then place a small drop on the specimen grid held by the forceps. By placing a piece of filter paper lightly against the grid withdraw almost all of the drop leaving a thin film. Then apply a drop of 2% (w/v) phosphotungstate (neutralized to pH 7·4 with NaOH) and similarly remove all but a thin film. Allow to dry. If the grids are found to be hydrophobic, dry a thin film of 1% (w/v) bovine serum albumen on them. This improves spreading at the expense of resolution. The phosphotungstate is electron-dense, so biological material which is not penetrated by the phosphotungstate will appear light against a dark background: hence the term negative-staining. Since the fractions are viewed from above rather than in section particles larger than about 100 nm cannot be adequately examined by this technique; they appear as large areas without structure. The appearance of synaptic vesicle fractions under negative-staining is shown by Whittaker & Sheridan (1965).

The other fractions require prefixation with buffered glutaraldehyde, and fixation in osmic acid. They are then dehydrated in a graded series of alcohols and embedded in epoxy resins such as Araldite or Epon. Sections are cut from all regions of the block to ensure representative sampling and they are then mounted on the specimen grids and stained with uranyl acetate and lead citrate. Fixation conditions suitable for brain subcellular fractions are described by Ross, Andreoli and Marchbanks (1971) and further details of electron microscopy and techniques of sample preparation are given by Pease (1964).

Chemical determinations. Total nitrogen can be determined by a micro-Kjeldahl method. Protein determination by colorimetric methods has been noted as subject to interference from lipids in some of the cerebral fractions; the use of organic solvents to avoid this is described by Aldridge (1957) and Aldridge & Johnson (1959). The Folin–Ciocalteau method as described by Lowry et al. (1951) or variants of it are suitable for protein determination in subcellular fractions.

In Table 10.2 reference is made to a variety of suitable enzymic markers for subcellular fractions and to methods of assay. The enzymes are often occluded in membrane bound organelles and for true estimates of their activity they must be released. For enzymes which are not affected by detergent the addition of a final concentration of 0·1% Triton X-100 is usually sufficient to destroy the occluding membrane. Alternatively a short period of ultrasonication can be used, though in both cases checks should be made that the treatment does not inactivate the enzyme. Hypo-osmotic treatment is less suitable because of the tendency of the membranes to reform after rupture and re-occlude some enzyme. This can be readily demonstrated in the case of lactate dehydrogenase in synaptosome fractions.

The occlusion of a cytoplasmic enzyme such as lactate dehydrogenase can give information about the state of the occluding membrane (see for example Marchbanks, 1967). The extent of occlusion can be measured by assaying the enzyme activity (under iso-osmotic conditions) before and after the addition of detergent. The occluded enzyme is the difference between these two estimates since only 'free' enzyme is measured initially and 'total' is measured after the addition of detergent which disrupts the membrane. Where an assay is done by continuous recording of spectrophotometer readings, determination of occlusion can be done rapidly and conveniently by measuring the 'free' rate initially and then adding detergent to the same cuvette to establish the 'total' rate. Further discussion of enzymology and the systematic study of brain enzymes *per se* is given in the next chapter.

REFERENCES

Abdel-Latif, A. A. (1966) *Biochem. biophys. Acta* **121**, 403.
Aldridge, W. N. (1957) *Biochem. J.* **67**, 423.
Aldridge, W. N., Emery, R. C. & Street, B. W. (1960) *Biochem. J.* **77**, 326.
Aldridge, W. N. & Johnson, M. K. (1959) *Biochem. J.* **73**, 270.
Anderson, N. G. (1968) *Q. Rev. Biophys.* **1**, 217.
Arnaiz, G. R. de L., Alberici, M. & De Robertis, E. (1967) *J. Neurochem.* **14**, 215.
Austoker, J., Cox, D. & Mathias, A. P. (1972) *Biochem. J.* **129**, 1139.
Autilio, L. A., Norton, W. T. & Terry, R. D. (1964) *J. Neurochem.* **11**, 17.
Bindler, E., Labella, F. S. & Sanwal, M. (1967) *J. Cell Biol.* **34**, 185.
Bisby, M. A. & Fillenz, M. (1971) *J. Physiol.* **215**, 163.
Bogdanski, D. F., Tissari, A. & Brodie, B. B. (1968) *Life Sci.* **7**, 419.
Bradford, H. F. (1969) *J. Neurochem.* **16**, 675.
Bradford, H. F. (1970) *Brain Res.* **19**, 239.
Cotman, C. W. (1972) In *Research Methods in Neurochemistry*, ed. Marks, N. & Rodnight, R, Vol. 1, p. 45. New York: Plenum Press.
Cotman, C. W., Herschman, H. & Taylor, D. (1970) *J. Neurobiol.* **2**, 169.
Croft, D. N. & Lubran, M. (1965) *Biochem. J.* **95**, 612.
Cuello, A. C., Hiley, R. & Iversen, L. L. (1973) *J. Neurochem.* **21**, 1337.
Dawson, R. M. C., Elliot, D. C., Elliot, W. H. & Jones, K. M. (1969) *Data for Biochemical Research*, 2nd edn, p. 609. Oxford Univ. Press.
Dean, C. R. & Hope, D. B. (1968) *Biochem. J.* **106**, 565.
De Belleroche, J. S. & Bradford, H. F. (1972) *J. Neurochem.* **19**, 585.
Del Castillo, J. & Katz, B. (1955) *J. Physiol.* **128**, 396.
De Robertis, E. (1967) *Science, N.Y.* **156**, 907.
De Robertis, E., Azcurra, J. M. & Fiszer, S. (1967) *Brain Res.* **5**, 45.
Dounce, A. L., Witter, R. F., Monty, K. J., Pate, S. & Cottone, M. A. (1955) *J. biophys. biochem. Cytol.* **1**, 139.
Eichberg, J., Whittaker, V. P. & Dawson, R. M. C. (1964) *Biochem. J.* **92**, 91.
Ellman, G. L., Courtney, K. D., Andres, V. Jun. & Featherstone, R. M. (1961) *Biochem. Pharmacol.* **7**, 88.
Emanuel, C. F. & Chaikoff, I. L. (1957) *Biochim. biophys. Acta* **24**, 254, 261.
Fiszer, S. & De Robertis, E. (1967) *Brain Res.* **5**, 31.
Fleck, A. & Munro, H. N. (1962) *Biochim. biophys. Acta* **55**, 571–583.
Gianetto, R. & deDuve, C. (1955) *Biochem. J.* **59**, 433–438.
Grahame-Smith, D. G. & Parfitt, A. G. (1970) *J. Neurochem.* **17**, 1339.
Gray, E. G. & Whittaker, V. P. (1962) *J. Anat.* **96**, 79.
Häggendal, J. (1963) *Acta physiol. scand.* **59**, 242.
Hanzon, V. & Toschi, G. (1960) *Exptl Cell Res.* **21**, 332.
Horne, R. W. & Whittaker, V. P. (1962) *Z. Zellforsch. Mikrosk. Anat.* **58**, 1.
Hosie, R. J. A. (1965) *Biochem. J.* **96**, 404.
Hubbard, J. I. (1970) *Prog. Biophys. & Mol. Biol.* **21**, 33.
Israël, M. & Frachon-Mastour, P. (1970) *Archs Anat. micros. Morph. exp.* **59**, 383.
Israël, M., Gautron, J. & Lesbats, B. (1970) *J. Neurochem.* **17**, 1441.
Israël, M. & Whittaker, V. P. (1965) *Experientia* **21**, 325.

Kanaseki, T. & Kadota, K. (1969) *J. Cell Biol.* **42**, 202.
Kasai, M. & Changeux, J. P. (1971) *J. Membrane Biol.* **6**, 1.
Kirkpatrick, J. B., Hyams, L., Thomas, V. L. & Howley, P. M. (1970) *J. Cell Biol.* **47**, 384.
Koenig, H., Gaines, D., McDonald, T., Gray, R. & Scott, J. (1964) *J. Neurochem.* **11**, 729.
Kornberg, A. (1955) *Meth. Enzym.* **1**, 441.
Kuff, E. L. & Schneider, W. C. (1954) *J. biol. Chem.* **206**, 681.
Kuroda, Y. & McIlwain, H. (1974) *J. Neurochem.* **22**, 691.
Labella, F. S. & Sanwal, H. (1965) *J. Cell Biol.* **25**, 179.
Lowry, O. H., Rosebrough, N. J., Farr, A. L. & Randall, R. J. (1951) *J. biol. Chem.* **193**, 265.
Marchbanks, R. M. (1967) *Biochem. J.* **104**, 148.
Marchbanks, R. M. (1968a) *Biochem. J.* **106**, 87.
Marchbanks, R. M. (1968b) *Biochem. J.* **110**, 533.
Marchbanks, R. M. (1975) In *Methods in Brain Research*, ed. Bradley, P. B., p. 114. Wiley:Chichester.
Marchbanks, R. M. & Israël, M. (1971) *J. Neurochem.* **18**, 439.
Marchbanks, R. M. & Israël, M. (1972) *Biochem. J.* **129**, 1049.
Martin, D. L. & Smith, A. A. (1972) *J. Neurochem.* **19**, 841.
McBride, W. J., Mahler, H. R., Moore, W. J. & White, F. P. (1970) *J. Neurobiol.* **2**, 73.
McIlwain, H. & Bachelard, H. S. (1971) *Biochemistry and the Central Nervous System*, 4th edn, p. 540. Churchill Livingstone: Edinburgh.
Michaelson, I. A., Richardson, K. C., Snyder, S. H. & Titus, E. O. (1964) *Life Sci.* **3**, 971.
Molinoff, P. B., Weinshilboum, R. & Axelrod, J. (1971) *J. Pharmac. exp. Ther.* **178**, 425.
Nicklas, W. J., Clark, J. B. & Williamson, J. R. (1971) *Biochem. J.* **123**, 83.
Olafson, R. W., Drummond, G. I. & Lee, J. F. (1969) *Can. J. Biochem.* **47**, 961.
Ozawa, K., Seta, K., Araki, H. & Handa, H. (1967) *J. Biochem., Tokyo* **61**, 352.
Pease, D. C. (1964) *Histological Techniques for Electron Microscopy*, 2nd edn. New York: Academic Press.
Porteus, J. W. & Clark, B. (1965) *Biochem. J.* **96**, 159.
Pull, I. & McIlwain, H. (1974) *Biochem. Soc. Trans.* **2**, 272.
Rassin, D. K. (1972) *J. Neurochem.* **19**, 139.
Rodnight, R., Weller, M. & Goldfarb, P. S. G. (1969) *J. Neurochem.* **16**, 1591.
Rodnight, R., Wynter, C. V. A., Cook, C. N. & Reeves, R. (1969) *J. Neurochem.* **16**, 1581.
Ross, L. L., Andreoli, V. M. & Marchbanks, R. M. (1971) *Brain Res.* **25**, 103.
Roughton, F. J. W. & Booth, V. H. (1946) *Biochem. J.* **40**, 319, 327.
Salo, T. & Kouns, D. M. (1965) *Analyt. Biochem.* **13**, 74.
Schmidt, G. (1955) *Meth. Enzym.* **11**, 523–530.
Sottocasa, G. L., Kuylenstierna, B., Ernster, L. & Bergstrand, A. (1967) *J. Cell Biol.* **32**, 415.
Stahl, W. L., Smith, J. C., Napolitano, L. M. & Basford, R. E. (1963) *J. Cell Biol.* **19**, 293.
Szerb, J. C. (1961) *J. Physiol.* **158**, 8P.
Tamir, H., Rapport, M. M. & Roisin, L. (1974) *J. Neurochem.* **23**, 943.
Thompson, R. J. (1973) *J. Neurochem.* **21**, 19.
Tolani, J. A. & Talwar, G. P. (1963) *Biochem. J.* **88**, 357.
Uchizono, K. (1965) *Nature, Lond.* **207**, 642.
Vanable, J. W. (1963) *Analyt. Biochem.* **6**, 393.
Verity, M. A. (1972) *J. Neurochem.* **19**, 1305.
Whittaker, V. P. (1959) *Biochem. J.* **72**, 694.
Whittaker, V. P. (1965) *Prog. Biophys. & Mol. Biol.* **15**, 39.

Whittaker, V. P. (1968) *Biochem. J.* **106**, 412.
Whittaker, V. P. & Barker, L. A. (1972) In *Methods of Neurochemistry*, ed. Fried, R. Vol. 2, pp. 1–52. New York: Marcel Dekker.
Whittaker, V. P. & Dowe, G. H. C. (1965) *Biochem. Pharmacol.* **14**, 194.
Whittaker, V. P., Michaelson, I. A. & Kirkland, R. J. A. (1964) *Biochem. J.* **90**, 293.
Whittaker, V. P. & Sheridan, M. N. (1965) *J. Neurochem.* **12**, 363.
Wilson, I. (1970) *Biochemistry* **9**, 4999.
Wilson, W. S. & Cooper, J. R. (1972) *J. Neurochem.* **19**, 2779.
Wilson, W. S., Schulz, R. A. & Cooper, J. R. (1973) *J. Neurochem.* **20**, 659.
Wolfe, L. S. & McIlwain, H. (1961) *Biochem. J.* **78**, 33.
Zomzely, C. E., Roberts, S. & Rapoport, D. (1964) *J. Neurochem.* **11**, 567.

11. Isolation and Study of Enzymes from Neural Systems

H. S. BACHELARD

Preparation of enzymes	244
Extraction of enzymes	246
Acetone powders	247
Purification of enzymes from neural tissues	250
Separation of isoenzymes	251
Kinetic studies of neural enzymes	253
References	258

Neural tissues provide the richest known sources of many enzymes, including some which are of widespread occurrence. Examples of these include enzymes of the glycolytic pathway, the two enzymes concerned with metabolism of cyclic AMP (adenylcyclase and phosphodiesterase) and the Na^+,K^+-ATPase involved in maintenance of cation distribution. Only very few enzymes have been characterized as occurring solely within neural systems and seem at present to be restricted either to enzymes of neurotransmitter metabolism, such as glutamate decarboxylase, or to those found in specialised parts of the system, such as the 2′,3′-nucleotidase of myelin (Chapter 10).

Other enzymes have been reported to be of negligible activity in the nervous system, which probably reflects its degree of metabolic specialization. Thus evidence that gluconeogenesis does not occur to any significant extent in the mammalian brain is supported by the absence of key enzymes required for reversal of glycolysis from pyruvate to glucose (McIlwain & Bachelard, 1971). Such features, together with the presence of cellular constituents not found outside the nervous system, have stimulated the development of specialised techniques for enzyme studies in neural systems. Neural tissues are characterised also by their high lipid content due to the preponderance of membrane lipoproteins (McIlwain & Bachelard, 1971); for this reason the initial stage of enzyme extraction and purification is frequently the preparation of an acetone powder of the tissue. The techniques involved in such preparations are therefore described in some detail below.

This chapter includes a description of the ways by which neural tissues are prepared for enzyme studies and some of the techniques which have been devised to examine their properties. It is not intended to include general aspects of enzymology (for these, see Dixon & Webb, 1964; Mahler & Cordes, 1971), nor are general methods of assay described. Abnormality or absence of enzymic activity is becoming increasingly important in understanding various brain disorders. Such abnormalities may remain undetected by routine screening for enzymic activity and their recognition may thus require kinetic analysis or examination for changes in isoenzyme patterns. Both of these aspects therefore receive some comment. A compilation of activities exhibited by neural systems has been given by Seiler (1969).

PREPARATION OF ENZYMES

Enzymes are frequently first detected and identified in tissue dispersions prepared with minimal treatment; this is always advisable, for inactivation may result from more extensive handling. It can be of particular importance during the initial stages of investigation, before sufficient is known of the individual properties of the enzyme to allow for adequate precautions to be taken. Simple dispersions may also be adequate for initial characterization of an enzyme and for gaining the first impression of the maximal rate of reaction of which a given weight of tissue is capable. Results so obtained should however be regarded as provisional until the enzyme has been further extracted or purified, matters which are discussed in a later section.

A simple dispersion is capable of exhibiting many of the individual enzyme potentialities of the tissue from which it is prepared, when appropriate substrates, coenzymes, buffers or inhibitors are supplied. Unusually labile enzymes, or enzyme activities which depend on specific association with structural elements, may however be lost or fail to be exhibited.

Neural as other tissues should be handled promptly for enzyme studies. Most cerebral enzymes examined seem to be stable in the brain *in situ* for an hour or more after death. Nevertheless this cannot be assumed to apply to enzymes not yet examined, and changes occurring in other neural constituents, though they may not affect enzyme activity directly, can make a process of fractionation non-reproducible. Generally cerebral enzymes have been found to be stable during unusual *in vivo* conditions, and also *in situ* after death (Robins *et al.*, 1958). Of some 15 enzymes examined, phosphofructokinase was the only one to lose significant activity in rabbit brain post mortem. All of the enzymes which have been examined after storage of small freeze-dried samples of rabbit brain for one year at $-20°$ proved to be completely stable (Strominger & Lowry, 1955).

Before removing the tissue from the animal, therefore, all apparatus and materials needed should be at hand and cooled to $0°$ where appropriate. With small laboratory animals, cerebral tissues can usually be cooled and dispersed in cold fluids within 3 min after death of the animal. Material transported from an operating theatre or slaughterhouse is usually best placed in a plastic container or bag which is then placed with ice in a vacuum can. It may also be frozen during transport if a method of extraction has been chosen which involves the frozen tissue. If delay in handling the tissue cannot be avoided, it is preferable to leave the intact tissue at or below 0° rather than to leave it after dispersion. It should be remembered, however, that storage in the frozen state can result in alterations in the ease with which an enzyme can be extracted or released from a cellular component. In the author's laboratory, freezing of ox or guinea pig cerebral cortex has resulted in increased difficulty in extracting the mitochondrial hexokinase activity, referred to below.

The most frequently used method of dispersing tissue samples of up to a few grams, is with a test tube homogenizer or cognate apparatus. These instruments and their use in preparing subcellular particles have already been described in Chapter 10; the major differences in their use for enzyme studies lies in the much wider range of fluid media in which the tissue may be ground. These range from water and organic solvents to various salt and non-electrolyte solutions.

Blenders or macerators are typically employed in dispersing some tens or hundreds of grams of material. This is placed in the cold instrument with 4 to 10 vol of cold fluid (2–50 vol have been used). and the blades run at maximal speed for 15 s to 5 min. This may be followed by a period of gentle stirring of about 1 h. The blenders or macerators differ from test-tube homogenizers in having a greater tendency to either damage or not completely release subcellular particles.

Rubbing in a mortar and pestle is now less frequently carried out as a method of preparation. The tissue, initially with no fluid or with a small volume only, was rubbed to a cream or sludge in the course of a few minutes, and this subsequently dispersed by addition of more fluid. An abrasive, as sand, was added to tissues with tough elements, such as ganglia or peripheral nerve. Peripheral nerve has been handled also by freezing it in liquid nitrogen and pulverising in a mortar, before dispersion in a glass homogenizer. Cerebral tissues also have been brought from a slaughterhouse at $-60°$ to $-80°$ and roughly broken, while solid, with a mortar and pestle before blending.

Changes in enzyme stability during purification can also occur, due to removal of small molecules, such as cofactors, or other factors which might act normally to retain the required conformation of the enzyme protein.

The fluids used for tissue dispersion vary widely. For detection of unknown enzymes, gentle dispersion in water or in isotonic aqueous sucrose solution is often used in preliminary studies, and, as described in Chapter 10, sucrose is still the most widely used medium for subcellular fractionation. When some of the characteristics of the enzyme are known, the dispersion medium chosen is frequently an aqueous hypotonic buffer system, designed for maximum release and stability, and often also so as to be compatible with the buffer system found to give optimal enzymic activity.

Extraction of enzymes

Even when dispersion of the tissue has been extremely thorough, it may still retain in subcellular particles other structures or otherwise masked enzymes which can become soluble or more active. Many additional procedures are therefore frequently adopted with the intention of revealing maximal enzyme potency. In any particular instance they may or may not do so, or may be deleterious; a variety of such procedures is therefore usually examined. The most common current first test for 'latent' or occluded enzyme activity is to compare initial velocities in the presence and absence of non-ionic detergents, such as Triton X-100. This has been shown to be very effective in revealing the occluded hexokinase and monoamine oxidase activities of cerebral mitochondria (Tipton & Dawson, 1968; Thompson & Bachelard, 1970) and the lactate dehydrogenase activity of synaptosomes (Chapter 10). Other surfactants have been employed and lysolecithin has been noted to have such action on slices or suspensions of cerebral tissues (Marples, Thompson & Webster, 1959). Lyoslecithin is made by the action of lecithinase; it is powerfully haemolytic. At about 1 mM, lysolecithin caused the liberation from sliced cerebral cortex of some 85% of its true and pseudocholinesterases during 30 min at 38°, whereas in its absence only 3-6% was liberated. Other techniques include the use of lipid solvents, particularly to prepare acetone powders (below) and ultrasound is often used to reveal latent or membrane-bound enzymic activities (see for example, Kaplan, 1955; Giuditta and Strecker, 1959).

Multiple sequential freezing and thawing can also result in increased activity, as was found for cerebral microsomal Na^+, K^+-ATPase (Schwartz, Bachelard & McIlwain, 1962; see also Samson & Quinn, 1967). Repeated freezing and thawing may also help in extraction as exemplified by facilitated solubilization of mitochondrial hexokinase activity (Bachelard, 1967). Such techniques should be used carefully as they may prove to be a disadvantage: there is always the possibility of releasing proteolytic enzymes from occluded sites such as lysosomes, with consequent inactivation of the enzyme being studied. This is frequently avoided if the subcellular localisation of the enzyme is known beforehand; extraction is then performed after separation of the appropriate subcellular fraction (Chapter 10). Inclusion of p-tosyl fluoride in the dispersion medium may also be effective in inhibiting proteolytic activity (White & Wu, 1973) although it is important to ensure that the inhibitor does not affect the activity of the enzyme under investigation.

It is also worth noting that addition of a chemical extractant may alter the enzymic properties and efforts should be made to remove the extractant before placing too much emphasis on properties such as kinetic constants (see, for example, Thompson & Bachelard, 1970). Such attention to the results of removal of extractant can provide useful information also on the criteria for solubilization since the enzyme may then revert to an insoluble state.

Acetone powders

Treatment of neural tissues with cold acetone has been widely and successfully employed as an initial step in enzyme extraction or purification. Rapid dispersion of the tissue in several volumes of pure acetone is necessary, followed by washing and removal of the acetone without permitting the tissue to absorb water from the air. The tissue forms a flocculent suspension, which dries to a light powder. One gram of cerebral tissue yields about 0·18 g of acetone-dried powder; the main constituent extracted by the acetone is water, but some lipids and constituents of small molecular weight are also dissolved.

For enzyme assay or further purification, the acetone powder is suspended in, or extracted with, aqueous fluids; their properties as extracting agents of acetone powders may differ considerably from their properties in extracting fresh tissue. Usually more enzyme activity is extracted from acetone-dried tissue (as from frozen-dried tissue) than from fresh tissue, but this is not universal.

(i) The following is an example of preparation of an acetone powder on a small scale, from a few grams of cerebral tissue (McIlwain & Tresize, 1957). It uses a homogenizer tube which can be centrifuged and which is provided with a relatively loosely fitting glass pestle.

In a vacuum-bowl of crushed ice are put the homogenizer and pestle, a stoppered flask with about 100 ml of acetone, and a stoppered 25 ml measuring cylinder. When these are cold, freshly excised tissue (2–3 g) is weighed ($\pm 0 \cdot 1$ g) and put in the tube, followed by 7 vol of the acetone and the pestle. The tissue is immediately homogenized with a motor (about 15 s at 2000 rev/min) and spun at about $1000\,g$ for 5 min in a refrigerated centrifuge at 0°. The acetone is decanted off, replaced by the same volume of fresh cold acetone, the deposit resuspended by running the pestle for a few seconds and the centrifuging repeated. The resuspension and centrifuging are again repeated, and on this occasion the deposit is quickly spread around the sides of the tube by sharp knocking on the hand, or by a thick glass rod. The tube is now put in a vacuum desiccator containing fresh $CaCl_2$ and shavings of paraffin wax, and evacuated (to about 0·03 mm Hg; an air-ballast oil-pump is advantageous in not needing traps). The pump is left running for about 0·5 h; the powder can be used after a few hours but is commonly left drying overnight.

Variants are to use the acetone at temperatures down to $-15°$; to use up to 100 vol acetone in a single treatment instead of successive treatments with smaller volumes; to disperse in a mortar and to collect by filtering instead of centrifuging, and to dry over sulphuric acid or phosphorus pentoxide. In the case of peripheral nerve the tissue has been frozen in liquid nitrogen and ground to a powder under liquid nitrogen, before extracting with acetone; this is to cope with its tough sheath. In some cases, acetone-dried material on a Buchner funnel has been washed with diethyl ether, so replacing the acetone by a more volatile solvent.

(ii) Larger scale preparation of acetone powders is carried out with blenders or macerators. These must be cooled to near $0°$ and kept cool, by working in a cold room or by appropriate ice-baths or jackets. An example follows. Acetone, 2 to 3 l, is kept overnight in a cold room together with a balance and a blender whose jar of about 1 l is calibrated at 500 ml. Sheep brain (50 g $\pm$ 1 g) is weighed and put in the blender with 500 ml acetone, covered with a tightly-fitting lid and run for 1 min and then intermittently for the next 5 min, being switched on when the larger particles settle round the blades. The suspension is transferred to two centrifuge cups, and these spun at 1000 g for 3 min at $0°$. The supernatants were decanted off, the tubes being inverted for about half a minute to drain. To each cup 150 ml of acetone was added, and the deposit re-suspended by vigorous mechanical stirring, followed by a few minutes gentle stirring; they were then centrifuged and drained as before. Re-suspension and centrifuging were repeated and after draining, the residue was spread round the cups and dried in a desiccator as described under (i) except that the pump was left running for 1–2 h. After leaving the evacuated desiccator at $0°$ overnight the dry powder (about 9 g) was transferred to a bottle which was left in a desiccator over P_2O_5 overnight. It was then stoppered and kept at $-20°$.

Variants in the larger scale preparation of acetone powders are to macerate the brain in a mortar with acetone at -8 to $-15°$, or to freeze in liquid air and pulverize in a steel mortar, before blending; to use up to 500 vol of acetone; and to collect the acetone-dried material on a Buchner filter instead of by centrifuging (Roberts & Frankel, 1950, 1951; Muntz, 1953; Berry & Whittaker, 1959). A commercial source of rabbit-brain acetone-powder is now available and has been used by Haubrich (1973) for studies on choline kinase activity (below).

Extraction from acetone powders

The methods of preparing acetone powders which have been described yield essentially the same product, provided that they have been carried out rapidly, in the cold, and without allowing re-entry of moisture. Different methods of extracting the powder, however, can give markedly different enzyme activities. Most of the extracting fluids used with intact tissues can be employed, and can be used at varying temperatures for varying lengths of time; examples follow.

Acetone powders have been used as source of enzymes by suspending them in fluids, without further separation: in 4 parts by weight of water, for phosphatases (Strickland *et al.*, 1956); in phosphate buffer, for glutamate decarboxylase (Roberts & Frankel, 1950, 1951). Usually, however, extracts have been made by suspending the powder in an aqueous fluid and centrifuging immediately or after brief standing. Thus to obtain enzymes concerned with ammonia formation, an acetone powder was suspended in water (1 g/10 ml) in a glass homogenizer tube, and after 2 min centrifuged at high speed for 30 min (Muntz, 1953). After a resin treatment, such extracts contained 1-8 mg of protein-N/ml. For glutamine-synthesizing enzymes, an acetone powder was stirred for 10 min with 10 parts of water, and centrifuged (Elliott, 1951); stages of acid precipitation, heating, and adsorption on calcium phosphate followed.

Extraction of an acetone powder of sheep brain with 20 mM-$KHCO_3$, followed by centrifuging at 1200 g, was used for choline acetylase (Berry & Whittaker, 1959). A similar enzyme, with acetic thiokinase, was extracted from an acetone powder of peripheral nerve with 50 mM-KCl and 7 mM-phosphate buffer (Berry & Rossiter, 1958). The properties of of rabbit brain choline kinase were studied after extraction of the acetone powder into 25 vol water. The aqueous supernatant fluid after centrifugation was adjusted to pH 5·0 by the slow addition, with stirring of 0·2 mM acetic acid. The supernatant obtained after centrifugation was adjusted to pH 7·5 with N-NaOH and the enzyme was purified further by ammonium sulphate fractionation and by column chromatography on DEAE–cellulose (Haubrich, 1973).

Extraction with either water or 150 mM-phosphates, pH 7, yielded glutathione reductase in maximal activity from acetone powders of guinea pig cerebral hemispheres (McIlwain & Tresize, 1957). Extraction with 50 mM-2-amino-2-hydroxymethylpropane-1, 3-diol, pH 7·6, for 20–30 min before centrifuging was employed to obtain a thiolesterase (Strecker *et al.*, 1955). The arylsulphatases of human brain were separated from acetone powders according to the following procedure (Perumal & Robins, 1973). The soluble sulphatases were removed from the acetone powder by repeated incubation with 12 vol 50 mM acetate buffer (pH 6·2) at 37° for 1 h and centifugation. The residual acetone powder was suspended in 0·5 M tris buffer (pH 8·0) and lysolecithin was added to a final concentration of 0·5% (w/v). After incubation of the suspension at 37° for 30 min and centrifugation, 82% of the initially insoluble arylsulphatase activity was recovered in the supernatant as solubilized activity. Further purification was achieved using ammonium sulphate fractionation. The choice of conditions for extraction probably often depends on the reaction mixture in which a chosen enzyme activity is to be determined, rather than being the result of trying a large variety of conditions.

PURIFICATION OF ENZYMES FROM NEURAL TISSUES

Once recognition and preliminary study of an enzyme from the entire tissue or organ has been carried out, a process which includes the establishment of a specific assay system, it is often advantageous to select the most active or most practical part of the tissue for further study. This may involve taking specified regions from the brain or spinal cord, as exemplified by the selection of the caudate nucleus as a starting point for studies on tyrosine hydroxylase (Nagatsu *et al.*, 1971; Poillon, 1973). Separation of a particular subcellular fraction of the whole organ or of a specified region from it may also prove useful as a first stage of purification. Synaptosome fractions are used increasingly as sources of enzymes (Chapter 10) and myelin had been isolated and part-purified as the preliminary stage in studies on non-specific esterases (Rumsby *et al.*, 1973). The application of rigorous criteria for the identity and purity of a subcellular fraction cannot be overemphasized and it is important to remember that the well-defined enzymic or chemical 'markers' for the fractions from non-neural tissues may differ from those for neural systems. Specific techniques for preparation of such fractions and the criteria for their identification are now well established (Chapter 10; see also McIlwain & Bachelard, 1971). There is no general recipe for purification of enzymes, whether of neural or of non-neural origin. This depends on the properties of the particular enzyme being studied and might include a selection from general techniques such as heat treatment, salt or solvent fractionation, chromatography, isoelectric focussing and electrophoresis. The recent technique of affinity chromatography (for a review, see Guilford, 1973) has been applied to the purification of tyrosine hydroxylase activity from the caudate nucleus; 3-iodotyrosine linked to Sepharose 4B was used (Poillon, 1973).

It is obviously an important first stage in enzyme purification to attempt to obtain it in a soluble form by the use of the techniques described above. However, it has already been noted that one of the characteristic properties of neural tissues is the high membrane content (McIlwain & Bachelard, 1971) and some success has been achieved in partial purification of membrane-bound enzymes, even though they remain insoluble. This has been of particular interest in studies on the membrane-bound Na^+,K^+-ATPase. This enzyme has resisted attempts to produce a truly soluble form and it may be that the architectural integrity of the lipoprotein matrix of the membrane is essential for its full biological activity. Evidence in support of this view comes from the effects on removal of extractants used to 'solubilize' the activity and also from the restoration by phospholipids of inactivated solubilized preparations produced by the use of various phospholipases and selected detergents. The restoration of activity has been shown to occur with the enzyme from the brain, peripheral nerve, and many non-neural sources (Tanaka & Abood, 1964; Bachelard & Silva, 1966; Taniguchi & Tonomura, 1971).

Significant enrichment of the specific activity (one of the criteria for partial purification) of the Na^+,K^+—ATPase followed treatment of membranous fractions with the sodium iodide reagent of Nakao et al. (1965). Although the enzyme remained insoluble, the treatment removed 70% of the protein and 95% of the unwanted Na^+-independent ATPase activity, giving Na^+,K^+-ATPase activity enriched some 3-fold in specific activity. Chemical analysis showed that the insoluble material which contained the enriched enzymic activity had an increased phospholipid: protein ratio (Pull & McIlwain, 1970). This enzyme has also been purified to a remarkable extent by selective treatment with chemical reagents. Hokin and coworkers (Uesugi et al., 1971) treated the insoluble NaI-preparation (noted above) from the brain with the detergent, Lubrol; the resultant soluble preparation was purified by a combination of zonal centrifugation and various precipitation techniques to give a final purification factor of 30–50. It is of interest that the final stage of purification (controlled precipitation with ammonium sulphate) caused the enzyme to revert to an insoluble state. This enzyme was judged to be about 50% pure, thus providing reason to hope for an eventual complete purification, though the enzyme may remain insoluble in aqueous media since all the evidence points to its existence as a lipoprotein complex.

Separation of isoenzymes

A given neural system, in common with other organs of the body, typically contains multiple forms (isoenzymes) of many of its constituent enzymes. Isoenzymes are defined as proteins having the same catalytic function but differing in many of their characteristics, including physicochemical and kinetic properties. The difference in physicochemical properties are useful in separation and the differences in kinetic properties may be of fundamental importance in understanding changes in pathological conditions.

Techniques. The most commonly applied techniques for separating cerebral isoenzymes have been electrophoretic methods on supports such as paper, cellulose acetate, starch and polyacrylamide sheets or gels. Chromatography using columns of ion exchange cellulose or Sephadex is also frequently used. Thus the five isoenzyme forms of cerebral aldolase have been separated qualitatively by starch-gel electrophoresis, and quantitatively by DEAE-cellulose column chromatography (Nicholas & Bachelard, 1969). The aldolase in sucrose extracts of guinea pig cerebral cortex was partially purified by fractional precipitation with ammonium sulphate. The activity which precipitated between 50% and 65% saturation was applied to columns of DEAE-cellulose which had been equilibrated with a buffer system consisting of 10 mM-tris HCl, 0·5 mM fructose diphosphate and 1mM-EDTA, pH 7·35. The five individual isoenzymes of aldolase present in the original extract were separated by elution using a stepwise gradient of NaCl up to 1·6 M.

Identification of the isoenzymes in the original extract and in the individual peaks of activity eluted from the column was carried out by starch gel electrophoresis in citrate–phosphate buffer. The gel was prepared using 1·8 mM-citric acid, 13 mM-Na_2HPO_4, pH 7·03, and the samples were applied to the gel by means of moistened paper wicks inserted into precut slits in the gel. After electrophoresis for 18 h at 2–3 V/cm and 35 mA, at 0°, in 6 mM-citric acid, 48 mM-Na_2HPO_4, pH 7·03, the gels were stained for aldolase activity by immersion in a specific reagent at 37° for 1 h. The reagent, pH 8·25, contained 3 mM-fructose 1,6-diphosphate, 40 mM-$NaHAsO_4$, 40 mM-$Na_4P_2O_7$, 0·2 mM-NAD^+, glyceraldehyde 3-phosphate dehydrogenase (0·2 unit/ml), phenazine methosulphate (0·01 mg/ml), nitro-blue tetrazolium (0·2 mg/ml) and glycine-NaOH buffer (60 mM).

Electrophoresis, which is rapid and requires small amounts of tissue extract, provides a useful method of detecting which forms are present and of monitoring qualitative changes in patterns in clinical investigation. Care must be exercised in quantitative measurement on the gels, since the extent of staining may not be related to the amount of enzyme present. Distribution and intensity of staining can be affected particularly by the extent of spread, by diffusion, of the enzyme during electrophoresis.

Applications. An example of the value of investigating cerebral isoenzymes is given by the studies of Sandler and coworkers on mono-amine oxidases. These are mitochonrial in occurrence so must first be extracted by the techniques described above. Extracts of rat and human brain have been shown to contain up to four separable forms of mono-amine oxidase activity. Mitochondria were prepared from sucrose homogenates by the methods described in Chapter 10. The activity was extracted from the mitochondria by sonication in 1·5% Triton X-100, which contained 1 mM-benzylamine to prevent inactivation of the enzymes. The activity was partially purified by a combination of ammonium sulphate fractionation (material precipitated between 30% and 55% saturation was used) and chromatography on columns of Sephadex G-200. The four forms of the activity were separated using polyacrylamide disc electrophoresis and detected with a reagent consisting of 0·1 M-phosphate buffer pH 7·4, Na_2SO_4 (0·2 mg/ml), nitro blue tetrazolium (0·8 mg/ml) and tryptamine–HCl (0·8 mg/ml) as substrate.

The four forms were found to differ in affinity for kynuramine as substrate, and when tested against various ^{14}C-labelled amines (tyramine, tryptamine, dopamine, kynuramine and benzylamine), showed considerable variation in substrate specificity. Their sensitivity to inhibitors of relevance to clinical therapy of depressive illnesses also varied

(Gorkin, 1966; Youdim et al., 1969, 1972; Collins et al., 1970). In view of the variation in substrate specificity there has been some doubt as to the validity of calling these multiple forms 'isoenzymes' but there is some evidence of immunological cross-reactivity among them (Hartman et al., 1969). The importance of such an approach to the investigation of cerebral isoenzymes is indicated by the possibility of designing inhibitors specifically for each individual form, with a view to increased specificity and effectiveness in drug treatment of depression (Youdim et al., 1969).

Multiple forms of enzymes have also been suggested to be implicated in other disorders of brain function. Two different forms of β-galactosidase (a lysosomal enzyme) have been separated from rabbit brain (Jungalwala & Robins, 1968). An acetone-powder was extracted with phosphate buffer (0·1 M, pH 7·0) and the two forms, measured against a synthetic substrate, were separated on DEAE-cellulose columns after partial purification by ammonium sulphate precipitation. When the extract was placed on DEAE-cellulose in tris-phosphate buffer, one form only bound to the column and so could be separated from the other which was not bound. Although there is some doubt that these two forms act on gangliosides *in vivo*, the possibility of multiple forms of β-galactosidase involved in the catabolism of gangliosides is of relevance to disordered ganglioside metabolism. Two genetic disorders which result in brain dysfunction have been shown to be due to defective breakdown of ganglioside GM_1, and to be due to deficient β-galactosidase activity. The properties of the deficient enzymes in the two cases differ. In 'type 1' the deficiency is more pronounced than in 'type 2': the enzyme has a lower pH optimum and is more resistant to heat inactivation than that of 'type 2' which behaves more like the enzyme in normal samples. Such observations have led to the suggestion that the two disorders, both inherited, may involve different forms of β-galactosidase activity (Okada & O'Brien, 1968; O'Brien, 1969; Pinsky et al., 1970; Wolfe et al., 1970).

Defects in one of the three cerebroside sulphatases (originally named arylsulphatases due to their action on synthetic substrates) are thought to be implicated in types of *metachromatic leucodystrophy*. Variations in content or activity of two identifiable forms of N-acetylhexosaminidase are considered to underlie two types of *Tay-Sachs disease*. The enzyme patterns in extracts of abnormal brains from patients suffering these disorders have been compared with those of extracts from normal tissues after separation by isoelectric focussing (Jatzkewitz, 1972).

KINETIC STUDIES OF NEURAL ENZYMES

The importance of kinetic studies of neural enzymes has only recently been recognized as an essential criterion for detecting enzyme abnormalities which might underlie dysfunction and disease. The descriptions of enzyme kinetics in current text books of biochemistry (with few exceptions, e.g. Mahler & Cordes, 1971) tend to be insufficiently comprehensive, giving only some selections of the various primary and none of the secondary graphical analyses that have been described. The excellent review articles that are available are often directed towards the specialist and may be too detailed, especially in mathematical treatment, for the general reader.

There follows, therefore, a short description of the kinetic approach and the graphical analysis appropriate to an examination of potential enzymic abnormality in samples from neural tissues. Further detail can be gained from the reviews of Cleland (1967, 1970). Recommendations of nomenclature and symbols for kinetic constants made by the Enzyme Commission (1973) seem limited; the following is based on that of Cleland (1963) and refers to the reaction (1) in the forward direction of the rate equation below, which is the general equation satisfying most bimolecular reaction mechanisms.

$$v = \frac{VAB}{(K_{ia}K_b + K_aB + K_bA + AB)} \quad (1)$$

where v is initial rate of reaction; A, B are concentrations of reactants A and B respectively; K_a, K_b are Michaelis constants for A and B respectively; K_{ia} is a dissociation constant, for dissociation of A from the enzyme, and V is the maximum rate at saturating concentrations of all reactants.

The graphical analyses which follow are mathematically derived from the general rate equation (Cleland, 1963, 1970). The classical primary plots of $1/v$ vs. $1/A$ (Fig. 11.1) or of A/v vs. A (Fig. 11.2) give the maximum rate (V) and the Michaelis constant (K), which is the substrate concentration required for half-maximal rate of enzymic conversion of that substrate. This was originally devised for enzyme reactions involving only one substrate, which is rare. Most reactions involve more than one substrate and often other reactants, such as coenzymes or activators. Therefore the above plots should be constructed from the results of experiments performed with saturating concentrations of all the reactants involved, except that of the substrate being varied, which should be maintained over a concentration range well below the concentration giving saturation.

Errors inherent in the construction of the plots render necessary some selection of the best type of plot to use, especially if the experimental conditions are likely to involve inaccuracies in determining the initial velocity, particularly at low substrate concentrations. Generally, the $1/v$ vs. $1/A$ plot (Fig. 11.1) is considered to be the least reliable because any error in the lower values for v will be exaggerated when plotted as $1/v$; 'weighted' regression analysis has been designed to overcome this (Wilkinson, 1961). Plots (Fig. 11.2) of A/v vs. A are more reliable, especially if the error in v remains small, but less reliable

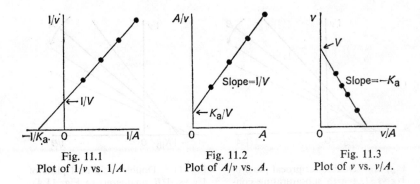

Fig. 11.1 Plot of $1/v$ vs. $1/A$.
Fig. 11.2 Plot of A/v vs. A.
Fig. 11.3 Plot of v vs. v/A.

than a third type (v vs v/A; Fig. 11.3) if the error in v is large (Dowd & Riggs, 1965). A plot of v vs. v/A is also useful in that any deviations from theory are likely to be exaggerated most. In contrast, the double-reciprocal plot of $1/v$ vs. $1/A$ tends to minimize the effect of unreliable experimental points.

Nevertheless the latter type is still the most commonly used, based on weighted regression lines, and forms the basis for further analysis by secondary plotting, or replotting, described below. Whether such primary plots give a measure of the actual Michaelis constant of the enzyme for the particular substrate under study may depend on the reaction mechanism and on the effects of interactions of the enzyme with other reactants (the second substrate in a two substrate reaction, a coenzyme or activator); one of these reactants could be inhibitory at high concentration. For more rigorous determination of kinetic constants, the construction of secondary plots is advised (Florini & Vestling, 1957; Cleland, 1963). In this approach the effect of other reactants is taken into account. In the treatment which follows, A is the substrate studied and B is another reactant (second substrate, coenzyme or activator; Frieden, 1957; Cleland, 1963). The initial velocities of the enzymic reaction are measured at various concentrations (all subsaturating) of A repeatedly at a range of fixed concentrations (also subsaturating) of B. Since this approach depends on initial velocity measurement, care must be taken to ensure that the measured rates are linear with time over the entire assay period.

Two double-reciprocal plots are drawn, using regression analysis, as depicted in Figs. 11.4 and 11.5. The plots will usually intersect at some common point which may be above, below or on the abscissa, depending on the kinetic constants, and therefore, the reaction mechanism. The dissociation constant for A, K_{ia}, is not necessarily the same as K_a (the Michaelis constant for A), depending on the mechanism. The same is true for K_{ib} and K_b, the analogous constants for B. It should be noted that K_{ib} may not exist as a meaningful constant, since there is no true

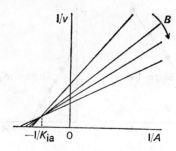

Fig. 11.4 Double reciprocal plots of $1/v$ vs. $1/A$ with subsaturating concentrations of substrate B. The arrow indicates the direction of increasing concentrations of B.

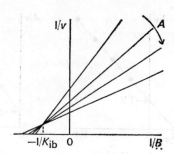

Fig. 11.5 Double reciprocal plots of $1/v$ vs. $1/B$, analogous to Fig. 11.4.

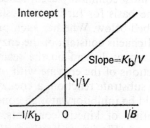

Fig. 11.6 Secondary plot of the ordinate intercepts of the lines of Fig. 11.4 vs. $1/B$.

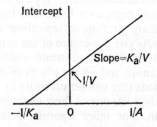

Fig. 11.7 Secondary plot of the ordinate intercepts of the lines of Fig. 11.5 vs. $1/A$.

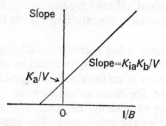

Fig. 11.8 Secondary plot of the slopes of the lines of Fig. 11.4 vs. $1/B$.

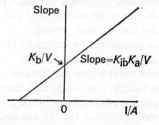

Fig. 11.9 Secondary plot of the slopes of the lines of Fig. 11.5 vs. $1/A$.

enzyme-B complex in certain types of reaction mechanism, such as ordered sequential (Frieden, 1957) or 'ping pong' (Cleland, 1963). The intercepts on the ordinates of the lines of the primary plots are measured and replotted as shown in Figs. 11.6 and 11.7. Figure 11.6 is the replot of the intercept of each line of Fig. 11.4 vs. the reciprocal of the concentration of B for that line (intercepts vs. $1/B$); Fig. 11.7 is the equivalent replot of intercepts vs. $1/A$ from Fig. 11.5. The Michaelis constants (K_b and K_a) are obtained from the intersection of the lines with the abscissa and V from the intersection of the lines with the ordinate. This should give the same value for V in both cases (Figs. 11.6 and 11.7) because it is the value when both A and B are at saturating concentrations (i.e. $1/A$ and $1/B$ are zero).

The slopes of the lines of the primary plots (Figs. 11.4 and 11.5) can provide further kinetic information. Figure 11.8 is the replot of the slopes from Fig. 11.4 vs. $1/B$; Fig. 11.9 is the replot of the slopes from Fig. 11.5 vs. $1/A$. The constants to be obtained are shown and K_{ia} comes from the ratio of the slopes of Figs. 11.6 and 11.8, and if appropriate, K_{ib} can be derived from Figs. 11.7 and 11.9. A similar approach, using replots of slopes and intercepts from primary plots produced in the presence of an inhibitor is regarded as essential in inhibition kinetic studies (Cleland, 1963) not described here.

Applications. A naturally-occurring enzyme abnormality may have an unchanged, or slightly modified maximum velocity (V) but a marked change in one of its Michaelis constants. In such a case, measurement of activity under 'optimal' conditions, which include saturating concentrations of reactants, may not detect the abnormality. An altered K_a value could cause a profound change in activity under conditions *in situ*, if the local intracellular concentration of the natural substrate is below that required for saturation, as is usually the case. This principle is illustrated diagrammatically in Fig. 11.10.

One illustration of this general principle is the altered kinetic properties of the defective enzyme in *citrullinaemia*, a rare inherited disease associated with mental retardation. It was the only enzymic abnormality to be detected in patients suffering the disorder. The enzyme, argininosuccinate synthetase, of fibroblasts cultured from a patient, was found to have a Michaelis constant for its natural substrate some 25-fold higher than normal (McMurray *et al.*, 1962; Tedesco & Mellman, 1967). Although direct evidence is lacking, this implies a much impaired ability to perform its normal catalytic function, if, as seems reasonable to assume, the concentration of substrate available to the enzyme is close to the normal value for the Michaelis constant.

Other examples of changed kinetic properties of enzymes have been reported recently. The Michaelis constant for glucose of hexokinase was found to be twice the normal value in brain biopsy samples taken during neurosurgical treatment of epilepsy (Bachelard *et al.*, 1975). A ten-fold increase in substrate Michaelis constants for hypoxanthine-

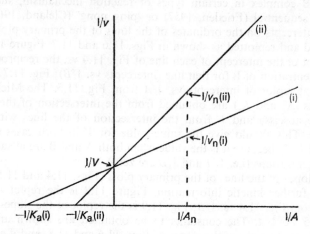

Fig. 11.10 Diagrammatic plots ($1/v$ vs. $1/A$) of a theoretical change in kinetic properties from normal (i) to an abnormal enzyme (ii) in which K_a rather than V is affected. The rate (v_n) at the intracellular substrate concentration (A_n) could be profoundly affected; the abnormality could be detected by accurate determination of the Michaelis constant.

guanine phosphoribosyl transferase has been detected in some of the patients suffering from the Lesch-Nyhan syndrome. This is a defect of purine metabolism which is associated with mental dysfunction and behavioural abnormality (Kelley & Arnold, 1973).

Potential enzymic dysfunction as a result of impaired ability of the enzyme protein to bind its coenzyme has been suggested to be one basis for defects in metabolism of sulphur amino acids, leading to the brain disorders of *homocystinuria* and *cystathioninuria* (Rodnight, 1968). This could be clarified by appropriate kinetic study but has proved difficult to achieve in view of the rarity of the disorders and limitations in obtaining brain samples.

Other examples of variations in the kinetic constants of enzymes, measured in tissue extracts, could be due to a change in the pattern of isoenzymes if one only of these were defective as described above. A kinetic analysis could contribute significantly to detection of such an abnormality, and perhaps to designing appropriate measures to overcome the resultant metabolic deficiency.

REFERENCES

Bachelard, H. S. (1967) *Biochem. J.* **104**, 286.
Bachelard, H. S. & Silva, G. D. (1966) *Archs Biochem. Biophys.* **117**, 98.
Bachelard, H. S., Thompson, M. F. & Polkey, C. E. (1975) *Epilepsia* (in press).
Berry, J. F. & Rossiter, R. J. (1958) *J. Neurochem.* **3**, 59.
Berry, J. F. & Whittaker, V. P. (1959) *Biochem. J.* **73**, 447.

Cleland, W. W. (1963) *Biochim. biophys. Acta* **67**, 104, 173, 188.
Cleland, W. W. (1967) *Annu. Rev. Biochem.* **36**, 77.
Cleland, W. W. (1970) In *The Enzymes*, 3rd edn, ed. Boyer, P. D., Vol. 2, p. 173. New York: Academic Press.
Collins, G. G. S., Sandler, M., Williams, E. D. & Youdim, M. B. H. (1970) *Nature, Lond.* **225**, 817.
Dixon, M. & Webb, E. C. (1964) *Enzymes*, 2nd edn. London: Longmans.
Dowd, J. E. & Riggs, D. S. (1965) *J. biol. Chem.* **240**, 863.
Elliott, W. H. (1951) *Biochem. J.* **49**, 106.
Enzyme Commission (1973) *Enzyme Nomenclature*, Amsterdam: Elsevier.
Frieden, C. (1957) *J. Am. Chem. Soc.* **79**, 1894.
Florini, J. R. & Vestling, C. S. (1957) *Biochim. biophys. Acta* **25**, 575.
Giuditta, A. & Strecker, H. J. (1959) *J. Neurochem.* **5**, 50.
Gorkin, V. Z. (1966) *Pharmacol. Rev.* **18**, 115.
Hartmann, B., Kloepter, H. & Yasunobi, K. (1969) *Fed. Proc. Fedn Am. Socs exp. Biol.* **28**, 857.
Haubrich, D. R. (1973) *J. Neurochem.* **21**, 315.
Jatzkewitz, H. (1972) *Biochem. Soc. Symp.* **35**, 141.
Jungalwala, F. B. & Robins, E. (1968) *J. biol. Chem.* **243**, 2458.
Kaplan, N. O. (1955) *Meth. Enzym.* **2**, 662.
Kelley, W. N. & Arnold, W. J. (1973) *Fed. Proc.* **32**, 1656.
Mahler, H. R. & Cordes, E. H. (1971) *Biological Chemistry*, 2nd edn. New York: Harper & Row.
Marples, E. A., Thompson, R. H. S. & Webster, G. R. (1959) *J. Neurochem.* **4**, 62.
McIlwain, H. & Bachelard, H. S. (1971) *Biochemistry and the Central Nervous System*, 4th edn. London: Churchill.
McIlwain, H. & Tresize, M. A. (1957) *Biochem. J.* **65**, 288.
McMurray, W. C., Mohyuddin, F., Rossiter, R. J., Rathburn, J. C., Valentine, G. H., Koegler, S. J. & Zarfas, D. E. (1962) *Lancet* i, 138.
Muntz, J. A. (1953) *J. Biol. Chem.* **201**, 221.
Nagatsu, T., Sudo, V. & Nagatsu, I. (1971) *J. Neurochem.* **18**, 2179.
Nakao, T., Tashima, Y., Nagano, K. & Nakas, M. (1965). *Biochem. biophys. Res. Commun.* **19**, 755.
Nicholas, P. C. & Bachelard, H. S. (1969) *Biochem. J.* **112**, 587.
O'Brien, J. S. (1969) *J. Pediat.* **75**, 167.
Okada, S. & O'Brien, J. S. (1968) *Science, N.Y.* **160**, 1002.
Perumal, A. S. & Robins, E. (1973) *J. Neurochem.* **21**, 459.
Pinsky, L., Powell, E. & Callahan, J. (1970) *Nature, Lond.* **228**, 1093.
Poillon, W. N. (1973) *J. Neurochem.* **21**, 729.
Pull, I. & McIlwain, H. (1970) *Biochem. J.* **119**, 367.
Roberts, E. & Frankel, S. (1950, 1951) *J. biol. Chem.* **187**, 55; **188**, 789.
Robins, E., Smith, D. E., Daesch, G. E. & Payne, K. E. (1958) *J. Neurochem.* **3**, 19.
Rodnight, R. (1968) In *Applied Neurochemistry*, ed. Davison, A. N. & Dobbin, J. p. 377. Oxford: Blackwell.
Rumsby, M. G., Getliffe, H. M. & Riekkinen, P. J. (1973) *J. Neurochem.* **21**, 959.
Samson, F. E. & Quinn, D. J. (1967) *J. Neurochem.* **14**, 421.
Schwartz, A., Bachelard, H. S. & McIlwain, H. (1962) *Biochem. J.* **84**, 225.
Seiler, N. (1969) *Handb. Neurochem.* **1**, 325.
Strecker, H. J., Mela, P. & Waelsch, H. (1955) *J. biol. Chem.* **212**, 223.
Strickland, K. P., Thompson, R. H. S. & Webster, G. R. (1956) *J. Neurol. Neurosurg. Psychiat.* **19**, 12.
Strominger, J. & Lowry, O. H. (1955) *J. biol. Chem.* **213**, 639.
Tanaka, R. & Abood, L. G. (1964) *Archs Biochem.* **108**, 47.
Taniguchi, K. & Tonomura, Y. (1971) *J. Biochem., Tokyo*, **69**, 543.

Tedesco, T. A. & Mellman, W. J. (1967) *Proc. natn. Acad. Sci. U.S.A.* **57,** 829.
Thompson, M. F. & Bachelard, H. S. (1970) *Biochem. J.* **118,** 25.
Tipton, K. F. & Dawson, A. P. (1968) *Biochem. J.* **108,** 95.
Uesugi, S., Dulak, N. C., Dixon, J. F., Hexum, T. D., Dahl, J. L., Perdue, J. F. & Hokin, L. E. (1971) *J. biol. Chem.* **246,** 531.
White, H. L. & Wu, J. C. (1973) *J. Neurochem.* **20,** 297.
Wilkinson, G. N. (1961) *Biochem. J.* **80,** 324.
Wolfe, L. S., Callahan, J., Fawcett, J. S., Andermann, F. & Scriver, C. R. (1970) *Neurology, Minneap.* **20,** 23.
Youdim, M. B. H., Collins, G. G. S. & Sandler, M. (1969) *Nature, Lond.* **223,** 626.
Youdim, M. B. H., Collins, G. G. S., Sandler, M., Jones, A. B. B., Pare, C. M. B. & Nicholson, W. J. (1972) *Nature, Lond.* **236,** 227.

12. Subsystems and Regions of the Mammalian Brain; The Retina

H. McIlwain and M. J. Voaden

The retina	261
Structure	262
The isolated retina	264
Examining the retina *in situ*	271
Specialized retinal systems	272
Isolated subsystems of the brain	275
Cerebral cortex: heterogeneity	276
Obtaining maximal yield of cerebral cortical slices	278
White matter	280
Lateral olfactory tract–piriform cortex	281
Optic tract–superior colliculus	282
Dentate gyrus and hippocampus	284
Dissection of the brain to regional blocks	284
Preparation of regional blocks	285
Preparations available from regional blocks	288
References	289

The very diverse types of neural structures in different animal species and in different parts of the animal body, present an attractive range of experimental opportunities. Physiological and zoological texts must be consulted for information on such systems; the few which are illustrated in this section, are selected through the distinctiveness of the type of chemical, metabolic or functional study in which they can be used rather than with the intention of giving an exhaustive account.

THE RETINA

The retina proper is composed of a pigmented epithelium and a neural matrix. It lines the inner posterior surface of the eyeball (Fig. 12.1) from which it may be isolated with relative ease, and is already formed at about the thickness (150–400 μm) of slices that are made artificially from other organs for metabolic studies. Both *in vivo* and *in vitro* it can be excited by its natural stimulus, light, and its functional state monitored by the electroretinogram.

The tissue contains photosensitive systems which initiate electrical impulses on illumination, and mechanisms which group and modify the impulses so generated. Apart from its photochemical portions, the neural retina resembles parts of the cerebral cortex in several respects, and in many studies of its general metabolism the photosensitive systems have been ignored. The photochemical elements are however in several species easily separated from the rest of the tissue by mechanical means, as is described at the end of this section, and they have been the subject of study in isolation.

Structure

Gross and fine retinal anatomy has been studied extensively. Excellent reviews are available, among them those by Walls (1942), Polyak (1957), Straatsma et al. (1969), Hogan et al. (1971), Cohen (1972), and Stell (1972). Classically 10 sublayers, as detailed in Fig. 12.1 are recognized. The broad classes of cells are the same in all or nearly all vertebrates and Fig. 12.1*B* summarizes also the distribution of the major cell types encountered. Müller cells form the main glial component and the external limiting membrane comprises the junctions of these with each other and with the visual cells. In cytoplasmic detail Müller fibres resemble astrocytes although they also have properties of ependymal and, in some species, oligodendroglial cells. Very few conventional astroglia, oligodendroglia or microglia have been found.

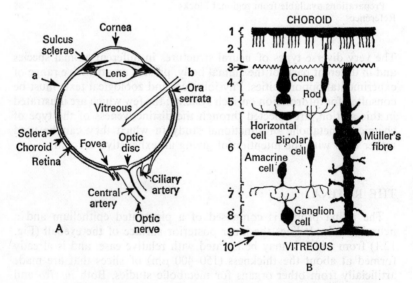

Fig. 12.1 *A:* Transverse section of an eye: *a*, position of supportive ring for an *in situ* eye-cup preparation; *b*, penetration point for intravitreal injection. *B:* Schematic diagram of the major constituents of the vertebrate retina. The layers are: *1*, the pigment epithelium; *2*, photoreceptor outer and inner limbs; *3*, outer limiting membrane; *4*, outer nuclear layer; *5*, outer plexiform (synaptic) layer; *6*, inner nuclear layer; *7*, Inner plexiform (synaptic) layer; *8*, ganglion cell layer; *9*, nerve fibre layer; *10*, inner limiting membrane.

The photoreceptor cells, lying closest to and enveloped by processes of the pigment epithelium, have classically been recognized as belonging to two functionally and morphologically distinct classes: rod cells which are active in scotopic or twilight conditions and cones which are concerned with daylight or photopic vision. The distinction is clear in mammals, teleosts and nocturnal birds but not in lower fishes, amphibians, reptiles and diurnal birds (Cohen, 1972; Stell, 1972).

Two main types of synaptic contact are seen in retinal tissue (Dowling, 1970; Stell, 1972). Conventional synapses are made by horizontal cell processes in the outer plexiform layer and by amacrine cell processes in the inner plexiform layer. In addition to these there are specialized contacts, made by receptors with both horizontal and bipolar cells as the postsynaptic elements and by bipolar cells with ganglion and/or amacrine cells postsynaptically. These are typified by a bar or 'ribbon' formation in the presynaptic element. Single cone terminals have been observed to make numerous contacts, the pedicles having a diameter of about 5–7 μm. In contrast mammalian rod spherules frequently make only single invaginated contacts and measure approximately 1–2 μm across. Bipolar terminals are also about this size, and diameters range from 0·5–1·5 μm, depending on the species.

Dramatic species differences exist in the number of the individual cell types that are present in a retina and this must be borne in mind when comparisons are attempted. The predominant photoreceptor populations in some of the most commonly studied species are shown in Table 12.1. Differences are known to exist between the mechanisms initiating rod and cone function (Tomita, 1972a) and also those main-

Table 12.1 Characteristics of some commonly studied eyes

Species	Approximate wet weight of retinae (mg)	Water (%)	Photoreceptor population* Centre	Periphery	Scleral texture†
Rat	12	85	R_M	R	+
Guinea pig	20	88	R_M	R	+
Rabbit, albino	80	84	M	R	+ +
Ox	700	92	M	M_R	+ + + +
Pig	250	88	M	M_R	+ +
Horse	700	90	M	M	+ + + +
Cat	190	91	M	R_M	+ + +
Dog	—	84	M	R_M	+ + +
Monkey	220	—	C	M	+
Pigeon	80	83	C‡	M	+ + +
Chicken	100	—	M	M_R	+ + +
Frog (R. temporaria)	10	89	M	M	+ + + +

*R, mainly rods; M, mixed photoreceptors; C, mainly cones. † Texture ranges from + + + +, firm, to +, flaccid. ‡Red Spot.

taining their homeostasis (Cohen, 1972). As regards the inner retinal layers, it is recognized that the behaviour of ganglion cells shows varying degrees of complexity and that this may be correlated with the number of amacrine cells present (Dowling, 1970; Stell, 1972). Thus there are more amacrine cells in retinae where the ganglion cells show complex reactions such as directionally selective responses to movement (e.g. frog, pigeon) and less where they respond predominantly to local contrast (e.g. primate, cat, rat).

The putative neurotransmitters dopamine, γ-aminobutyric acid, glycine and acetylcholine have been implicated in various aspects of amacrine activity. Species differences exist and have been reviewed by Graham (1974).

In addition to species variation in cell numbers and retinal organisation, area specialization may exist also within a single retina. Species comparisons of this aspect are, as yet, limited but metabolic differences have been detected between defined areas of the pig retina by Weiss & Kosmath (1971). Possibly correlating with this is the general observation that cones, when present, are more numerous in the central portions of the tissue.

Primate retinae have a specialized area, the fovea, in which the inner retinal layers are absent and only a population of photoreceptors, specialized cones, exists. The axonal portion of these is extended to synapse with bipolar cells in the surround and may exceed 125 μm in length. Foveal areas with varying characteristics are also found in lower phyla and are discussed by Walls (1942).

Apart from differences in general anatomy, variation occurs also in the extent to which retinae are vascularized (François & Neetens, 1974). In all chordates the retina is supported, at least in part by the choriocapillaris at the base of the pigment epithelium. In addition, in most mammals the inner retinal layers are supplied to greater or less depth by branches of the central retinal artery. An exception is the rabbit where the vessels lie on the surface of the tissue over two wing-shaped areas of medullated nerve fibres. The retinae of amphibians, birds and reptiles are avascular. The former have a vascular network on the surface of the tissue whereas the latter are maintained in part by special intravitreal structures, the pecten and conus capillaris respectively.

The isolated retina

Practical aspects of handling retinae and maintaining them *in vitro* have been surveyed by Ames (1965), Graymore (1970a, b) and Sickel (1972).

The ease with which an eye can be enucleated depends on the species. In rats the eyeball is proptosed by separating the eyelids with thumb and forefinger; a single cut at the base with fine, sharp scissors will then free it. In other species, for example, cats and pigeons, it is necessary to free the globe of attached muscular and other tissue by dissecting into the socket with blunt, curved excision scissors. Care must be taken not to cut branches of the ciliary blood vessels present in the orbit. The optic nerve is cut at the final stage and the eyeball thus freed.

The strongest areas of attachment of the retina to the eyeball occur at the ora serrata (Fig. 12.1A), and at the point at which the optic nerve leaves the eye. Methods of removing it depend, therefore, on sectioning the globe just posterior to the ora serrata. The position of the latter must be ascertained for each species as the distance between it and the sulcus sclerae (the meeting point of the cornea and sclera) varies. In rabbit and rat the two are very close. With larger eyes sectioning can be accomplished by holding the eyeball, with forceps, at fragments of muscle, starting an initial equatorial cut with a sharp scalpel or razor blade and then continuing with scissors. When an eye is firm it is often possible to complete the separation with one combined sideways and downwards movement of a razor blade. It is suitable for this procedure if the eye is held in a large Petri dish that has been lined with a wad of filter paper, soaked in the experimental buffer.

When smaller eyes are being studied finer precision is needed and various blocks have been designed to aid the process (e.g. Ames, 1965). One of the simplest is that described by Graymore (1958) (see Fig. 12.2A). Hemispherical pits are drilled in the surface of a block of Perspex and fine drainage canals run from each well to the base of the block. The aim is for the eye to position in the well with the ora serrata exposed just above the surface of the Perspex. It is held steady by placing the corneal surface into a suitably proportioned hole drilled in one end of a plastic strip (Fig. 12.2B and C). A sharp flat razor blade applied with a combined forward and sideways movement along the surface of the block will then section the eye immediately behind the ora serrata. Scissors are used to free any residual attachment.

Larger eye-cups may be cut into portions and the sections of retina subsequently freed, whilst smaller retinae are often used intact. It is necessary first to remove vitreous from the surface of the preparation and the ease of this will depend on the species. Where a large amount of gel vitreous remains in the eye-cup it can frequently be removed intact by gripping into it with forceps, care being taken not to contact the neural tissue. Alternatively the method detailed diagramatically in Fig. 12.2D can be used. In man and sheep the procedure is complicated by a greater overall attachment between the vitreous body and the retina. In frog, also, a residual layer of vitreous remains attached to the retina: it can be reduced by gently mopping with a tip of filter paper.

Albumen (mol wt 68,000) moves through the vitreous body of the rabbit at a rate approaching free diffusion (Maurice, 1959). Other species may differ as vitreous hyaluronic acid levels are very low in rabbit (see Berman and Voaden, 1970) but it does suggest that the presence of this tissue on the retinal surface will not substantially alter experimental results. Complications arise in the weighing of the retina,

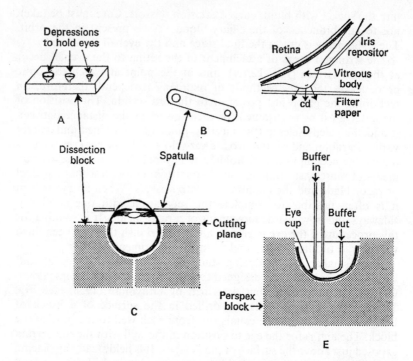

Fig. 12.2 Handling the retina for metabolic studies. *A:* Plastic block to aid the sectioning of a small eye. *B:* Plastic spatula to hold the eye steady. *C:* Position of the eye in the block. *D:* Removal of vitreous body from an eye-cup. The eye-cup is inverted and the vitreous touched lightly onto filter paper. Contact between retinal tissue and the filter paper must be avoided; if the filter paper is slightly moistened capillary drag to points c, d will be lessened. *E:* Arrangement for superfusion of the inner retinal surface of an isolated eye-cup.

however, and where vitreous contamination is an unknown factor more accurate results will be obtained if the experiments are based on dry rather than wet weights. As the vitreous body is 99% water it will barely contribute to the tissue weight in the dry state.

Various instruments can be used to free a retina from the eyeball. A small spatula or an iris repositor are often of help, particularly when portions of tissue are being freed from sections of a larger eye-cup. In addition camel-hair brushes have been employed successfully. A suitable procedure with an intact eye-cup is to gently invert it into a dish of chilled medium; a round ended glass rod is often used to aid this process. The scleral edge may then be gripped with one pair of fine forceps whilst another pair is run along this same edge underneath the retina.

Gentle movement in the medium should then parachute the retina exposing the main attachment at the optic nerve, which can be cut.

A factor to be considered when working with isolated preparations of retina is the extent to which a preparation is 'contaminated' with pigment epithelium. This is of importance not only because of the need to maintain a uniform experimental preparation but also because a major permeability 'barrier' into the tissue exists in the pigment epithelial layer (see Berman & Voaden, 1970). This may not be of great concern where glucose is the exogenous substrate as it penetrates readily through the blood-retinal barrier, but could cause erroneous results where substances normally excluded, e.g. neuroactive amino acids, are being studied. In the teleost retina lowered metabolic rates *in vitro*, in the presence of the pigment epithelium, have been attributed to a lack of free gaseous exchange (Santamaria *et al.*, 1971).

Processes of the pigment epithelium interdigitate with the receptors in all retinae and in most systems in which the two tissues are separated they will be broken off and isolated with the neural retina. Also, some of the receptor tips will remain with the pigment epithelium. In pigmented eyes the amount of epithelium remaining attached to the retina will be readily seen and often the main portion can be removed by gently shaking the tissue in the dissection medium or brushing lightly with a camel-hair brush. Care must be exercised, however, since it is very easy to break off the outer limbs of the photoreceptor cells as well (see below).

In some diurnal frogs, reptiles, birds and teleost fish pigment flows into the epithelial processes in response to light and out of them in the dark. Whether movement of the processes also occurs is uncertain but the two tissues do separate much more readily in a dark-adapted eye and 'clean' preparations of both may potentially be isolated. As with the neural retina, the pigment epithelium is also being investigated *in vitro* and experiments in which it has been isolated with attached choroid and studied as a membrane are reported by Lasansky & De Fisch (1966). A detailed description of this region of the retina has been given by Moyer (1969).

Retinal weight. Freed retinal tissue can be weighed on a rider such as that shown in Chapter 6, excess fluid being removed by touching the retina lightly on to glass. This will often lead to tissue damage and/or loss, however, and an alternative method, described in detail by Graymore (1958), has proved convenient for general use. A small filter disc (Whatman No. 50 filter paper, cellulose acetate electrophoresis paper or collodion) is soaked in a buffer, drained briefly on a pad of filter papers to remove excess moisture and weighed on a torsion balance. The retina is then floated onto the disc and the whole redrained. Residual vitreous on the preparation can be seen readily at this stage and may be removed gently with a torn edge of slightly moistened filter paper. The disc plus retina is then reweighed and the retinal wet weights determined by difference. Alternatively discs may be dried to constant weight and, with consideration to changes in weight of control papers treated in the same way, the retinal dry weights obtained. As this method of weighing is applied at the end of an experiment, changes in, for example, the water content of the tissue during an incubation period must be determined separately and corrections made to the experimental values. In addition the possibility of a reduction in tissue weight because of the metabolism of endogenous constituents or the mechanical loss of photoreceptor outer limbs must also be considered. The average wet weights and water contents of some of the most commonly studied retinae (neural portions) are shown in Table 12.1.

Retinal metabolism. The metabolic characteristics of isolated retinae have been surveyed by Lolley (1969), Lolley & Schmidt (1974), Graymore (1969, 1970b) and Voaden (1974). Extremely high activity is a general finding and in mammalian retinae respiration is approximately twice as great as in brain and aerobic glycolysis six to seven times as high. The 'excess' activity is associated with photoreceptor function— considerably lower levels being found in retinae without photoreceptors (Graymore, 1969, 1970b; Santamaria *et al.*, 1971) and also in the immature tissue. The effects of buffer composition on metabolic activity have been discussed by Graymore (1969, 1970b); both respiration and glycolysis are increased by bicarbonate.

Most investigations of retinal metabolism have been performed on the bleached tissue. It is important to bear in mind that in this state the visual cycle will be non-functional and that pathways normally associated with its working, e.g. the hexose monophosphate shunt (Futterman *et al.*, 1970) will not be maximally stimulated. Indeed, the electron acceptor phenazine methosulphate will almost double the observed activity of the shunt in an isolated retina. Studies designed to show the effects of light on retinal metabolism are yielding apparently conflicting results (Lolley, 1969; Sickel, 1972 but cf. Voaden, 1974) perhaps because of differing stimulus conditions (see below).

When experiments are performed on dark-adapted retinae the problem is for the observer to 'see' without affecting the retinal preparation. In these circumstances it has become convention to use low-intensity red light as, theoretically, this will least stimulate rod photoreceptors. For predominantly cone retinae the aim must be to work at the lowest intensities commensurate with the experimental manipulations. A preferable alternative, however, is infrared illumination plus an infrared image converter.

The retina offers the unique possibility of being easily controlled by its natural stimulus, light, and of producing a conveniently measured electrical response. This response, the electroretinogram has been studied and analysed extensively and the conclusions summarized and discussed by Granit (1963), Brown (1968) and Tomita (1972b). Practical details for measuring the electroretinogram have been given by Granit.

In addition, studies with microelectrodes and 'spots' of light of defined diameter, both moving and stationary, are revealing an interplay of both excitatory and inhibitory nervous activity, particularly in the inner layers of the retina (Dowling, 1970; Stell, 1972).

In all experiments in which the retina is stimulated, control of intensity and timing of the exposure to light are of the utmost importance as the tissue is capable of responding to a wide range of stimulus parameters; covering 10 log units of light intensity and also discriminating wavelength. Techniques for maintaining the retina in a functional state *in vitro* have been developed and described by Ames (1965), Arden & Ernst (1970) and Sickel (1972) and centre around double-sided perfusion chambers in which the retina is mounted as a central membrane. Standard Krebs-Ringer bicarbonate and phosphate media preserve the morphology of the mammalian retina but from a functional point of view modifications have been made. These systems have been used to monitor ionic fluxes in the retina (see also Ames *et al.*, 1967) and correlations observed between metabolic and electrical activity.

Superfusion systems can be used to study both the uptake and efflux of exogenously applied labelled compounds and the release of endogenous constituents of the retina. The calculation and analysis of rate constants for efflux have been discussed by Brading (1971) and specifically for brain and retina by Cohen (1973) and Kennedy & Voaden (1974) respectively. It is a general finding in the brain and retina that two main components are present plus an initial rapid efflux. The observed efflux pattern will be the sum of the individual exponential components. It is important, therefore, that efflux is continued for sufficient time to allow the slowest of these to be resolved i.e. when the semi-logarithmic plot of the desaturation curve reaches a linear phase (approximately 2–4 h). The linear component can then be extrapolated back to zero time and, by subtraction from the original curve the first main component obtained. The process is then repeated to obtain the initial component. The size of this will depend on the extent to which the eye-cup or retina is washed before the efflux is monitored as it is formed by the loss of adhering loading medium. The first main component is probably intracellular but may have a contribution from extracellular sources; the second one is intracellular. The rates for the first and second main components depend on the compound being studied.

Example: efflux of γ-aminobutyrate. This is an illustrative analysis giving the main rate-constants for efflux of [^{3}H] γ-aminobutyric acid from the frog retina in the absence of calcium (see Kennedy & Voaden, 1974). The isolated frog retina was preloaded with [^{3}H] γ-aminobutyrate for 1 h in a Ca^{2+}-free amphibian Ringer solution, containing 0·1 mM-amino-oxyacetic acid to inhibit the metabolism of the added amino acid. At the end of the incubation period, the retina was washed for 2–3 min in an excess of fresh Ringer, to remove adhering loading medium and then superfused with Ca^{2+}-free medium containing 0·1 mM-amino-oxyacetic acid and 2 mM-EDTA, in a closed perfusion chamber of volume 1·5 ml. The flow rate was 1 ml/min. Ten-minute perfusate samples were collected and the levels of radioactivity measured on a scintillation spectrometer, with results shown in Fig. 12.3. Extrapolation

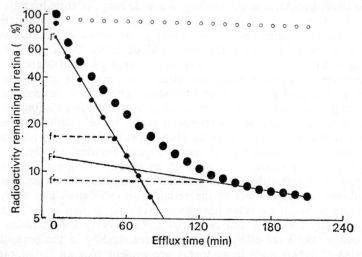

Fig. 12.3 Efflux [^{3}H]γ-aminobutyric acid from the frog retina *in vitro*. The large closed circles show loss of [^{3}H]γ-aminobutyric acid from frog retina, loaded and superfused in the absence of calcium ions. The small closed circles show the first main component of efflux, derived as described in the text. For comparison, the efflux of γ-aminobutyric acid from frog retinae loaded and superfused in the presence of calcium (open circles) has been included.

of the terminal linear component of the curve to zero time gives the percentage of the total initial radioactivity released in the second component (F'). In the present experiment F' is 12.2%; the percentage residual activity of the component (f') after a known time interval is then ascertained. In the present example 130 min after the start of superfusion, f' is 8·86%. Therefore, substituting into the rate equation: $f' = F' e^{-k't}$ where k' is the rate constant for the second main component, gives $e^{-130k'} = 0·726$. Reference to a standard e^{-x} table shows that $0·726 = e^{-0·32}$. Therefore $130k' = 0·32$ and $k' = 0·00247$. The half-time efflux of this component is $0·693/k$, or 280 min.

The first main component for efflux is obtained by a similar analysis. Thus $F = 70·8\%$ and, after a time interval of 50 min, $f = 16·6\%$. Therefore $16·6/70·8, = 0·2343 = e^{-50k}$ and $k = 1·4537/50$ or $0·0291$; the half-time for efflux is 24 min. The remainder of the radioactivity that was present in the retina at the start of perfusion (seen as a displacement of the zero point of the 1st main component) is most probably due to adherent loading medium not removed during the washing phase.

Examining the retina *in situ*

In addition to studies with the isolated tissue, useful information can be obtained if the retina is maintained within the eye and the posterior portion of the organ investigated as an eye-cup preparation. This may be done with the eye either isolated or left *in situ*. In this latter situation the retinal vasculature can be maintained in a functional state.

The ease of such eye-cup preparations depends not only on the size of the eye-ball but also on the degree of scleral rigidity. This ranges from the relatively firm composition seen, e.g. in the frog and pigeon to the flaccid skin-like structure found in the rat and monkey. Comparative scleral textures are listed in Table 12.1. When a firm composition is present the eye-cup can be positioned in a hemispherical depression in a block of Perspex, as shown in Fig. 12.2E and fine-gauge steel tubing used for superfusion. For work with mammalian retinae the buffer must be prewarmed before it enters the eye.

A preparation in which the eye-cup has been studied *in situ* and the electroretinogram monitored is described for cat by Kramer (1971). More flaccid eyes may be investigated *in situ* using the following procedure which has been developed by Dr M. S. Starr and is described in detail for the rabbit. When the animal has been anaesthetized to a depth at which corneal reflexes are abolished, the eye-ball is proptosed with loose fitting Spencer Wells forceps; the blood supply must not be hindered. A rigid stainless-steel wire loop, formed to a size that will fit snugly over the ora serrata of the eye is placed on and sutured to the sclera. When this is in position the cornea is removed–the first penetration into the eye being made with a fine scalpel or razor blade and the operation completed with scissors. All cuts are made as close to the ora serrata as possible. The lens of the eye may detach with the anterior structures or be left *in situ* when it can be removed with the vitreous body. To do this, it is necessary to free the attachment of the vitreous body to the retina, that occurs at the ora serrata, by cutting into the gel in this region with fine, curved scissors. The vitreous body can then be removed *en masse* with forceps and any residual gel mopped up carefully with filter paper. The retina adheres strongly to filter paper and must not be touched with it. The proptosing forceps are then removed, the eye-cup returned into the orbital socket and the inner retinal surface superfused with a suitable medium. In situations in which the blood supply to the ocular structures is intact the medium in the eye-cup does not have to be continuously exchanged. However, with an isolated eye-cup preparation, buffer flow-rates of about 1 ml/min are utilized and where the isolated retina is studied as a membrane in enclosed chambers, flow rates of 2–5 ml/min have proved necessary to preserve normal electrical activity (Sickel, 1972).

The retina in vivo. A number of studies on retina, e.g. involving autoradiography or permeability analysis, have been based on direct injection of compounds into the vitreous body *in vivo*; see Fig. 12.1. Many of these have been very successful but a word of caution is needed as Maurice (1957) has shown that the injection of volumes greater than 0·05–0·1 ml into the rabbit eye can cause irreversible damage to the retina—due, most probably, to the ischaemia produced by the increased intraocular pressure. In addition the rate of uptake of substances from the vitreous may be diffusion limited as heavier incorporation of compounds into the isolated retina is a common observation. Once a suitable depth for intravitreal injection has been established, small bore tubing can be placed on the needle to act as a 'stop' or guide in subsequent experiments.

Specialized retinal systems

The retina, having the ordered, layered structure shown in Fig. 12.1, lends itself to autoradiographic, histochemical and microchemical studies. In the latter case the elegant techniques of O. H. Lowry have been applied to tangential sections obtained from frozen retinae. It is beyond the scope of the present survey to detail these studies but descriptions and modifications of the techniques and instruments involved have been given by Lowry *et al.* (1961), Burt (1966) and Lowry & Passonneau (1972). The results from histochemical and microchemical analysis of various retinal constituents and enzymes have been reviewed by Lowry (1964), Lolley (1969) and Graymore (1969, 1970b). A technique for processing the retina for autoradiography has been described by Ehinger & Falck (1971) and others by Marshall & Voaden (1974 a, b).

Useful information on retinal functioning can be obtained with the judicious use of chemical fractionation. Various means exist by which specific layers of the retina can be made to degenerate (Graymore 1969, 1970b; Lolley 1969). For example, photoreceptor cells are affected adversely by sodium iodoacetate, light of specific wavelengths (see also Harwerth & Sperling, 1971), hyperoxia (in rabbits) and, indirectly, by iodate which directly affects the pigment epithelium. Inner retinal layers can be destroyed by the parenteral administration of sodium glutamate to immature animals and, in vascularised retinae, by anoxia. The hereditary condition of retinitis pigmentosa is also of interest in this context as it occurs in a variety of animals and is characterized by a specific disappearance of the retinal photoreceptors, the inner retinal layers apparently remaining normal.

Apart from studies with the aim of investigating the outer limbs of photoreceptor cells (see below), the classic techniques of subcellular fractionation have not been extensively applied to retinal tissue. Nevertheless the tissue does yield detached nerve endings on homogenisation and, in addition, four specific regions of photoreceptor cells (the synaptic pedicles, nuclei and the inner and outer limbs) can be isolated (Neal & Atterwill, 1974; Marshall, Medford & Voaden, 1974). Individual neuronal cells have been isolated near intact from retinal tissue by treating it with hyaluronidase followed by papain (Drujan and Svaetichin, 1972).

Visual pigments and photoreceptor outer segments. The photosensitive pigments of vertebrate eyes are present within the outer segments of the photoreceptor cells (Fig. 12.1*B*). Rhodopsin, the visual pigment in rods, has been extensively purified after solubilizing with the aid of detergents (Hubbard *et al.*, 1971), whereas cone pigments have proved more difficult to extract and have been characterized mainly by single cone microspectrophotometry and, in man, by reflection densitometry (Liebman, 1972).

The dimensions of rod outer segments show species variation but in rat they are about 25 μm long and 1–3 μm in diameter. They may be separated fairly easily from the rest of the tissue by shaking in buffer and when isolated by this means are intact and behave as sensitive and reversible osmometers (Korenbrot, 1973). This method has formed the basis for bulk preparations of outer limbs (Collins, Love & Morton, 1952) but is time consuming. More recently, subcellular fractionation has been applied to homogenized retinal tissue and preparations of fragmented outer limbs obtained and used for biochemical studies (Futterman *et al.*, 1970; Daeman *et al.*, 1970). In the method of Futterman and coworkers, 80 dark-adapted fresh bovine retinas are homogenized in 40 ml of 0·154 M-KCl. The mixture is then diluted to 240 ml with more of the KCl solution, the nuclei sedimented by centrifugation at 600–800 g for 5 min and a crude photoreceptor outer-segment pellet obtained by centrifuging the supernatant at 1600 g for 10 min. This can be used immediately or further purified on a nine-step discontinuous sucrose gradient (24–32% w/v sucrose) by centrifuging at 2300 g for 70 min at 4°C. The bulk of the photoreceptor fragments are recovered in a discrete red-purple band. A main contaminant of outer limb preparations is mitochondria from the inner segment of the cell. A photoreceptor membrane fraction free from these can be obtained using the sucrose flotation procedure detailed by Futterman & Rollins (1973).

Rhodopsin consists of a lipoprotein, opsin, combined with a chromophore, 11-*cis*-retinal (vitamin A aldehyde) which on illumination is isomerized and converted, via a series of intermediates, to all-*trans*-retinol (vitamin A). The methodology of work on visual pigments and their chromophores has been surveyed and detailed by Hubbard *et al.*, (1971). Both retinal and retinol can be estimated separately or in a mixture of both by reaction with antimony trichloride ($SbCl_3$)— reading the absorption spectrums of the resultant blue-coloured products on a recording spectrophotometer in the scanning mode. $SbCl_3$ is poisonous, however, and extreme care is needed during its use An alternative method for measuring concentrations of retinal is that devised by Futterman and Saslaw (1961) (see also Daemon et al., 1970). It is based on the reaction of the aldehyde with thiobarbituric acid to yield a red-coloured product which shows maximal absorbance at 530 nm. Retinal isomers, 11-*cis* and all-*trans*, show equal chromogenicity in this method which can be used as a basis for measuring rhodopsin content and retinol dehydrogenase activity of the retina.

Illustrative experiment: measuring retinol dehydrogenase. This shows the coenzyme dependence of retinol dehydrogenase in the retina of the adult albino rat. The reagents needed are: (1) 0·25 M-sucrose; (2) 0·1M-glycine buffer, pH 9·6 (Pearse, 1968, p. 584); (3) 2% w/v NAD in glycine buffer; (4) 2·3% w/v NADP in glycine buffer; (5) 0·75% w/v retinol in 10% w/v Tween 80 (a polyoxyethylene derivative of sorbitan monooleate) in acetone; (6) 4% w/v thiourea in glacial acetic acid (filtered through glass wool); (7) 0·6% w/v thiobarbituric acid in absolute ethanol (filtered and stored at 4°C); (8), 0·01% w/v all-*trans*-retinal in absolute ethanol (shielded from light and stored at 4°C); (9), 2 ml reagent (8) and 2·5 ml water diluted to 25 ml with 90% w/v ethanol; this solution is prepared immediately before use.

A standard curve is established as follows. Aliquots of reagent (9) are diluted to provide a series of standard solutions containing 2 to 20 μg of retinal in 3 ml 90% ethanol. Reagent (6), 1 ml, and 1 ml of reagent (7) are added. The solutions are mixed and the tubes then stood in the dark for 30 min at room temperature. The developed colour is light sensitive and the tubes must be kept shielded at all times. The absorbencies are measured on a spectrophotometer at 530 nm and a standard curve constructed.

Retinal extracts are then prepared, using eight retinae for each experiment. They are obtained as described above and are touched lightly onto glass to remove excess moisture. The wet weights can be determined directly on a torsion balance or else the homogenizing tube into which the tissue is pooled may be preweighed and the total weight of the retinae obtained by difference. Sucrose, 1 ml of 0·25 M, is then added to each tube and the contents briefly homogenized. The homogenate is transferred to either a vial or test-tube, to which is added 0·1ml reagent (3) or reagent (4), 0·1ml of reagent (5) and 1·8 ml of reagent (2). The solutions are mixed thoroughly and the vial or tube stoppered and incubated, with shaking, at 37°C. Portions of 0·5 ml are withdrawn after 60 min and are added to 2·5 ml of n-propanol. The solutions are mixed, stood in the dark, for 30 min, centrifuged and the supernatants transferred to glass test-tubes (stoppered). The thiourea reagent (6), 1 ml, is added to each tube followed, after mixing, by 1 ml of thiobarbituric acid reagent (7). The mixtures are then left in the dark for 30 min and the absorbencies measured on a spectrophotometer at 530 nm. The quantities of retinal produced are ascertained by reference to the standard curve.

Typical values for retinol dehydrogenase activity in the retina of

SUBSYSTEMS AND REGIONS OF THE MAMMALIAN BRAIN

the adult albino rat (Fernando, Graymore & Kissun, unpublished observations) are, with NAD as coenzyme, 591 ± 184 µg retinal produced/h.g wet wt of retina $(n=10)$ and, with NADP, 1494 ± 340 µg. retinal/h.g wet wt of retina $(n=5)$. The activity is most probably due to at least two enzymes with differing coenzyme dependence (cf. Kissun *et al.*, 1972).

ISOLATED SUBSYSTEMS OF THE BRAIN

Several parts of the mammalian brain which have been successfully examined as isolated subsystems are listed in Table 12.2, together with examples of the information available about them. The studies listed have been carried out using simple fluid media such as glucose-bicarbonate Krebs-Ringer solutions. They have lasted about 0·5–2·5 h: that is, they are investigations of surviving tissues, and not of cultured tissues which would necessitate aseptic conditions and specialized media.

Table 12.2 Responses of isolated cerbral subsystems to electrical stimulation

Part of the brain (references)	Electrical responses	Other observations
Subcortical white matter; corpus callosum (Bollard & McIlwain, 1957; Kurokawa, 1960; Yamamoto & McIlwain, 1966)	Conducted impulses	Rodent and human tissues examined Tract-section interrupts response
Lateral olfactory tract (Yamamoto & McIlwain, 1966)	Conduction: 10–12m/s	
Optic tract (Kawai & Yamamoto, 1969)	Conduction: brief biphasic action potential: short latency and refractory period	
Lateral olfactory tract–piriform cortex (Yamamoto & McIlwain, 1966; Campbell *et al.*, 1967; Richards & Sercombe, 1968; McIlwain & Snyder, 1970 Richards & Smaje, 1974)	Pre- and postsynaptic in cortex; negative wave and unit discharges. Frequency-dependent attenuation; post-tetanic potentiation	Actions of butobarbitone, halothane, ether, scopamine, γ-aminobutyrate, chlorpromazine

(*Table continued overleaf*)

Table 12.2 *continued*

Part of the brain (references)	Electrical responses	Other observations
Optic tract-superior colliculus (Kawai & Yamamoto, 1969; Kawai, 1970)	Pre- and postsynaptic in brachium and colliculus; negative spike and wave discharges with relatively long and variable latency and refractory period	Some spontaneous and glutamate-induced firing; action of lysergic acid diethylamide
Medulla (Bollard & McIlwain, 1957)		Metabolic responses
Hypothalamus (Chase et al., 1969)		Actions of Li^+, ouabain, chlorpromazine
Corpus striatum (Katz & Kopin, 1969)		Tetrodotoxin; Li^+, ouabain action
Caudate nucleus (Yamamoto, 1973)	Action potentials modified by previous stimulation; by Ca^{2+}	Dopamine modified action potentials and cell-firing frequency. Actions of phenoxybenzamine and methamphetamine
Hippocampus: dentate gyrus and associated structures (Yamamoto & Kawai, 1968; Yamamoto, 1970, 1972; Jeffreys, 1975)	Negative spike and wave discharges. Rapid response from an electrically-coupled synapse	Low-Cl- media induced seizure discharges, propagated; Mg, Mn, tetrodotoxin and dendritic propagation
Neocortex (McIlwain, 1951, 1953; Rowsell, 1954; Greengard & McIlwain, 1955; Brierley & McIlwain, 1956; Li & McIlwain, 1957; Cummins & McIlwain, 1961; Srinivasan et al. 1969; Yamamoto & Kawai, 1967; Pumphrey, 1969; McIlwain & Pull, 1972)	Displacement and recovery of membrane potential; spike discharges; transmitted direct cortical response	With human tissues and neoplasms; induced gliosis; numerous drug interactions and release of neurohumoral agents

Preparations derived from the rat or guinea pig unless otherwise specified. For further metabolic responses by the systems. see Table 8.5.

Cerebral cortex: heterogeneity

The mammalian cerebral cortex has already featured much in other parts of this book, and has been regarded on a macroscopic scale as largely uniform from a chemical and metabolic point of view. This is often approximately so, but the cellular make-up of the cortex differs in different cortical areas and at different depths from the surface (see Fortuyn, 1914; Krieg, 1946; Hubel & Wiesel, 1959).

An appraisal of cellular elements in slices cut from the parietal cortex of the guinea pig in the fashion described below (Li & McIlwain, 1957; Hillman & McIlwain, 1961) suggested the first slice of 0·35 mm to contain layers I, II, and the upper portion of layer III, with horizontal cells and medium-sized pyramidal neurons as the largest elements. A second slice, taking the cortex between 0·35 and 0·7 mm from the surface, contained layer IV with parts of layers III and V, with Golgi type II cells likely to be intact. It must be further emphasized that the slices cut first, from the outer convexity of the hemispheres, carry the natural outer surface of the brain, and that it is likely that this surface has specific functional relationships with the cerebrospinal fluid which normally bathes it. Apart from such relationships, it is also important to note that this slice has one artificially-cut surface while successive slices, cut from greater depths in the cortex, have two such surfaces.

Both chemical composition and metabolic properties have been observed to differ in slices cut at different depths in the cerbral cortex. Glycogen was found at greatest concentration in the outer slices (Le Baron, 1955; McIlwain & Tresize, 1956) as also was chloride (Thomas & McIlwain, 1956). Resynthesis of glycogen in the tissue *in vitro* also yielded higher values in the outer slices; and outer slices were most active in anaerobic glycolysis (Dixon, 1953). Loss of glucose from incubation fluids was smaller with outer slices of cerebral cortex than with inner ones, but the loss of glucose caused by outer slices proved susceptible to being increased by insulin (Raefelson, 1958). This was attributed to the intact outer surface of the first slices; the action of insulin was not seen with inner slices. Lipid content and lipid metabolism have been found to be greater in the inner slices (Majno & Karnovsky, 1958): those cut from a depth of about 0·4 mm incorporated isotopic carbon from acetate into lipids, at twice the rate observed with outer slices.

A calibrated plastic template was used by Bennett *et al.* (1958) to obtain defined areas of cerebral cortex from the rat for study of cholinesterase. Histochemical techniques of sectioning frozen cerebral tissues give good opportunities for examining enzyme activities at different depths in the cortex, and these have now been applied to a number of enzyme systems (Lowry *et al.*, 1954; Robins *et al.*, 1956; Adams, 1969). Hess & Pope (1959) note especially that the different adenosine triphosphatases activated by calcium and by magnesium salts each differ in potency in tissue taken from different depths in the cortex, but independently: one being maximal in layer I and the other in layers II–IV.

Electrical activity observed in neocortical samples from the guinea pig (outer surface, various areas) has included the restoration and maintenance of resting cell-membrane potentials, with their dependence on supply of oxygen and other media constituents. Cell-firing has been observed by intracellular and by extracellular electrode placements, usually in reponse to applied stimuli (Chapter 10). Numerous metabolic changes are associated with such excitation (Table 12.2 and Chapter 8). The cutting of quite thin slices from the guinea pig neocortex, comprising only the outer 0·2 mm, has been employed in analysing the electrical responses to surface stimulation (Yamamoto & Kawai, 1967). These preparations were shown histologically to include only the molecular layer, and the response observed by small ball-tipped silver wires consisted of two successive surface-negative waves, which were attributed to the distal portions of apical dendrites.

Obtaining maximal yield of cerebral cortical slices

It is often valuable to obtain from a small laboratory animal, as the rat or guinea pig, the maximal yield of cerebral cortex in the form of a few slices of as large a size as possible. For this purpose the following procedure has been found suitable. It can be expected to yield from the adult guinea pig six slices 0·35 mm thick and weighing 110–140 mg, with two lesser ones; and from the rat four to six slices of 80–100 mg.

Preparation employs the cutting table, the blade or bow-cutter and cutting guide, and the spatulae of Chapter 6. The cerebrum, removed as previously described, is divided to the two hemispheres. One is left in its Petri dish and the other placed with its cortical surface downwards on the saline-moistened filter paper of the cutting table. Here the brain stem and midbrain which are uppermost, are removed with the spatula, cutting from the posterior towards the anterior portion, which is momentarily supported by the left forefinger so that the spatula is pressed towards the finger (Fig. 12.4). The lateral ventircle is then entered with the spatula from its junction with the third ventricle, which has been opened on separating the hemispheres. The spatula is then used to extend the ventricle, cutting a tract of white matter which is part of the fornix-system. The unfolded medial temporal-parietal part of the cortex is then unrolled on to the moist filter paper of the table, the anterior and posterior angles being nicked with the spatula to facilitate this (Fig. 12.4B). The remainder of the basal ganglia now projects up from a sheet of cortex, about 26 × 18 mm in the guinea pig; about half this area is occupied with subcortical white matter. The projecting ganglia are now removed by a horizontal cut with the spatula in the plane of the white matter (at this stage the subcortical white matter may be sampled: see below).

The sheet of moist filter paper carrying the cortex is picked up from the table and another sheet of paper dipped in saline put in its place. The first sheet with the cortex is inverted and put on the new filter paper, and then the first paper removed, leaving the cortex with its outer surface uppermost. Slices are now cut with the blade and guide. Before cutting, it is especially necessary to ensure that the guide is completely in contact with the tissue, for any air bubbles trapped between them will cause holes in the slices cut. The trapping is more likely with the flat sheet of tissue than with the normal, convex surface of the cortex.

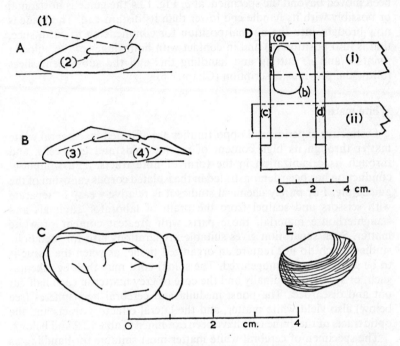

Fig. 12.4 Cutting large slices of cerebral cortex, and a slice of subcortical white matter, from the guinea pig brain. A hemisphere is laid on moist filter paper on a cutting block. *A* and *B*: stages in opening the hemisphere to give a sheet of cortex; successive operations are (1) cutting away the basal ganglia and associated structures; (2) inserting a spatula into the lateral ventricle; (3) and (4) extending the lateral ventricle, which enables most of the tissue remaining above the plane of the ventricle, to be unfolded on to the filter paper as a sheet. *C*: the resulting sheet. *D*: Positions of blade and guide in sweeping out air bubbles before cutting (i, *a* and *b*); and (ii) in subsequent cutting, with the blade on the ridges of the guide at points *c* and *d*. A bow-cutter may be used in place of the blade illustrated. *E*: direction of fibre tracts in the sheet of subcortical white matter.

The blade and guide are accordingly dipped in saline, the excess saline removed by a rapid jerk, and the blade first laid on the cortex at the point farthest from the operator. The guide is then rested on the blade so that both touch the cortex at *a*, Fig. 12.4. Blade and guide slope away from the operator so that at the part of the cortex nearest to him, the blade is about 0·5 mm above the specimen and the guide is about 0·5 mm above the blade. The blade is now drawn towards the operator, its tip *a* still remaining in contact with the guide; the guide is allowed to rest on the cortex as this is done. With the removal of the blade, the guide must be brought gradually nearer to a horizontal position in order to lie on the cortex, and when the tip of the blade has been moved beyond the specimen, at *b*, Fig. 12.4 the guide is horizontal or possibly with its handle end lower than its distant end. The blade is now brought into the normal position for commencing to cut a slice, that is into position (ii) and in contact with both ridges of the guide at points *c* and *d;* cutting and handling this and the subsequent slices follow the previous description (Chapter 6).

White matter

Distinctive experimental opportunities are afforded by cerebral white matter through its high content of myelin lipids and glial cells, and through its organization in the form of fibre tracts. Thus, impulse-conduction has been demonstrated in the isolated corpus callosum of the guinea pig. For purely chemical studies it is relatively easy to separate with scissors and scalpel from the brain of laboratory animals and slaughterhouse material, those parts with greatest content of white matter. Such separation gives suitable preparations also for metabolic studies which do not require an organized tissue, as when the tissue is to be chopped or homogenized. The spinal cord may then be taken as such, or split longitudinally and the core of grey matter in each half cut out and discarded. The pons, medulla and cerebral hemispheres (see below) also yield white matter, and the lateral olfactory tracts and the optic tracts of the guinea pig have been examined (Table 12.2 and below).

The specimen of cerebral white matter most suitable for handling as a tissue slice has proved to be subcortical white matter (Bollard & McIlwain, 1957; Kurokawa, 1960). In preparing the cerebral hemispheres of guinea pigs or rats to obtain a maximum yield of tissue, a point is noted (p. 278) when the outer wall of the lateral ventricle is exposed as the upper surface of a sheet of cortex. At this point, a slice can be cut with the bow-cutter and guide from the whole of the exposed surface, using a guide giving a slice thickness of $0.35-0.4$ mm. The white matter is softer and more easily "smeared" and distorted than is grey matter; the preparation described is carried out with cutter and guide dipped in the saline of a subsequent experiment, and the slice washed from them to a dish of saline. After cutting the slice, some white matter should still be seen to remain on the block of tissue left behind on the cutting table; this gives assurance that the slice cut has sections in which white matter extends throughout its thickness. The slice still however has attached areas of grey matter, readily visible, and these are cut off with a scalpel or sharpened spatula used as described in Chapter 6. In this way, each hemisphere of the guinea pig yields a kidney-shaped piece of white matter about 8×15 mm in area.

These slices have been employed in experiments on the spread of electrical stimulation from the localized application of pulses, and in this connection the direction of fibre-tracts in the sample was determined by Kurokawa (1960) and is noted in Fig. 12.4. The white-matter slices have been subdivided in some experiments so that four specimens are obtained from one guinea pig. After cutting these slices, the remainder of the hemisphere may be inverted on to a second sheet of filter paper on the cutting table, and at least two large slices of cerebral cortex obtained from each hemisphere.

Lateral olfactory tract–piriform cortex

This system, prepared from the guinea pig or rat, was introduced for combined chemical and electrical studies by Yamamoto & McIlwain (1966) and has proved versatile and dependable. In the small rodents and the rabbit the lateral olfactory tract is fused to the surface of the piriform lobe; detailed anatomical and histological studies of the system have been made by Cajal (1955 republication) and by Valverde (1965). Many electrophysiological and pharmacological studies have been made of the relevant parts of the brain *in vivo* (see for example Legge *et al.*, 1966; Biedenbach & Stevens, 1969; Phillips *et al.*, 1963). An electron-microscopic study has been made in association with observation of electrical responses (Yamamoto *et al.*, 1970).

To prepare the l.o.t.–piriform cortical system, the piriform lobe is first obtained as a block of tissue. A guinea pig or rat is struck on the back of the neck, exsanguinated and the skull opened as described in Chapters 1 and 6, but removal of the frontal bones is extended forward to expose the olfactory bulbs. This is aided by cutting the bone with a clipper termed an ingrown nail cutter (Scholl, London; the cutter has a 9 mm cutting blade which closes on a thin lower plate, horizontal in use and inserted between the dura and the relevant part of the skull). With a scalpel held vertically two transverse cuts are now made to separate the brain from the olfactory bulbs and the spinal cord. The narrow sharp blade of the spatula of Fig. 6.1 held axially, is inserted below the brain and with side-to-side movements along the base of the skull is brought forward to free the brain and to lift it from the skull. The brain is divided with the spatula to the two hemispheres of which one is placed on the newly-cut surface on the saline-moist filter paper of a cutting table (Fig. 6.3). The lateral olfactory tract shows clearly as a white streak extending on the piriform cortex from the anterior, cut, end of the specimen for about 12 mm; it runs posteriorly approximately parallel to the rhinal fissure which separates the piriform cortex from the neocortex. With a fine (e.g. ophthalmic) scalpel held perpendicular to the surface of the cortex near the fissure, a cut is made from the anterior end of the hemisphere parallel to the rhinal fissure and about 0·5 mm on its neocortical side, and so placed that the tract is about at the centre of the block so separated. It is necessary to alter the angle of the scalpel to the vertical during this cutting, as the fissure is not straight and the cortex is piriform or pear-shaped.

The block of tissue so separated is placed on its scalpel-cut surface on the moist filter paper on the cutting table. The declevity previously present at the piriformity of the lobe, which made difficult the use of a guide in slicing, has now largely been lost. The lateral olfactory tract should be uppermost and running axially along the specimen near the centre of its surface. A slice cut now with a bow-cutter or blade (Fig. 6.4, 6.3), and a guide giving a tissue-thickness of 0·3–0·35 mm, can be expected to weigh 30–40 mg. Its appearance is shown in Fig. 9.4. The preparation of the two piriform specimens from an excised brain needs only 1–2 min and the remainder of each hemisphere yields samples of neocortex if these are also required. For electrophysiological work a single piriform sample only may be needed, but it can be advantageous to prepare the two such tissues available from an animal, to preincubate them together and transfer one to an auxiliary vessel when experiments with the other are commenced. In metabolic work the two piriform and two superficial neocortical samples can valuably be examined in different vessels of the same experiment (see Chapter 8 and Table 8·5). The lateral olfactory tract itself was dissected free from the remainder of the tissue in interpretative experiments (Yamamoto & McIlwain, 1966).

A major value of this preparation as it is obtained from a guinea pig is the facility it offers for stimulation at the anterior portion of the lateral olfactory tract, and observation of electrical responses at points 15 mm distant. Conduction in the lateral olfactory tract and postsynaptic events in the surrounding cortex were characterized by Yamamoto & McIlwain (1966) and Richards & Sercombe (1968), using surface electrodes initially and subsequently microelectrodes which penetrated the tissue; see Chapter 9 and Fig. 9.4. The microelectrodes detected unit cell-firing and contributed to understanding the surface responses. The preparation is suceptible to a number of neurohumoral agents and drugs. Metabolic responses were obtained by stimulating the lateral olfactory tract, and metabolic responses greater in magnitude obtained by field stimulation of the whole preparation; this enabled the proportion of tissue activated from the tract, to be evaluated.

Optic tract–superior colliculus

This preparation gives a means of following the input from an optic tract of the guinea pig through the lateral geniculate body to the superior colliculus and has been employed in metabolic and electrophysiological studies (Kawai & Yamamoto, 1969; Heller & McIlwain, 1973).

With an animal prone, and after opening the skull as described in Chapters 1 and 6, the frontal lobes were freed by a cut posterior to the olfactory bulbs, made with a scalpel held vertically. The frontal lobes were then retracted and the optic nerves cut at their point of entry to the skull. The spinal cord was cut across just posterior to the cerebellum and a narrow sharp spatula (Fig. 6.1) used to remove the brain, when it appears as shown in Fig. 12.5. It was then inverted on a cutting table to present its ventral surface uppermost. The optic chiasma was divided at the midline and the left optic tract preserved. The left hemisphere was then removed, and the brain stem preserved; this carried the posterior hypothalamus, cerebral peduncle, the lateral geniculate body and superior colliculus (*B*, Fig. 12.5). The optic tract was freed from the peduncle with fine scissors, the lower part of the peduncle cut away and the specimen placed on the cut surface just made. With a recessed glass guide and bow-cutter, a slice 0·35–0·4 mm thick was cut from the superior colliculus and adjacent lateral geniculate body leaving intact the connection to the optic tract (*C*, Fig. 12.5).

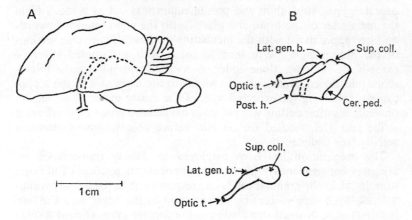

Fig. 12.5 Preparation of an optic tract–superior collicular subsystem from a guinea pig brain. *A:* The brain as removed from the skull; interrupted lines show relevant parts which are not visible from the surface. *B:* The dissected block (see text) carrying the left optic tract, lateral geniculate body, superior colliculus, cerebral peduncle and posterior hypothalamus. The optic tract is partly dissected free, and its former position is shown by interrupted lines. *C:* The preparation used. *Lat. gen. b.*, lateral geniculate body; *Sup. coll.*, superior colliculus; *Optic t.*, optic tract; *Cer. ped.*, cerebral peduncle; *Post. h.*, posterior hypothalamus.

One such preparation, weighing 20–25 mg, is available from an animal and 2–3 min are taken in obtaining it; much of the brain remains available for other use. Response to its stimulation at the optic tract while maintained *in vitro*, has been observed at the lateral geniculate body and at the superior colliculus, employing surface electrodes. Responses have also been observed by micropipette electrodes within the superior colliculus. A postsynaptic field potential was suppressed by 0·1–1 μM-serotonin (Kawai & Yamamoto, 1969; for a related effect *in vivo* see Phillis *et al.*, 1967) and this action was potentiated by lysergic acid diethylamide. Output of serotonin and of adenine derivatives have been found to occur on stimulation of the preparation (Tables 8.5 and 12.2); the output of serotonin was inhibited by lysergic acid diethylamide (Kawai, 1970).

Dentate gyrus and hippocampus

Isolated preparations from this part of the guinea pig brain have been maintained and examined *in vitro* by Yamamoto (1972b) and Yamamoto & Kawai (1967, 1968). The brain was removed from an animal in the fashion described in Chapter 1 and divided sagitally. By removing the brain stem, the hippocampus and dentate gyrus were exposed and sections 0·25 to 0·35 mm thick cut from the gyrus as shown at *A*, Fig. 9.5. For some investigations the hippocampus, including the dentate gyrus, subiculum and presubiculum was cut as a block from the remainder of the brain and placed with the regio inferior upwards on filter paper moist with the incubating medium. Transverse sections of the hippocampus were then prepared with a mounted sliver of razor-blade. This was done under observation with a binocular microscope and the cuts made at right angles to the main axis of the hippocampus following the expected path of the mossy fibres. Each section immediately after cutting was placed in oxygenated glucose–bicarbonate saline and preincubated for 30 min before observation of electrical activity (see Chapter 9).

The preparation has been employed to display transmission by synapses between mossy fibres and hippocampal neurones following stimulation at the granular cell layer; responses observed at the pyramidal cell layer were susceptible to variation in the Mg^{2+} and Ca^{2+} of bathing fluids. Normal stimulation at the dentate gyrus elicited a spike response and negative wave, but similar stimulation in chloride-free media produced a train of afterdischarges of large amplitude, corresponding to a seizure discharge (Fig. 9.5; Table 12.2).

DISSECTION OF THE RAT BRAIN TO YIELD SEPARATED REGIONAL BLOCKS

For examining tissue constituents and specifying their region of occurrence, dissection of the brain is frequently performed in a fashion characteristically different from that exemplified by the preceding descriptions. Those procedures usually aimed at retaining at least two regions in functional connection. With the procedures now to be described, connecting tracts are deliberately severed to obtain small regions, some of which represent distinct divisions, subdivisions or 'nuclei' recognized anatomically. These regions, for example the caudate nucleus, are frequently aggregates of cell bodies with synaptic structures carrying characteristic neurotransmitters; the transmitters indeed may themselves contribute to staining characteristics by which the nuclei have been recognized. The present dissection procedures are applicable when it is wished to account for the total quantity of cerebral constituent or activity, and therefore begin by dividing the brain to specified 'blocks' with little or no discarded material.

Preparation of regional blocks

A procedure in which chosen blocks of tissue were separated from the rat or guinea pig brain, and subsequently chopped mechanically to slices of uniform size, has been used in a number of laboratories (see McIlwain & Buddle, 1953; McIlwain, 1966; Sattin, 1966). The following preparation is based on the dissections of Bayoumi & Smith (1973), Glowinski & Iversen (1966) and Goodman et al. (1973). The brain was exposed in the skull as described in Chapter 1 and from the prone animal was separated by transverse vertical cuts immediately anterior to the cerebral hemispheres and posterior to the cerebrellum, and by horizontal cuts along the base of the skull. It was then placed on the cutting table (Fig. 6.3) and divided according to Fig. 12.6. Cut (i) passes through the optic chiasma, and cut (ii) is defined by the outline of the cerebellum. The posterior portion is subsequently separated by cut (iii) to a *cerebellar* block, and a block comprising the *medulla oblongata* plus *pons*.

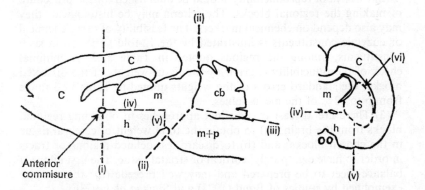

Fig. 12.6 Division of the rat brain to regional blocks. Interrupted lines show the positions of scalpel-cuts designated (*i*)–(*vi*) and described further in the text. Left: as would be seen in saggital section between the hemispheres; right, as would be seen in coronal section at the level of cut (*i*). C, cerebral cortical block; cb, cerebellar block, h, hypothalamic block; m, midbrain block (with thalamic and subthalamic tissue); m+p, block with medulla oblongata plus pons; s, striatal block with tissue of caudate nucleus, globus pallidus and putamen.

The central section was inverted on the cutting table and divided as follows. A *hypothalamic* block was separated by a horizontal cut (iv) at the level of the anterior commisure, plus vertical cuts one of which is shown at (v) and which follow the outline of the hypothalmus (for variants at this stage of the procedure, see Table 12.3, legend). To obtain a *striatal* block the cut (vi) was made, starting from a lateral ventricle, continuing just central to the corpus callosum and subcortical white matter to meet the horizontal cut (iv), and then, again, to the lateral ventricle (Fig. 12.6). Striatal tissue was so obtained from each hemisphere. The striatum extends anteriorly to cut (i) and was cut from that region also; for some purposes the four pieces of striatal tissue were handled together. A *hippocampal* block was obtained by making a cut between the two hemispheres, through the corpus callosum to reach one of the lateral ventricles. This revealed the hippocampus as an arc of tissue at the rim of the infolded hemisphere. The hippocampus with the attached dentate gyrus was cut free, from each side of the brain. The remaining material between the sites of the hypothalamic and hippocampal blocks was then taken as a *midbrain* block. Blocks of *cortical* tissue were available on either side of cut (i).

Comments on preparing regional blocks. (1) It is often appropriate to make further subdivisions of the blocks obtained by a process such as that described. Simple examples are quoted in Table 12.3 and will be evident also from preceding sections of this Chapter. Kuhar *et al.* (1973) examined opiate-binding to 42 regions of the brain of *Macaca mulatta*, including five parts of the hypothalamus, and found the binding capacity of the regions to vary between 1·4 to 65 fmol/mg protein.

(2) Criteria of reproducibility should be established for the procedure of making the regional blocks. The criteria may be histological; they may also depend on chemical markers. The feasibility of using chemical or enzymic constituents is illustrated by the 12-fold variation in such constituents, among the regions quoted in Table 12.3. A minimal check of reproducibility is given by the tissue-weights of the dissected blocks; the standard errors of the weights quoted in Table 12.3 range from 6 to 12 % of the mean values.

(3) There are at least two distinct approaches in obtaining regional blocks from the brain: (a) to obtain the total weight of cerebral tissue in the chosen blocks: and (b) to discard less-defined regions or tracts in order to have e.g. 'purely' cortical or striatal tissue. The first allows a balance-sheet to be prepared and may well precede (b); the two are exemplified by studies of Bond (1973) and Sims *et al.* (1973).

(4) Tissue temperatures between 0° and room temperature have been used in preparing regional blocks. Room temperature, if it is not above about 15°C, is to be recommended in the interest of speed and simplicity. Any cooling should not be below 0° if cell-containing tissues are desired, and should not be done by immersion in fluid. Isotonic and hypotonic fluids are taken up by tissue blocks between 0° and 30°C

Table 12.3 Distribution of tissue components in regional blocks from rat brain

Regional block	(1) Weight (mg ± S.D.)	(2) Extracellular space (ml/100 g)	(3) Noradrenaline content (μg/g ± s.e.)	(4) Dihydroxyphenylalanine decarboxylase (μmoles/g h)	(5) High-affinity uptake of γ-aminobutyrate μmoles/g min
Cerebellum	267 ± 22	19·5	0·17 ± 0·02	1·6 ± 0·1	3·4
Medulla and pons	252 ± 24	—	0·72 ± 0·02	3·8 ; 5·4	2·9
Hypothalamus	110 ± 14	—	1·79 ± 0·01	8·6 ± 0·5	16·7
Striatum	102 ± 17	20·8	0·25 ± 0·03	18·7 ± 1	7·0
Midbrain	146 ± 10	—	0·37 ± 0·03	5·2 ± 0·3	10·8
Hippocampus	151 ± 22	19	0·20 ± 0·01	2·4 ± 0·2	13·8
Cortex	804 ± 58	12·8	0·24 ± 0·02	1·9 ; 3·1	15·8

Values (1) and (3), Glowinski & Iversen (1966); (2), Goodman et al. (1973); (4), Sims et al. (1973); (5), Bond (1973). Dissections by the different authors (q.v.) differed in detail; thus in the dissection for data (5), the outline of the hypothalamus was more closely followed than was the case for that for data (3), and yielded 50–60mg of tissue. In (4) the values for cortex are from occipital and frontal regions, and for medulla-pons are from lateral and midline regions. An analogous dissection has been carried out by Sun et al. (1975) in investigating the distribution of adenosine and of AMP deaminases, 5'-nucleotidase and β-glycerophosphatase; a hypothalamic block showed greatest deaminase activities.

more extensively than at 37°, and thus invalidate measurements based on tissue weight. Hypertonic solutions necessary to diminish change in tissue weight, alter the distribution of fluid between intra- and extracellular phases of the tissue.

Preparations available from regional blocks

The regional blocks are handled in characteristically different fashions for different types of study.

(1) *Retaining maximal cell integrity and relatively specified location.* The regional blocks yielded by dissection are usually not suitably shaped for metabolic studies, nor for other studies which depend on metabolic maintenance of tissue integrity. In this they differ from the cerebral subsystems described previously. The blocks usually require cutting to thinner sections, and this can be done in several fashions according to the needs of subsequent experiments. To prepare a few samples, dissection of the block may be carried out with scalpel or razor blade under observation with a binocular microscope (Yamamoto, 1972b).

A mechanical tissue chopper such as that described in Chapter 6 can be used to convert the whole of the chosen block into portions suitable for maintenance as surviving tissues. For maximal retention of cell-integrity, minimal chopping should be performed: for example, one series of parallel cuts at 0·3–0·5 mm intervals, in a plane chosen in relation to the histological structure of the block. Grey-matter regions require the thinner slices; this will yield 8–25 cross-sections from several of the blocks of Table 12.3, each section of some 5–10 mg fresh weight. These are still adherent and the whole may be transferred to one experimental vessel for prompt use as described in Chapter 6 (examples: McIlwain & Buddle, 1953; Sattin, 1966; Bustos & Roth, 1972). After chopping, the tissues of the adherent sections are still in their original positions in relation to each other, and may be placed individually in separate vessels of oxygenated media for incubation. The narrow and relatively sharp spatula of Fig. 6.1, after dipping in the medium, is used for this separation and transfer. When so handled, a subsequent determination, of for example protein content, is needed as a measure of sample-size. Samples separated in this fashion from, for example the hypothalamic block are not replicates, for the hypothalamic nuclei and tracts are differently distributed among the chopped samples. The samples will however be identifiable according to their position, being more or less lateral or central, anterior or posterior.

Chosen areas may also be separated after tissues have been chopped, though this is more tedious. Thus from a guinea pig hemisphere a block was isolated by cuts similar to (i) and (ii), Fig. 12.6. The block was chopped mechanically at 0·35 mm intervals and the sections teased apart in a shallow black-glazed dish of incubation medium, when cross-sections similar to that of Fig. 12.6 were seen and cuts made around the striatum with a scalpel while the sections rested on the bottom of the dish. By accumulating these parts from 12 sections, about 40 mg of striatal tissue was obtained.

(2) *Mixed samples from a regional block.* An alternative manner of using a tissue-chopper deliberately randomizes the differences within a selected block of tissue and yields replicate samples for distribution among a number of control and experimental vessels. A regional block is chopped relatively finely, e.g. at 0·15–0·25 mm intervals in two vertical planes at 45 to 90° to each other. This gives approx. 1000–3000 fragments from the regional blocks of Table 12.3. These may be suspended in a few ml of incubation-medium using a wide-tipped pipette and subsequently distributed among 5–10 vessels by rapid pipetting. For metabolic experiments this preparation, and the preparation of the preceding section (1), require to be handled as cell-containing samples: that is, incubated in media of composition resembling extracellular fluids. For examples of their use see McIlwain & Buddle (1953) and Bond (1973).

(3) *For yielding subcellular fractions.* The regional block is dispersed and fractionated as described in Chapter 10. It is to be noted that any subcellular fractionation of cerebral tissue is best carried out with a defined region of the brain. In particular, synaptosomal fractions differ markedly in chemical and enzymic properties when prepared from regions which are distinct in their neurotransmitter content. Relevant examples are the studies of Sattin (1966), Kuhar *et al.* (1973) and Sims *et al.* (1973); see Table 12.3.

(4) *For extraction of chemical constituents or enzyme activities.* For these studies the whole brain or the regional blocks may be frozen with solid CO_2 mixtures or with liquid nitrogen. Handling the regions for maximal extraction of materials or for maximal exhibition of enzyme activity applies the methods and precautions of earlier chapters. A distinct method of dissecting cold rat brain was applied by Sethy *et al.* (1973) in determining components of the acetylcholine system following cerebral lesions. In examining glutamic acid decarboxylase in 7 regions from normal and pyridoxine-deficient rats, Bayoumi & Smith (1973) placed the dissected regions in liquid nitrogen and accumulated material from 5–6 brains before homogenizing the samples in water or 0·25% w/v triton X-100.

REFERENCES

Adams, C. W. M. (1969) *Handb. Neurochem.* **2**, 525.
Ames, A. (1965) In *Biochemistry of the Retina*. ed. Graymore, C. N. pp. 22–25. New York: Academic Press.
Ames, A. Tsukada, Y. & Nesbett, F. B. (1967) *J. Neurochem.* **14**, 145.
Arden, G. B. & Ernst, W. (1970) *J. Physiol. Lond.* **211**, 311.

Bayoumi, R. A. & Smith, W. R. D. (1973). *J. Neurochem.* **21**, 603.
Bennett, E. L., Krech, D., Rosenzweig, M. R., Karlsson, H., Dye, N. & Ohlander, A. (1958) *J. Neurochem* **3**, 144.
Berman, E. R. & Voaden, M. J. (1970). In *The Biochemistry of the Eye*. ed. Graymore, C. N. pp. 373–471. London: Academic Press.
Biedenbach, M. A. & Stevens, C. F. (1969) *J. Neurophysiol.* **32**, 195.
Bollard, B. M. & McIlwain, H. (1957) *Biochem. J.* **66**, 651.
Bond, P. A. (1973) *J. Neurochem.* **20**, 511.
Brading, A. F. (1971) *J. Physiol. Lond.* **214**, 393.
Brierley, J. B. & McIlwain, H. (1956). *J. Neurochem.* **1**, 109.
Brown, K. T. (1968) *Vision Res.* **8**, 633.
Burt, A. M. (1966) *Microchem. J.* **11**, 18.
Bustos, G. & Roth, R. H. (1972) *Br. J. Pharmacol.* **46**, 101.
Cajal, S. R. y (1955) *Studies on the Cerebral Cortex (Limbic Structures)*, trans. Kraft, L. M. London: Lloyd-Luke.
Campbell, W. J., McIlwain, H., Richards, C. D. & Somerville, A. R. (1967) *J. Neurochem.* **14**, 937.
Chase, T. N., Katz, R. I. & Kopin, I. J. (1969) *J. Neurochem.* **16**, 607.
Cohen, A. I. (1972) In *Handbook of Sensory Physiology*. Vol. VII/2: *Physiology of Photoreceptor Organs*, ed. Fuortes. pp. 63–110. Berlin: Springer.
Cohen, S. R. (1973) *Brain Res.* **52**, 309.
Collins, F. D., Love, R. M. & Morton, R. A. (1952) *Biochem. J.* **51**, 292.
Cummins, J. T. & McIlwain, H. (1961) *Biochem. J.* **79**, 330.
Daemen, F. J. M., Borggreven, J. M. P. M. & Bonting, S. L. (1970) *Nature, Lond.* **227**, 1259.
Dixon, K. C. (1953) *J. Physiol., Lond.* **120**, 267.
Dowling, J. E. (1970) *Invest. Ophthal.* **9**, 655.
Drujan, B. D. & Svaetichin, G. (1972) *Vision Res.* **12**, 1777.
Ehinger, B. & Falck, B. (1971) *Brain Res.* **33**, 157.
Fortuyn, A. B. D. (1914) *Archs Neurol. Psychiat., Chicago* **6**, 221.
François, J & Neetens, A. (1974) In *The Eye*. Vol. 5: *Comparative Anatomy and Physiology of the Eye*, ed. Davson. Part 1. London: Academic Press.
Futterman, S. & Saslaw, L. D. (1961) *J. biol. Chem.* **236**, 1652.
Futterman, S., Henrickson, A., Bishop, P. E., Rollins, M. H. & Vacano, E. (1970) *J. Neurochem.* **17**, 149.
Futterman, S. & Rollins, M. H. (1973) *J. biol. Chem.* **248**, 7773.
Glowinski, J. & Iversen, L. L. (1966) *J. Neurochem.* **13**, 655.
Goodman, F. R., Weiss, G. B. & Alderdice, M. T. (1973) *Neuropharmacology* **12**, 867.
Graham, L. T. Jr. (1974) In *The Eye*. Vol. 6: *Comparative Anatomy and Physiology of the Eye*, Edit. ed. Davson Part 2. London: Academic Press.
Granit, R. (1963) *Sensory Mechanisms of the Retina (with an Appendix on Electroretinography)*. New York: Hafner.
Graymore, C. N. (1958) *Br. J. Ophthal.* **42**, 348.
Graymore, C. N. (1969) In *The Eye*, Vol. 1: *Vegetative Physiology and Biochemistry* 2nd edn. ed. Davson. pp. 601–645. London: Academic Press.
Graymore, C. N. (1970a) *Br. Med. Bull.* **26**, 130.
Graymore, C. N. (1970b) In *Biochemistry of the Eye*. ed. Graymore, C. N. pp. 645–735. London: Academic Press.
Greengard, O. & McIlwain, H. (1955) *Biochem. J.* **61**, 61.
Harwerth, R. S. & Sperling, H. G. (1971) *Science, N.Y.* **174**, 520.
Heller, I. H. & McIlwain, H. (1973) *Brain Res.* **53**, 105.
Hess, H. H. & Pope, A. (1959) *J. Neurochem.* **3**, 287.
Hillman, H. H. & McIlwain, H. (1961) *J. Physiol.* **157**, 263.
Hogan, M. J., Alvaredo, J. A. & Weddell, J. E. (1971) *Histology of the Human Eye. An Atlas and Textbook*. Philadelphia: Saunders.

Hubbard, R., Brown, P. K. & Bownds, D. (1971) Meth. Enzym. **18**, 615.
Hubel, D. H. & Wiesel, T. N. (1969) *Nature Lond.* **221**, 747.
Jeffreys, J. G. R. (1975) *J. Physiol.* in press.
Kakiuchi, S., Rall, T. W. & McIlwain, H. (1969) *J. Neurochem.* **16**, 485.
Katz, R. I. & Kopin, I. J. (1969) *Biochem. Pharmacol.* **18**, 1935.
Kawai, N. (1970) *Neuropharmacology* **9**, 395.
Kennedy, A J. & Voaden, M. J. (1974). *J. Neurochem.* **22**, 63.
Kawai, N. & Yamamoto, C. (1969). *Int. J. Neuropharmacol.* **8**, 437.
Kissun, R. D., Graymore, C. N. & Newhouse, P. J. (1972) *Expl. Eye Res.* **14**, 150.
Korenbrot, J. I. (1973) *Expl Eye Res.* **16**, 343.
Kramer, S. G. (1971) *Invest. Ophthal.* **10**, 438.
Krieg, W. J. S. (1946) *J. comp. Neurol.* **84**, 277.
Kuhar, M. J., Pert, C. B. & Snyder, S. H. (1973) *Nature, Lond.* **245**, 447.
Kurokawa, M. (1960) *J. Neurochem.* **5**, 283.
Lasansky, A. & de Fisch, F. W. (1966) *J. Gen. Physiol.* **49**, 913.
Le Baron, F. N. (1955) *Biochem. J.* **61**, 80.
Legge, K. F., Randic, M. & Straughan, D. W. (1966) *Br. J. Pharmacol.* **26**, 87.
Liebman, P. A. (1972) In *Handbook of Sensory Physiology*. Vol. VII/1 *Photochemistry of Vision*, ed. Dartnall, pp. 481–528. Berlin: Springer.
Li, C.-L. & McIlwain, H. (1957) *J. Physiol., Lond.* **139**, 178.
Lolley, R. N. (1969). In *Handbook of Neurochemistry*. Vol. 2: *Structural Neurochemistry*, ed. Lajtha, A, pp. 473–504. New York: Plenum.
Lolley, R. N. & Schmidt, S. Y. (1974) In *The Eye*. Vol. 6: *Comparative Physiology of the Eye*, ed. Davson & Graham, pp. 343–378. London: Academic Press.
Lowry, O. H. (1964) In *Morphological and Biochemical Correlates of Neural Activity*. ed. Cohen & Snider, pp. 178–191. New York; Hoeber.
Lowry, O. H. & Passonneau, J. V. (1972) *A Flexible System of Enzymatic Analysis*. New York: Academic Press.
Lowry, O. H., Roberts, N. R., Schulz, D. W., Clow, J. E., & Clark, J. R. (1961) *J. biol. Chem.* **236**, 2813.
Lowry, O. H., Roberts, N. R., Leiner, K.Y., Wu, M.-L., Farr, A. L. & Albers, R. W. (1954) *J. biol. Chem.* **207**, 39.
Majno, G. & Karnovsky, M. L. (1958) *J. exp. Med.* **107**, 475.
Majno, G., Gesteiger, E. L., La Gattuta, M. & Karnovsky, M. L. (1958) *J. Neurochem.* **3**, 127.
Marshall, J., Medford, P. A. & Voaden, M. J. (1974) *Exp. Eye Res.* **19**, 559.
Marshall, J. & Voaden, M. J. (1974a) *Expl Eye Res.* **18**, 367.
Marshall, J. & Voaden, M. J. (1974b) *Invest. Ophthal.* **13**, 602.
Maurice, D. M. (1957) *J. Physiol., Lond.* **137**, 110.
Maurice, D. M. (1959) *Am. J. Ophthal.* **47**, 361.
McIlwain, H. (1951) *Biochem. J.* **49**, 382.
McIlwain, H. (1953) *J. Neurol. Neurosurg. Psychiat.* **16**, 257.
McIlwain, H. (1966) *Proc. 3rd Int. Congr. neurol. Surg.* 1965 (Excerpta Med. Congr. Ser. 110), 442.
McIlwain, H. & Buddle, H. L. (1953) *Biochem. J.* **53**, 412.
McIlwain, H. & Tresize, M. A. (1956) *Biochem. J.* **63**, 250.
Moyer, F. H. (1969). In *The Retina. Morphology, Function and Clinical Characteristics*. ed. Straatsma, Hall, Allen, & Crescitelli. pp. 1–30. Univ. California Press.
Neal, M. J. & Atterwill, C. K. (1974) *Nature* **251**, 331.
Pearse, A. G. E. (1968) *Histochemistry. Theoretical and Applied* .London: Churchill
Phillips, C. G., Powell, T. P. S. & Shepherd, G. M. (1963). *J. Physiol. Lond.*, **168**, 65.
Phillis, J. W., Tebecis, A. K. & York, D. H. (1967) *J. Physiol., Lond.* **190**, 563.
Polyak, S. (1957) *The Vertebrate Visual System*. Univ. Chicago Press.
Pull, I. & McIlwain, H. (1972) *Biochem. J.* **130**, 975.

Pumphrey, A. M. (1969) *Biochem. J.* **112**, 61.
Rafaelson, O. J. (1958) *Lancet*, ii, 941.
Richards, C. D. & Sercombe, R. (1968) *J. Physiol., Lond.* **197**, 667.
Richards, C. D. & Smaje, J. C. (1974) *J. Physiol., Lond.* **239**, 103P.
Robins, E., Smith, D. E. & Eydt, K. M. (1956). *J. Neurochem.* **1**, 54, 77.
Rowsell, E. V. (1954) *Biochem. J.* **57**, 666.
Santamaria, L., Drujan, B. D., Svaetichin, G. & Negishi, K. (1971) *Vision Res.* **11**, 877.
Sattin, A. (1966). *J. Neurochem.* **13**, 515.
Sethy, V. H., Roth, R. H., Kuhar, M. J. & van Woert, M. H. (1973) *Neuropharmacology* **12**, 819.
Sickel, W. (1972) In *Handbook of Sensory Physiology*, Vol. VII/2: *Physiology of Photoreceptor Organs*, edit. Fuortes, M. G. F. pp. 667–727. Berlin: Springer.
Sims, K. L., Davis, G. A. & Bloom, F. E. (1973) *J. Neurochem.* **20**, 449.
Srinivasan, V., Neal, M. J. & Mitchell, J. F. (1969) *J. Neurochem.* **16**, 1235.
Stell, W. K. (1972) In *Handbook of Sensory Physiology*, Vol. VII/2: *Physiology of Photoreceptor Organs*, ed. Fuortes, M. G. F. pp. 111–213. Berlin: Springer.
Straatsma, B. R., Hall, M. O., Allen, R. A. & Crescitelli, F. (Eds.) (1969) *The Retina. Morphology Function and Clinical Characteristics.* Univ. California Press.
Sun, M. C., McIlwain, H. & Pull, I. (1975) *J. Neurobiol.*, in press.
Thomas, J. & McIlwain, H. (1966) *J. Neurochem.* **1**, 1.
Tomita, T. (1972a) In *Handbook of Sensory Physiology*. Vol. VII/2: *Physiology of Photoreceptor organs.* ed. Fuortes, M. G. F. pp. 483–511. Berlin: Springer.
Tomita, T. (1972b) ibid. pp. 635–665.
Valverde, F. (1965) *Studies of the Piriform Lobe.* Cambridge, Mass.: Harvard Univ. Press.
Voaden, M. J. (1974) *Biochem. Soc. Trans.* **2**, 1224.
Walls, G. L. (1942) *The Vertebrate Eye.* Cranbrook Institute of Science. Bulletin 19. The Cranbrook Press, U.S.A.
Weiss, H. & Kosmath, B. (1971) *Albrecht V. Graefes Arch. Ophthal.* **181**, 329.
Yamamoto, C. (1972a). *Expl Neurol.* **35**, 154.
Yamamoto, C. (1972b) *Expl Brain Res.* **14**, 423.
Yamamoto, C. (1973) *Adv. Neurol. Sci.* **17**, 64.
Yamamoto, C. & Kawai, N. (1967) *Expl Neurol.* **19**, 176.
Yamamoto, C. & Kawai, N. (1968) *Jap. J. Physiol.* **18**, 620.
Yamamoto, C. & McIlwain, H. (1966) *J. Neurochem.* **13**, 1333.
Yamamoto, C., Bak, I. J. & Kurokawa, M. (1970) *Expl Brain Res.* **11**, 360.

13. Bodily Metabolites and Drugs

R. RODNIGHT

Collection and treatment of specimens	294
Urine: general aspects	294
Preservation of urine specimens	294
Collection of urine specimens	295
Blood: general aspects	297
Collection and treatment of blood specimens	297
Cerebrospinal fluid	298
Experimental control in investigation of body fluids	299
Applications to clinical conditions	301
Phenylketonuria	303
Urinary phenylpyruvic acid	303
Plasma phenylalanine	304
Recommended method	304
Homocystinuria	305
Urinary homocystine	306
Hartnup disease	308
Histidinaemia	311
Histidine α-deaminase activity in skin	312
Plasma histidine	312
Argininosuccinic aciduria	313
Galactosaemia	314
Drugs in body fluids	314
Amphetamine detection in urine	316
References	318

Mutual influences between the nervous system and other parts of the animal body are manifold; their investigation through the analysis of the organs and fluids of the body is thus part of neurochemistry in its widest sense. Many aspects of research in this field are however parts of distinct disciplines, such as endocrinology, which possess their own extensive literature. Further, with the growth of clinical chemistry over the past decade there is now no shortage in the literature of detailed descriptions of methods for investigating metabolic abnormalities affecting the nervous system. The size of this literature precludes any attempt to review it comprehensively here; rather the aim of this chapter is to introduce to students of neurochemistry and clinicians interested in entering the field some of the problems involved and to illustrate them by discussing some of the available methods for studying selected disorders. Certain analytical procedures are described in detail, also for illustrative purposes. The approach is biased towards problems encountered in psychiatric rather than neurological hospitals and towards some of the commoner inborn errors of metabolism seen in hospitals for the mentally retarded and occasionally in psychiatric practice. A short section on analysis of body fluids for drugs commonly used for the treatment of nervous disorders is also included. A variety of books and reviews concerning investigations into bodily chemistry in the mentally ill and mentally subnormal are available; as examples Crome & Stern (1967), Himwich (1967; 1970), Rodnight (1968), Allan & Raine (1969), Davis (1970) and Weil-Malherbe & Szara (1971) may be consulted.

Investigations of this nature are dependent on successful co-operation between laboratory workers and medical and nursing staff in matters such as the collection of specimens, arranging of diets and making clinical observations. Co-operation is of great importance in studies of the mentally ill, since to obtain meaningful results the most careful control of experimental conditions is required. Whenever possible, it is desirable for chemist and clinician to take equal responsibility at the start of an experiment for its planning.

COLLECTION AND TREATMENT OF SPECIMENS

Urine: general aspects

The various body fluids differ in their availability and in the nature of the information given by their analysis. Urine is almost always obtainable in large quantity and may normally be collected with minimal disturbance of the subject. Its chemical composition tends to give an integrated picture of the body's metabolic activities over a period of time and metabolites occurring in very small amounts in the tissues or blood are often readily detected in urine. On the other hand urine composition is regulated by renal function and also it reflects to a greater extent than does that of blood, the diet and fluid intake of the subject.

Preservation of urine specimens

(a) *During collection.* If a urine specimen cannot be immediately sent to the laboratory, as for example when collecting pooled specimens, it is important to discourage by suitable means chemical change and bacterial growth in the specimen. If refrigeration of specimens is possible it should always be adopted, even if chemical preservatives are also used. If this is not possible, the effectiveness of various preservatives at room temperature must be considered in relation to the proposed analyses; for further discussion of these problems see Varley (1968). For most biochemical purposes and when storage at room temperature for longer than 24 h is not required, the incorporation of about 1% acetic acid and 0·5% toluene or chloroform is satisfactory. The reagents are added to the specimen container issued to the ward or to the subject, and after each addition of urine the bottle is shaken to saturate the specimen with preservative. The acid maintains the urine at pH 3–4, at which reaction many urinary constituents are more stable than they are at higher pH values. The use of strong mineral acids is rarely necessary and may lead to loss of labile constituents. Toluene and chloroform are ineffective in inhibiting bacterial growth if the specimen is already heavily infected when they are added.

In this case the employment of a powerful bacteriostatic agent, such as sodium ethylmercurithiosalicylate (Merthiolate, Eli Lilly & Co. Ltd.) may be necessary, although the possibility of mercury interfering with subsequent analyses must be borne in mind. In special circumstances other reagents may be considered: for example ethylenediaminetetraacetic acid may help to preserve constituents sensitive to heavy metals.

(b) *Pending analysis.* If a urine specimen cannot be analysed immediately it is received in the laboratory, it should be preserved at 0° or better at $-20°$. However, the stability of the constituents of urine, even at $-20°$, cannot be assured, particularly in the case of labile constituents present in very small concentrations, such as indole- and catecholamines. Even creatinine has been shown to decline on long storage at 0° (Bollard *et al.*, 1960). If delay in analysing a urine specimen is inevitable it is good practice to divide the specimen, immediately on its reception in the laboratory, into two samples and to add to one sample a known amount of the substance or substances to be subsequently determined. Any losses occurring during storage will then be readily detectable by analysis of both samples.

Stout polythene containers are widely used for routine urine collection. They are more convenient and safer than glass bottles, but if light-sensitive constituents are to be analysed it may be preferable to use amber glass bottles.

Collection of urine specimens

Single specimens. In general, single specimens are adequate only for the detection of gross metabolic abnormalities, as for example a severe aminoaciduria. A preprandial specimen, collected in the early morning on waking when urinary specific gravity is highest, should be obtained if possible. If the subject can be trusted to empty the bladder and if total volume of the specimen and time interval from the previous specimen are known, a measure may be obtained of excretion rates over that period. However, quantitative or even semi-quantitative deductions from analysis of single specimens are of limited value, even when these data are known. If the urinary excretion rate is not known analytical results from single specimens should always be related to other urinary constituents rather than urinary volume; for instance individual amino-acids are conveniently related to total nitrogen, creatinine or amino-nitrogen in the specimen. For further comments on the employment of creatinine as standard see below.

Pooled specimens. The many factors that influence urine composition make it advisable, whenever possible, to minimize variations in renal excretion occurring over short time periods by analysing urine collected continuously for periods of 24 h or longer. In arranging a 24 h urine specimen it is usual to request the subject to discard the first morning specimen and then to pool all the urine obtained for the next 24 h, including the first specimen on the following day, in a suitable container containing preservative. This timing ensures that the completed specimen reaches the laboratory with the minimum of delay.

Accuracy of pooled samples. In the case of mentally disturbed subjects collection of urine over periods of time may be impossible owing to incontinence, or the completeness of the specimen may be in doubt because of the subject's inability to co-operate. In these circumstances it is a common practice to estimate the accuracy of 24 h urine collections by measuring urinary creatinine, the daily excretion of which has often been considered to be a constant for any individual and independent of urinary volume, physical activity or average dietary variation (Beard, 1943; Peters & Van Slyke, 1946). The expected value is usually given as 20 mg of creatinine/kg body wt/24 h; when the daily output falls below this value the accuracy of a 24 h specimen is suspected. Recent work, however, has shown this assumption to be misleading. In a group of normal subjects Vestergaard & Leverett (1958) found that in a number of individuals creatinine output may vary greatly from day to day. Thus it was observed that 5 out of 18 subjects had a coefficient of variation in creatinine excretion of 10% or greater, and only in 5 other subjects was the variation of the means less than $\pm 4\%$. Over shorter periods (2–6 h) much greater variability was found. Several workers have since shown that variability in creatinine excretion may be even greater in mentally disturbed subjects (Pscheidt *et al.*, 1966; Chattaway *et al.*, 1969; Price, 1971). In the study of Price completeness of collection was independently evaluated in normal and schizophrenic subjects by determining the recovery of a pharmacologically inert and rapidly cleared compound (protocatechuic acid) administered at the onset of the collection period. Several schizophrenic subjects gave weight-corrected creatinine excretion rates in the urine which were statistically strikingly above or below the means for the normal subjects, despite satisfactory recoveries of protocatechuic acid. The measurement of creatinine output as a guide to the accuracy of urine collection should therefore be used with reservation. It may be noted that urinary creatinine output is not affected by the consumption of fresh meat, but is significantly raised if meat extracts or canned meat products are added to the diet (Vestergaard & Leverett, 1958). It should be emphasised that quantitative chemical studies on severely mentally disturbed subjects can be satisfactorily carried out only in a specially equipped and staffed metabolic ward with facilities for ensuring the accurate collection of urine specimens.

Blood: general aspects

In the fasting state the level of a blood constituent at any moment reflects the balance between its release and/or its utilization by the tissues, and its excretion. Blood analyses can therefore give a more immediate picture of chemical change in the tissues than can urine analysis, for the composition of blood tends to reflect tissue activities at the time of its withdrawal. Blood, moreover, contains more substances directly participating in tissue metabolism and a lower concentration of excretory products than does urine. Against these factors must be set the influence of homeostatic mechanisms tending to maintain the internal environment constant and thus diminishing the chance of detecting by blood analysis abnormal tissue metabolism. The additional analytical problems involved in blood analysis such as the presence of proteins and cellular elements, must also be considered.

Blood may be regarded as a tissue as well as a body fluid and blood cells afford means of studying enzymic activities in cellular material from human subjects (see, for example, Yunis, 1969). There are now numerous examples of this approach in the study of inherited disorders of metabolism, some of which are mentioned later in this chapter.

Collection and treatment of blood specimens

Technique of venipuncture. Techniques for withdrawing venous blood from the medial cubital or other cutaneous veins of the arm are described in detail elsewhere (e.g. Varley, 1968).

For some subjects, and particularly for the mentally ill, venipuncture may prove an unduly disturbing experience. In research investigations, therefore, where no immediate benefit to the patient is expected to result, the ethical aspects of the venipuncture procedure must be considered; advice from the appropriate ethical committee may be needed. It is further possible that the apprehension associated with venipuncture may be reflected in changes in blood composition. This is most likely to apply to hormonal constituents, the blood concentrations of which are well known to be sensitive to environmental stress. Adrenaline release was probably the explanation of the observations of Coppen & Mezey (1960) who reported that venipuncture caused an 11% increase in basal metabolic rate in a group of hospital patients. It may be advisable for clinician and chemist to discuss possbile means of reducing the emotional disturbance involved, such as: (a) the most suitable person to actually perform the venipuncture, and environmental factors such as the subject's surroundings and the presence of other personnel or patients; (b) habituation of the subject by withdrawal of blood specimens on successive days (analysis of these may give some indication of the extent to which stress is affecting the results); and (c) the use of a local anaesthetic to abolish the pain involved in introducing the needle through the skin.

Containers for blood specimens. Containers should be chemically clean and dry, and when necessary should contain the correct amount of preservative or anticoagulant. Containers of plastic or glass coated with a silicone preparation should be used if preservation of the integrity of cellular elements, particularly platelets, is important.

Prevention of haemolysis. The use of dry apparatus and the correct amount of anticoagulant are important factors. Also blood should not be forcibly ejected from syringe needles; it is preferable to dismount the needle before transferring the specimen to the container.

Anticoagulants. The most widely used anticoagulant is heparin (0·2 mg/ml of blood), which is satisfactory for most purposes. Platelet clumping, however, may be less marked in blood collected with sodium citrate (5 mg/ml of blood) as anticoagulant than with heparin. EDTA is also used; 0·1 vol. of a 1 % solution of the disodium salt in 0·9 % NaCl is effective. A method for collecting unclotted blood without the use of anticoagulants is as follows (Chandler, 1960). The blood is allowed to flow from a silicone-coated hypodermic needle inserted in the vein through an attached length of polyvinyl chloride tubing; after some 2 ml of blood has run out of the farther end of the tubing, the latter is closed in two positions by screw clamps and the enclosed blood, which has not been in contact with air, fails to clot for some hours.

Sodium fluoride (1 mg/ml of blood) inhibits glycolysis and preserves blood glucose for two or three days. Its use is important in studies of carbohydrate metabolism.

Preparation for analysis. Blood should not be stored, even at 4°, for longer than strictly necessary before separation of cells or preliminary extractions are carried out. Excessive centrifugal fields can cause haemolysis, as does storage below 0°. Many of the tissue-extracting agents described in Chapter 1 are applicable to blood.

Cerebrospinal fluid

The close proximity of the cerebrospinal fluid to the brain makes its analysis important in relation to cerebral function. Studies of its composition have been most extensive in neurological conditions, where changes are often of diagnostic importance. Pathological changes in cerebrospinal fluid composition in disease are reviewed by Prockop (1973). For valuable discussion of the significance of cerebral metabolites in cerebrospinal fluid in relation to disease see Moir *et al.* (1970).

The withdrawal of cerebrospinal fluid requires full clinical facilities; for descriptions of technique see Lups & Haan (1954); DeJong (1958). The quantity obtainable varies, but if pressure is normal it will usually be not greater than 15 ml. A major problem involved in conducting investigations of cerebrospinal fluid composition in abnormal conditions is that of obtaining normal specimens for comparison. For some purposes, spinal fluid withdrawn from subjects with suspected organic or vascular disease of the brain as a preliminary to air encephalography may be used. Cerebrospinal fluid from these sources cannot be considered as normal, but enables comparison to be made with a radically different population, as, for example, subjects suffering psychosis.

A number of factors other than disease processes are known to affect the composition of the cerebrospinal fluid. Amongst these are physical activity, age, sex and transport processes between the cerebrospinal fluid and blood stream. Physical activity is important because cerebrospinal fluid drawn from the lumbar region, which is the normal site for sampling, differs in composition from the cisternal and ventricular fluids. The gradient is due, in part at least, to differences in the site of origin in the central nervous system for metabolites appearing in the cerebrospinal fluid (Post *et al.*, 1973a; Garelis & Sourkes, 1973). Exercise has been shown to increase the rate of mixing of fluid from the upper and lower regions of the cord and may therefore alter the composition of the lumbar fluid (Fotherby *et al.*, 1963). In another study the concentration of 5-hydroxyindolyl acetic acid and homovanillic acid in cerebrospinal fluid from a group of depressed patients was significantly raised as the result of simulated manic activity on the part of the patients (Post *et al.*, 1973b). With regard to age a tendency for the concentration of acid monoamine metabolites in cerebrospinal fluid to increase with this factor was noted by Bowers & Gerbode (1968) and Gottfries *et al.* (1971). Males tend to have lower concentrations in the cerebrospinal fluid of these substances than females (Sjostrom & Roos, 1972). The rate at which metabolites are transported from the cerebrospinal fluid to the blood stream is another important variable affecting its composition. Administration of the drug probenecid to man blocks the transport process for a number of metabolites, such as those involved in amine metabolism, thus permitting their accumulation in the cerebrospinal fluid. The rate of accumulation under these conditions is sometimes taken as a measure of the rate at which the particular metabolite is being formed by cerebral metabolism. For description and application of this technique see Sjostrom & Roos (1972) and Goodwin *et al.* (1973).

EXPERIMENTAL CONTROL IN INVESTIGATION OF BODY FLUIDS

The problem of controlling experimental variables in investigations of body fluids in man is very complex, but is clearly one of the utmost importance if valid conclusions are subsequently to be drawn.

In mental illness particularly, the biological characteristics distinguishing normal from abnormal may be slight and readily obscured by uncontrolled variations in metabolic status: conversely, and perhaps more frequently, apparent metabolic differences between normal and abnormal subjects may in fact be due to differences in diet, body weight, nutritional status, physical activity, or simply in age or sex. It has already been pointed out that rigorous experimental control is possible only in a hospital ward, equipped with facilities and trained staff for metabolic investigations, and where normal and abnormal subjects live under precisely the same regime. Whilst such conditions are rarely attainable in practice they should always be considered as a desirable objective. With more limited facilities the investigator must apply the maximum control possible, bearing in mind that a small well-controlled investigation is often of more value than uncontrolled observations on many subjects. Full details of the extent of experimental control achieved should always be stated when reporting results.

The following section considers some of the main variables involved in body fluid investigations. Some general principles involved in experimentation on man are discussed by Beecher (1958); for statistical aspects of experimental design see, for example Snedecor (1956) and Maxwell (1958).

Nutritional status of subjects. As a result of disturbed eating habits, mentally ill subjects often suffer from nutritional deficiencies that are not always clinically apparent but may be sufficient to influence the chemical constitution of the body fluids. This applies both to chronically ill patients of long stay in hospital and to the acutely ill new admission in whom a long period of emotional suffering may have profoundly affected food intake. The re-establishment of a normal nutritional status after a change to an adequately supplemented and supervised diet, may be a matter of weeks or even months of attentive nursing (Horwitt, 1963).

Diet. Variation in dietary intake can be readily detected in the body fluids, especially in urine. For example many, if not the great majority, of the numerous phenolic derivatives commonly detectable in urine by chromatography (Smith *et al.*, 1959) can be shown to be derived from dietary constituents (Armstrong & Shaw, 1956; Smith, 1960). Other examples include the influence of items of diet on urinary derivatives of the indole and catechol amines (Udenfriend *et al.*, 1959) and urinary purines (Bollard *et al.*, 1960). The most satisfactory way to control this factor is to ensure that all subjects receive the same diet. This may be a normal, a simplified or a synthetic diet. A normal diet is free from the possible emotional consequences of monotony, but unless supervision is very strict it may be difficult to avoid the omission of minor items by a proportion of the subjects (for an example of this in relation to urinary phenols see Mann & LaBrosse, 1959). A simplified diet consists of fresh meat protein, carbohydrate and fat from defined sources together with supplements of essential factors, but omits all preserved foodstuffs, beverages, fruits and vegetables. Urinary excretion patterns for a variety of substances are greatly simplified by this diet. Synthetic diets of wholly defined composition are available commercially (e.g. Complan, Glaxo

Ltd.) but are monotonous and consequently limited in application; for an early example of their employment in the study of mental illness see Folin (1904). In circumstances where it proves impossible to maintain the subjects on the same diet, they or their attendants should be requested to record in detail the various items of their diet. In the event of chemical differences being detected it will then be necessary to evaluate these through more stringent dietary control.

A number of substances found in the body fluids are derived from the action of the intestinal flora on dietary constituents. Thus apart from a direct influence on body fluid composition the diet may exert an indirect influence by affecting the intestinal climate and the predominant type of organism present.

In a well-controlled investigation the subjects should receive no drugs for at least a week before making the initial observations.

Intestinal function. As indicated above, products of bacterial metabolism in the intestinal tract are found in the body fluids. The extent to which this occurs will depend on many factors which probably include, as well as the diet mentioned above, intestinal motility, mucous secretion and absorption, and possibly gastric acidity. Consideration of the contribution of bacterial metabolism is particularly important in studies of nervous and mental disorder, as in these conditions disturbed intestinal function is frequently observed. For example urinary indole excretion is particularly influenced by bacterial metabolism in the gut and to a great extent this accounts for the raised excretion of certain indoles often noted in mentally ill subjects (Rodnight, 1961; 1968). The intestinal origin of a body fluid constituent can usually be confirmed by administering an antibiotic: for instance 1–2 g of Terramycin (Pfizer Ltd) daily for 72 h normally will diminish bacterial action appreciably but more prolonged and extensive treatment is required to approach sterility of the intestinal contents. Abnormal amounts of an intestinal product in the body fluids may indicate an abnormality in tissue metabolism as does the indicanuria of Hartnup disease (Jepson, 1966; Rodnight, 1968).

Control subjects. As far as possible these should be matched for age, sex and body weight with the abnormal population. For example, physical characteristics, especially body weight and sex have been observed to affect purine excretion (Bollard *et al.*, 1960). Control subjects are always necessary except perhaps in the investigation of a periodic illness where the subject in remission may be used as his or her own control.

Applications to clinical conditions

Screening for metabolic disorders. Screening procedures are valuable for surveying large populations for the occurrence of individuals exhibiting abnormality in body fluid composition. They were first developed for studies of aminoaciduria (which comprise the largest group of inborn errors involving the nervous system), but now have a wider application, for example in the field of disorders of carbohydrate metabolism. They give only qualitative or at best semi-quantitative information and do no more than indicate subjects worth investigating further by more precise techniques. Screening procedures have a particularly important application in surveying neonatal populations since many inherited metabolic disorders respond to dietary treatment only if this is started in the first months of life. Most of the available methods employ chromatography or electrophoresis of urine or of blood extracts and for these purposes thin layers of cellulose powder or silica gel have largely replaced paper as supporting medium. The two-dimensional procedures give more information; many of them combine solvent chromatography with high voltage electrophoresis since with this technique resolution of amino acids in body fluids is superior to that attainable with different solvent pairs. Examples of procedures capable of resolving nearly all the important amino acids in urine and plasma are described by Troughton et al. (1966) and Walker & Bark (1966); a very rapid micro-procedure of similar design but with less resolving power is due to Samuels (1966). Other techniques are described by Crome & Stern (1967). An elegant one-dimensional procedure for plasma amino acids is described by Ireland & Read (1972). Blood (20 μl) is collected on filter paper and dried as in the Guthrie procedure (see below) for detecting hyper-phenylalaninemia by microbiological assay. The amino acids are eluted from the dried spot with 70% ethanol, the extract concentrated and then fractionated on thin layers of cellulose using n-butanol-acetone-acetic acid-water (33:35:10:20) as solvent. Some of the problems involved in setting up and interpreting screening procedures are discussed by Raine (1969) and Seakins et al. (1972).

Amino acid analysis. Disorders of amino acid metabolism constitute a significant proportion of the known metabolic diseases affecting the nervous system (Efron, 1965; Nyhan, 1967; Mozziconacci et al., 1968; Woolf, 1972). New cases are usually detected by a screening procedure as indicated above, but will then require further investigation by quantitative analysis. Quantitative techniques are also important in the study of the heterozygote carriers of these conditions. With the increasing availability of a variety of commercial analyzers and great improvements in speed and resolution, automatic amino acid analysis on ion exchange columns has become a routine procedure for many laboratories. Detailed consideration of this area would therefore be superfluous here. Most manufacturers give information on the applicability of their equipment to the analysis of body fluids; in studying specific disorders elution programmes can often be varied to permit a good resolution of the amino acid under study at the expense of resolving others. Elution buffers of lithium instead of sodium citrate are often recommended as they are superior in resolving glutamine and asparagine, which are high in plasma, from serine (Benson et al., 1967; Peters et al., 1968; Kedenburg, 1971). For further information see, for example, Efron (1966a), Perry (1967), Samyn et al., (1970), Ertingshausen & Adler (1970) and Scott (1972).

PHENYLKETONURIA

The disordered phenylalanine metabolism of this condition involves an abnormal urinary excretion of phenylpyruvic acid and of other related phenols, and a high phenylalanine content in the blood, where levels some fifty times the normal 10 μg/ml may be found (Knox, 1966, Crome & Stern, 1967; Woolf, 1967, 1972). Diagnosis is made by detection of excess phenylpyruvic acid or o-hydroxyphenylacetic acid in the urine and by determination of plasma phenylalanine. Plasma phenylalanine determinations are of value not only in diagnosis, but also (a) in controlling treatment of the condition by a low phenylalanine diet, which when instituted early in life, gives good results (see McBean & Stephenson, 1968; Fuller & Schuman, 1969; Woolf, 1972 for recent assessment) and (b) in the detection of the heterozygous carriers of the condition. Plasma phenylalanine concentrations in the latter subjects are only slightly higher than normal, but on administration of an oral dose of phenylalanine the concentrations increase to values about twice as high as observed in control subjects similarly dosed.

Urinary phenylpyruvic acid

The original ferric chloride procedure is performed by acidifying 5 ml of fresh urine with not more than 0·25 ml of 5 N-H_2SO_4 and adding 1–2 drops of 10% (w/v) $FeCl_3$. A green colour fading within about 15 min indicates the presence of phenylpyruvic acid; the sensitivity of the reaction depends on the concentrations of H^+ and Fe^{3+} ions added. The more commonly used procedure, the Phenistix test, is based on the work of Rupe & Free (1959). The Phenistix (Ames Co., Indiana, U.S.A.) consists of a piece of thick filter paper impregnated with ferric ammonium sulphate, Mg^{2+} and an organic buffer at pH 2·3. The paper is dipped in the urine specimen or pressed against the wet diaper from the suspected child and 30 s later any colour developed is compared with a scale of colours provided by the manufacturers; a green colour indicates the possible occurrence of phenylpyruvic acid in the urine. Neither of these tests is specific for phenylketonuria; in histidinaemia for example, the urine may give a green colour with $FeCl_3$ owing to the presence of imidazole compounds. The specificity of the $FeCl_3$ test is discussed by Crome & Stern (1967).

Phenylpyruvic acid is unstable in urine, making it uncertain of detection in specimens that are not fresh. Woolf (1967) therefore devised a screening procedure for another more stable excretion product in the condition, *o*-hydroxyphenylacetic acid. A drop of urine is dried on filter paper (enabling it to be sent through the post if necessary), which is subsequently stapled to another sheet of chromatography paper and examined for phenols by one-dimensional chromatography. The procedure also detects the presence of excess *p*-hydroxyphenylacetic acid (raised in tyrosyluria) and the imidazole derivatives excreted in histidinaemia.

Plasma phenylalanine

Several methods are available, but not all are suitable for routine determinations in studying phenylketonuria. Paper or thin layer chromatography has the advantage of high specificity and of giving information on other amino acids but lacks precision; it will readily detect the severe untreated case, but may give equivocal results in others. Full amino acid analysis of plasma by ion exchange chromatography is the preferred procedure in confirming a tentative diagnosis by urine analysis, but is unnecessary for the control of treatment. For this latter purpose and for plasma phenylalanine determinations in tolerance tests used in detecting heterozygotes three methods are available; the L-amino acid oxidase method of La Du & Michael (1960), the fluorometric method of McCaman & Robins (1962) and the microbiological assay of Guthrie & Susi (1963). Of these the fluorometric procedure combines great sensitivity with simplicity and is recommended for most purposes. It is a little less specific than the L-amino acid oxidase method, but requires less blood. The microbiological assay of Guthrie is also sensitive and simple to use once established, but is less accurate; it has been widely applied in the screening of new born infants, particularly in the U.S.A. It was also recommended for this purpose by the U.K. Medical Research Council Working Party on Phenylketonuria (1968). However in a large scale comparative study of the Guthrie procedure with an automated version of the fluorometric method Holton (1972) considered the latter superior.

Recommended method

The method depends on the enhancement given by certain dipeptides of the fluorescence of a phenylalanine-ninhydrin product. It needs as apparatus a fluorometer, 12×75 mm test tubes, 10 μl constriction pipettes and 0·2 ml micropipettes. The reagents used are (*a*) 0·3 M-sodium succinate buffer pH 5·8; (*b*) 30 mM-ninhydrin (A.R. grade); (*c*) 5 mM-L-leucyl-L-alanine; (*d*) 0·3 and 0·6 M-trichloroacetic acid; (*e*) copper reagent: 1·6 g of Na_2CO_3, 65 mg of sodium potassium tartrate, 60 mg of $CuSO_4.5H_2O$ per litre of water (see original paper for comment on preparation); (*f*) phenylalanine standards: 0·05 mM and 0·5 mM in 0·3 N-trichloroacetic acid. The phenylalanine reagent is made up freshly for each determination by mixing (a) (b) and (c) in the proportions 10:4:2.

Procedure. Plasma (0·25 ml) is added to 0·25 ml of 0·6 M-trichloroacetic acid and the precipitated protein centrifuged out in the cold. Into 5 test tubes 0·16 ml of the phenylalanine reagent is pipetted. To 2 of these tubes 0·01 ml of plasma extract is added, to another 2 tubes 0·01 ml of high and low phenylalanine standards and to the last tube 0·01 ml of 0·3 M-trichloroacetic acid. The tubes are placed in a water bath at 60°C for 2 h, cooled and 1 ml of the copper reagent (e) is added. Fluorescence is measured after 10 min with an activation wavelength of 380 nm and emission wavelength of 470 nm. If a spectrophotofluorometer is not available an instrument equipped with filters may be used. The filters recommended by McCaman & Robins (1962) are Corning 5860 (365 nm) in the primary and a combination of Corning 4303 and 3384 (505–530 nm) in the secondary pathway. Linearity between the high and low standards may be assumed, but should be checked when the method is first established.

HOMOCYSTINURIA

In homocystinuria the affected subjects are unable to convert homocysteine to cystathionine efficiently due to a deficiency in the activity of the enzyme cystathionine synthetase (Carson *et al.*, 1963; Gerritsen & Waisman, 1964, 1966; Cusworth, 1969; Woolf, 1972). Homocysteine and homocystine therefore appear in the urine and the plasma; plasma methionine concentrations may be also raised. In some cases the block appears complete (e.g. Shih & Efron, 1970): in others it is partial and in a proportion of cases reactivation of the pathway by feeding pyridoxine in high dosage has proved possible (Kaeser *et al.*, 1969; Mudd *et al.*, 1970). The clinical syndrome is complex; bony and ocular abnormalities are usual and mental retardation a common feature; depression and psychosis have also been observed (Kaeser *et al.*, 1969). In frequency of occurrence in the population the condition is second only to phenylketonuria.

Liver biopsy specimens have been used to demonstrate the absence of cystathionine synthetase activity in liver from affected subjects (Mudd *et al.*, 1967). Fibroblasts cultured from normal human skin were found to exhibit cystathionine synthetase activity, but fibroblasts derived from the skin of homocystinuric patients had low or undetectable activity (Uhlendorf & Mudd, 1968). The enzyme was also induced with phytohemoagglutinin in cultures of lymphocytes obtained from normal human blood, but was not inducible or severely deficient in lymphocytes from patients with homocystinuria (Goldstein *et al.*, 1972).

Laboratory diagnosis involves in the first instance the detection of homocystine in the urine. An electrophoretic procedure due to Bessman *et al.*, (1967) as modified by Harvey (1968) is recommended. For confirmation plasma amino acid analysis before and after an oral dose of methionine (50 mg/kg body wt) is required (Brenton *et al.*, 1965; Kaeser *et al.*, 1969).

Urinary homocystine; recommended procedure

Homocystine is oxidized to homocysteic acid with performic acid. Homocysteic and cysteic acids are separated from other amino acids by ion-exchange chromatography followed by high voltage electrophoresis. The paper is stained with a cadmium–ninhydrin reagent and the colour due to homocysteic acid eluted and measured spectrophotometrically. The procedure determines both free homocystine and any homocystine present as the mixed disulphide with cystine.

Procedure. The urine sample (5 ml) is freeze dried or concentrated to less than 1 ml on a rotary evaporator. Performic acid is prepared by addition of 1 ml of 30% hydrogen peroxide to 9 ml of 90% formic acid and allowing the mixture to stand for 1 h; to oxidise the sample 12·5 ml of the reagent at 0°C is added to the concentrated or freeze dried specimen and the mixture kept at 0°C for 4 h. Of the oxidised urine 1 ml is filtered through a column (0·5 cm × 2·5 cm) of Amberlite CG 120 (or equivalent ion exchange resin) in the H$^+$-form made up in 0·05 M-HCl. Cysteic and homocysteic acids emerge with the effluent and are washed off the column with 9 ml of water. The combined effluent and washing is freeze fried and the residue redissolved in 0·5 ml of water.

The sample is fractionated by high voltage electrophoresis at 50V/cm for 3·5 h on Whatman no. 3 MM paper (or equivalent) in 2 M-formic acid adjusted to pH 1·9 with acetic acid. As a screening procedure 25 μl is a suitable quantity to examine, but this quantity might prove too high in an extract from a homocystinuric subject. Standards in the range 5–15 μg of homocysteic acid are put up in parallel and at least one internal standard should be incorporated. Homocysteic and cysteic acids move towards the anode, the respective distances being approximately 13 cm and 20 cm from origin depending on the equipment. Normal urines frequently contain an unknown ninhydrin positive spot which migrates about 11 cm from origin, but is clearly separated from added homocysteic acid.

After electrophoresis the paper is dried in an oven at 80°C and dipped in the cadmium–ninhydrin reagent of Atfield & Morris (1961). (This is prepared by dissolving 0·2 g of cadium acetate in 20 ml of water and 4 ml of acetic acid. Acetone, 200 ml, is added and the mixture shaken until any precipitate is dissolved; finally 2 g of ninhydrin are added). After allowing the solvent to evaporate the paper is placed in the dark in an ammonia-free atmosphere for 18–20 h. The pink-coloured areas corresponding to homocysteic acid are cut out, shredded and shaken in a stoppered tube with 4 ml of methanol for 20 min in a mechanical shaker.

After filtering through sintered glass, extinctions are measured at 500 nm in 1 cm light-path.

Comment. The original procedure of Bessman *et al.* (1967) omits the ion-exchange step. This is not essential if the homocystine content of the urine is high, in which case it may also be possible to use unconcentrated urine. However, if the urinary homocystine is only moderately raised, or in assessing progress of treatment with pyridoxine when only traces of the amino acid may be present, the modified procedure of Harvey (1968) is superior. Two-dimensional paper chromatography of oxidised urine was used in the original study (Carson *et al.*, 1963) of homocystinuria, but is less satisfactory than electrophoresis because lack of a suitable solvent pair to separate homocysteic acid from cysteic acid.

A comparison of unoxidised and oxidised urine by one-dimensional chromatography in *n*-butanol-acetic acid–water (60:15:15) may in fact give as much information as two-dimensional chromatography. The appearance of a major spot near the origin in the oxidised sample would suggest a high content of homocystine or cystine in the urine. If facilities for chromatography or electrophoresis are not available the cyanide–nitroprusside reaction for SH-compounds as performed by Gerritsen & Waisman (1964) may be used as a screening procedure. This does not, however, distinguish between excess cystine and homocystine and may give equivocal results unless excretion of the abnormal amino acid is particularly high.

An ion-exchange procedure for urinary homocystine is due to Gerritsen & Waisman (1964). The urine was oxidised with performic acid as described above and then fractionated (extract equivalent to 0·5 ml of urine) on a 0·9 cm × 20 cm column of an anion exchange resin (Biorad AG 2 × 10, 140–325 mesh, Cl-form) with 0·1 M-chloroacetic acid as elutant. The column was operated at 30 ml/h and ninhydrin added to the eluate after 20 min; homocysteic acid emerged between 45 and 90 ml and cysteic acid between 65 and 90 ml, the peaks being sharp and well separated.

Another qualitative screening test for homocystinuria has been described by Rosenthal & Yaseen (1969). Diagnosis by gas chromatography has been reported (Greer & Williams, 1967). Analysis of urine for *S*-amino acids using conventional autoanalyser equipment is discussed by Perry *et al.* (1967).

Analysis of S-amino acids in plasma in homocystinuria. Problems arise because of the instability of the reduced (SH) forms of the *S*-amino acids during chromatography on columns of ion exchange resins. This is usually overcome by oxidising cysteine and homocysteine to their respective disulphides by allowing plasma extracts to stand at room temperature and neutral pH for about 4 h. Alternatively the SH-forms may be converted to their stable *S*-carboxymethyl derivatives by reaction with iodoacetate (Brigham *et al.*, 1960), but this is unnecessary unless determination of the reduced forms is required. In most standard procedures for amino acid analysis in body fluids using sodium citrate buffers (e.g. the Technicon system or those based on Hamilton, 1963; 1968) homocystine emerges between the phenylalanine and ammonia peaks; in normal plasma there are no other peaks in this area, but in urine the pattern is more complex and careful indentification by internal standards may be necessary. Plasma from homocystinurics often contains the mixed disulphide of cysteine and homocysteine; in standard systems this normally appears between leucine and isoleucine. Perry (1967) describes a modification of a standard procedure which is capable of resolving methionine, methionine sulphoxide, cystine, homocystine and the mixed sulphide (see also *Perry et al.*, 1968).

In determining homocystine and cystine in plasma it is important to deproteinize the specimens rapidly after venipuncture since delay can result in the binding of the amino acids to plasma proteins.

HARTNUP DISEASE

This recessive hereditary condition is characterized by a gross and constant renal tubular aminoaciduria, abnormalities in urinary indole excretion and by intermittent clinical manifestations often resembling pellagra (Rodnight, 1961; Jepson, 1966; Milne, 1969; Woolf, 1972). Non-specific psychiatric complaints and cerebellar ataxia are commonly seen. The clinical picture is apparently caused by a defect in the transport of tryptophan in the small intestine and renal tubules. Dietary tryptophan is therefore poorly absorbed and less is available for conversion to nicotinamide in the liver; unabsorbed tryptophan enters the large intestine where it is degraded to indolic products which are absorbed by the blood stream, detoxicated in the liver and excreted in the urine. Affected subjects therefore convert an oral dose (70 mg/kg body wt) of tryptophan less efficiently to kynurenine and nicotinamide than do normal subjects, while proportionally more of the amino acid is converted to indolic products in the urine (Wong & Pillai, 1966; Milne, 1969). By contrast intravenously administered tryptophan is normally metabolized (De Laey, 1964; Wong & Pillai, 1966). The transport defect also involves a wide range of other amino acids thus explaining the gross aminoaciduria. Plasma amino acids tend to be lower than normal; inefficient absorption of amino acids and excessive loss in the urine probably accounts for the smaller stature achieved by Hartnup subjects (Milne, 1969). It has not proved possible to detect the transport defect in cultured fibroblasts from Hartnup patients (Groth & Rosenberg, 1972).

Hartnup disease is diagnosed by demonstration of a typical pattern of amino acids in the urine. The pattern is more significant than the total amino acid excretion. The 24 h excretion of the following amino acids is usually increased five- to tenfold: alanine, asparagine, glutamine, histidine, isoleucine, leucine, phenylalanine, serine, threonine, tryptophan, tyrosine and valine; glycine, aspartate and glutamate may also be raised, but there is no increase in the excretion of proline, hydroxyproline, methionine or arginine. The latter point is a useful diagnostic pointer which distinguishes the aminoaciduria in this condition from the non-specific aminoaciduria of galactosaemia or Wilson's Disease, where proline is raised. Typical values for two Hartnup patients are given in Table 13.1. Procedures for examining urine specimens for the characteristic amino acid pattern and for studying the indoluria of the condition are given below.

Table 13.1 Excretion of amino acids in Hartnup disease

Amino acid	Urinary output mg/24 h)	
	Subject A	Subject B
Alanine	650	500
Aspara gine	500	330
Aspartic acid	Not detected	160
Glutamic acid	430	40
Glutamine	3200	2500
Glycine	650	1000
Histidine	650	500
Leucine } Isoleucine	550	330
Lysine	430	500
Phenylalanine	210	250
Serine	1300	1000
Threonine	500	500
Tryptophan	40	40
Tyrosine	500	330
Valine	650	330
Total amino acid N	1290	1099

From Hersov & Rodnight (1960). The subjects were sisters, aged 29 (A) and 27 (B).

Urinary amino acids. For screening purposes chromatography in one dimension of a volume of urine equivalent to 2 seconds excretion is adequate. The urine is applied to paper or thin layers of cellulose and developed with n-butanol–acetic acid–water (12:3:5, by vol) for 4–6 h. The paper or plate is dried and the amino acids revealed by spraying with 0·2% ninhydrin in acetone and heating at 80°C for 3 min. Chromatograms from normal urine show a few purple or red spots with R_F values below 0·5 and only traces of material with higher R_F, whereas urine from cases of Hartnup disease yields chromatograms grossly overloaded with ninhydrin-positive material. To confirm a tentative diagnosis of Hartnup disease made in this way it is necessary to investigate the pattern of the urinary amino acids by semiquantitative two-dimensional chromatography or quantitative analysis in an amino acid analyzer. The two dimensional techniques mentioned earlier (Troughton et al., 1966; Walker & Bark, 1966) are suitable; for discussion of procedures for automatic analysis of urinary amino acids see Hamilton (1968).

Urinary indoles. In addition to tryptophanuria the indoluria of Hartnup's disease is characterized by a raised excretion of indican, and of indolylacetic acid and its conjugates (Jepson, 1968). These urinary indoles are, however, neither consistently raised in the condition nor are they sufficiently characteristic to be of much diagnostic value. Their origin in the gut through bacterial action has been demonstrated by treatment of subjects with antibiotics such as neomycin (Jepson, 1968).

The indoluria has been studied by two-dimensional chromatography on paper or thin layers of cellulose or silica (Jepson, 1960; Hansen & Crawford, 1966). Fresh untreated urine is developed with the solvent pair (*a*) *n*-propanol–ammonia solution (sp. gr. 0·88)–water (16:1:3, by vol); and (*b*) *n*-butanol–acetic acid–water (12:3:5, by vol). As an alternative solvent for the first run, *iso*propanol–ammonia solution (sp. gr. 0·88)–water (20:1:2, by vol) may be used. For paper chromatography 0·005% of a 24 h specimen should be examined, but the thin layer techniques are considerably more sensitive. Indoles are located by dipping or spraying the dried paper or plate with Ehrlich's reagent (2 g of *p*-dimethylaminobenzaldehyde dissolved in 15 ml of conc. HCl and made to 100 ml with pure acetone). Indican appears as a brown spot, urea as a large yellow spot and the remaining indoles as purple spots; approximate R_F values and other characteristics are given in Table 13.2.

Table 13.2 R_F values (approximate) and staining characteristics of urinary indoles which are increased in Hartnup disease

Compound	R_F in solvent		Colour and time of reaction with Ehrlich's reagent
	(*a*)	(*b*)	
Tryptophan	0·44	0·37	Purple; 5–10 min
Indican	0·58	0·48	Brown; 2–5 min
Indoxylglucuronide	0·29	0·28	Purple; develops immediately and fades within 1 min
Indolylacetic acid	0·43	0·93	Purple; 2–5 min
Indolylacetylglutamine	0·43	0·56	Purple; 5–10 min
Urea (data given for comparison)	0·43	0·41	Yellow; 5 min

For details of solvents, see text.

A more sensitive reagent is based on *p*-dimethylaminocinnamaldehyde. Of an 0·2% solution of the substance in ethanol, 4 ml is acidified with 1 ml of conc. HCl just before application to the paper or thin layer plate.

Tryptophan gives a grey-violet colour, indican a brown colour and indolylacetic acid a brown violet colour. As little as 50 ng of the latter can be detected on thin layer plates of silica gel; for further information Kaldewey (1968).

The quantitative output of indican in the condition may be determined by the method of Sharlit (1933; see also Rodnight & McIlwain, 1955). This is not of diagnostic value, but subjects excreting more than 200 mg of indican/24 h should be examined for the typical amino-aciduria of Hartnup disease. Severe indicanuria is occasionally seen in other conditions, especially in severe constipation (Berlin, 1957). Quantitative determination of urinary indoles after oral loading with tryptophan has contributed to the elucidation of the nature of the biochemical defect in the condition (Jepson, 1966; Wong & Pillai, 1966; Milne, 1969). A dose of L-tryptophan of 20 mg/kg body wt to affected subjects results in the elevation of urinary indican, indolylacetic acid and indolylacryloylglycine for 24 h for longer; in normal subjects any rise in indole excretion that may occur is short-lived.

HISTIDINAEMIA

This inherited disorder of histidine metabolism (La Du, 1966; Seakins & Holton, 1969) is characterized by a deficiency of the enzyme L-histidine ammonia-lyase (histidine α-deaminase, histidase, EC 4.3.1.3) which converts histidine to urocanic acid. As a result plasma histidine concentrations are high and the urine contains as degradation products imidazole pyruvic and lactic acids. The symptoms, which appear in childhood, are generally milder than those seen in phenylketonuria; in many cases the block appears only partial (Woody *et al.*, 1965). The inheritance by autosomal recessive gene and the heterozygote carriers may be detected by histidine loading tests.

The characteristic biochemical abnormality in the condition is the marked elevation in plasma histidine concentration. A high urinary histidine is of little diagnostic value unless accompanied by the presence of excess imidazole pyruvic and lactic acids. As indicated above, urine containing the latter compounds often gives a green colour with $FeCl_3$ or in the Phenistix test; in such screening tests confusion with phenylketonuria may therefore result. The imidazole acids may be detected on paper or thin layer chromatograms with diazolised sulphanilic acid (Pauly's reagent) (e.g. Woody *et al.*, 1965) or imidazole pyruvic acid may be determined quantitatively as its enol–borate complex (Cain & Holton, 1968).

The diagnosis of histidinaemia may often be confirmed by determination of the activity of histidine α-deaminase in the samples of the stratum corneum from skin where it normally occurs; also urocanic acid is normally found in skin extracts (or in sweat) but is typically absent or very low in the disorder. Techniques for determining plasma histidine concentration and histidine α-deaminase activity and urocanic acid in skin are considered below.

Histidine α-deaminase activity and urocanic acid content of skin: recommended procedure

The procedure recommended is due to Whitfield & Shepherd (1970). Reagents required are: (a) 0·08 M-sodium pyrophosphate (pH 9·2); (b) 0·24 M-L-histidine (pH 9·2); (c) 0·08 M-reduced glutathione (GSH) in 0·08 M-sodium pyrophosphate; (d) 7% (v/v) perchloric acid; (e) standard urocanic acid dihydrate (30 μM in 0·041 M-pyrophosphate buffer, pH 9·2); (f) 10 M-HCl and Zn powder. An ultraviolet spectrophotometer is also needed.

Procedure. About 20 mg of skin is removed from the thumb and finger pads with curved scissors. Sampling instructions are given in detail by Whitfield & Shepherd (1970); provided the stratum corneum only is taken the procedure is painless. The larger pieces of skin are cut up with scissors and weighed. They are then transferred to a homogenizer tube and dispersed in 1·8 ml of ice cold water. The clear supernatant is transferred to a 20 ml volumetric flask and the deposit washed with 1 ml ice cold water, the washing being added to the flask. The deposit of insoluble skin is then suspended in cold water in a volume of no more than 0·8 ml.

To the combined supernatants in the volumetric flask, 1 ml of 10 M-HCl is added and the extract made up to 20 ml. The extinction of the extract is then read at 277 nm. Then, to a 2 ml sample of the extract approximately 30 mg of zinc powder is added and the mixture incubated at 37°C for 1 h. After centrifuging the extinction of the supernatant is read again at 277 nm. Nascent hydrogen from the zinc and HCl reduces the urocanic acid to imidazole propionic acid, which does not absorb at 277 nm. The urocanic acid content is given by the difference between the two readings as compared with the reading given by the standard solution of urocanic acid.

To determine skin histidine deaminase activity, 0·2 ml of pyrophosphate buffer (a) is pipetted into four test tubes. To tubes 1 and 3 are added 20 μl of 0·24 M-histidine; to tubes 2 and 4 are added 20 μl of water. To all tubes are then added 20 μl of GSH solution, followed by 0·15 ml of skin suspension. Immediately following addition of the tissue 0·8 ml of 7% perchloric acid is added to tubes 2 and 4 (zero time controls). Tubes 1 and 3 are incubated at 37°C for 75 min with occasional mixing. At the end of this period all tubes are centrifuged and extinction of the supernatant read at 277 nm. Enzyme activity is calculated as μmol of urocanic acid formed/g of skin/h, making appropriate correction for the control.

Plasma histidine

There are no particular problems in determining plasma histidine by automatic amino acid analysis, as the amino acid is well separated from other peaks. If an amino acid analyser is not available the procedure of Baldridge & Greenberg (1963) is available. This is derived from the L-amino acid oxidase procedure of La Du & Michael (1960) for determining phenylalanine in plasma. Histidine reacts with the enzyme in the presence of borate ions at pH 7·8 to yield an enol-borate complex of imidazole pyruvic acid, the absorption of which is measured at 292 nm. There is little oxidation of histidine by L-amino acid oxidase at pH 6·5, in contrast to the rapid reaction of the enzyme with phenylalanine and tyrosine at this pH. By observing oxidation of plasma extracts at both pH values a correction factor for the phenylalanine and tyrosine content of the sample could be derived. The method gave excellent agreement with results for histidine obtained with the amino acid analyser.

ARGININOSUCCINIC ACIDURIA AND RELATED CONDITIONS

This is the commonest of a rare group of hereditary disorders involving enzymes concerned in the production of urea from ammonia (Efron, 1966b; Moser et al., 1967, Crome & Stein, 1967; Woolf, 1972). The subjects are usually severely retarded and exhibit a variety of neurological symptoms including generalised convulsions; however, cases with milder symptomatology have been described (Armstrong et al., 1964). The symptoms resemble those seen in ammonia intoxication and hyperammonaemia has been observed in affected subjects post-prandially or after an ammonia load (Moser et al., 1967). The subjects have a decreased ability to convert argininosuccinic acid to arginine and fumaric acid; consequently argininosuccinic acid appears in the urine, where the excretion rate may be as high as 3–9 g/24 h, depending on protein intake. Typically the subjects excrete 14–20% of the urinary nitrogen as the conjugated amino acid. Argininosuccinic acid also occurs in the plasma and in higher concentrations in the cerebrospinal fluid. Plasma citrulline concentration is raised. It is of interest to note that blood urea concentrations are normal in the condition, suggesting that the enzyme block is partial. However, the activity of argininosuccinase in erythrocytes obtained from affected subjects was either absent or very low compared to normal values (Tomlinson & Westhall, 1964); no difference was found between normal and abnormal subjects in arginase activity in erythrocytes in this study.

Argininosuccinic aciduria is detected by chromatography or electrophoresis of urine. Fresh specimens should be analysed as the compound readily forms stable anhydrides in acid solution (Westhall, 1960). Some workers prefer to complete the process of anhydride formation by boiling the acidified urine for 2·5 h or letting it stand at pH2 for 48 h (see Cusworth & Westhall, 1961) before analysis. A satisfactory screening procedure (Moser et al., 1967) utilizes high voltage electrophoresis.

About 10 μl of urine is subjected to a gradient of 100 V/cm for 45 min at pH 5·4; free argininosuccinic acid moves towards the anode, while the anhydrides remain close to the origin; all the compounds react with ninhydrin. Column chromatography is available for quantitative determination; on a 0·9 cm × 50 cm column of Dowex 50 operated at pH 4·19 and 30°C, free argininosuccinic acid emerged at 25 ml of eluate and the two anhydrides at 50 and 60 ml (Efron, 1966).

A gas chromatographic method for analysis of argininosuccinic acid in urine has been described (Sprinkle *et al.*, 1969). The amino acid is precipitated from urine by $Ba(OH)_2$, cleaved at alkaline pH to aspartic acid and ornithine and the latter determined in the chromatograph as their *N*-trifluoro-acetyl-*N*-butyl esters.

GALACTOSAEMIA

In galactosaemia (Isselbacher, 1966; Wool, 1972; Kalckar *et al.*, 1973) the enzyme galactose-l-phosphate uridyl transferase (EC2.7.7.12) which converts galactose-l-phosphate to glucose-l-phosphate, is inactive or defective. Consequently galactose and its phosphate accumulate in body fluids and tissues. The condition is inborn and appears early in life. In untreated subjects widespread pathological changes develop including liver and kidney damage (and consequently aminoaciduria) and eventually severe mental retardation. This can be minimized by promptly instituting a galactose-free diet.

Detection of galactose in the body fluids involves its differentiation from glucose, since both sugars give the usual reducing tests. In suspected cases the use of glucose oxidase will indicate whether the mellituria is due to glucose or some other sugar; for final identification chromatographic methods are used.

Treatment has been found most effective when the galactose-free diet is instituted at birth. At this stage galactose cannot be detected in the body fluids, but diagnosis of infants suspected of galactosaemia from previous family history may be made by assay of galactose 1-phosphate uridyl transferase in erythrocytes from cord blood; the enzyme is absent or greatly diminished in galactosaemia. A simple spectrophotometric method for its determination in erythrocytes is described by Ellis & Goldberg (1969); other procedures are reviewed by Beutler (1969).

DRUGS IN BODY FLUIDS

The greatly increased incidence of drug abuse over the past few years together with the growth in drug therapy for mental disease, have stimulated the development of a whole new range of rapid, sensitive and reasonably specific methods for detecting centrally acting drugs and their metabolites in the body fluids. A number of reviews and monographs on the subject are avialable, of which the manual of Clarke (1969) is particularly useful as a source book; other works recommended include a textbook by Curry (1969) and reviews on more specialised aspects by Cochin (1966) and Hammar et al. (1969) and papers in Psychopharmacology Bulletin (1970). Key research papers which form the basis of a number of important procedures are one by Beckett et al. (1967) on the detection of stimulants in urine by gas chromatography and thin layer techniques and another by Davidow et al. (1966) on thin layer chromatographic methods for narcotics in urine (see also Berry & Grove, 1971, 1973; Davidow & Fastlich, 1974).

The scope of analytical problems met by a specialized clinical laboratory undertaking body fluid analysis for drugs is illustrated by Table 13.3, which lists the principal drugs for which detection tests are available and in routine use at the Drug Division of the Chemical Pathology Laboratory of the Bethlem Royal and Maudsley Hospitals. Urine specimens from all new inpatients and from outpatients on each visit, in the associated Drug Dependence Research Unit are routinely examined for amphetamines, narcotics (morphine, methadone and cocaine) and barbiturates by the procedures indicated in the Table. Special instructions are issued to wards and clinics concerning collection specimens. For instance in investigating cases of suspected drug abuse or overdose, separate collection of all the urine specimens passed over a period is recommended rather than the pooling of specimens for 12 or 24 h periods. Each specimen is saved in a 250 ml screw-capped plastic bottle specially washed in the laboratory; it is important to avoid contamination with phenol antiseptics or detergents. In suspected cases of overdose the importance of saving all the urine passed on each occasion is stressed. Staff are encouraged take the initiative and collect specimens as soon as possible after the ingestion of drugs is suspected as many drugs are rapidly excreted.

Most of the current methodology used for analysis of body fluids for drugs is based on either thin-layer chromatography or gas–liquid chromatography, or a combination of both methods; only a few simple colour reactions in test tubes retain their usefulness (e.g. for salicylates). Neither method gives absolute identification, but for most drugs specificity is very high compared to the older procedures and it is rarely necessary to resort to derivative formation or more advanced techniques such as infrared analysis. Thin-layer procedures do not require expensive equipment and are technically less demanding than gas chromatography, but they are often less specific. To illustrate a typical procedure in this field a gas chromatographic method for the detection of amphetamine in urine is described below; we are grateful to Dr M. Buckell for details of this procedure, which is based on that of Beckett et al. (1967).

Amphetamine abuse is commonly encountered in addiction research; it can result in severe psychotic reactions often resembling schizophrenia (Connell, 1958).

Mass spectrometry and immunoassay are powerful research tools that may ultimately have application to large scale routine screening.

Amphetamine detection in urine

Equipment. A suitable gas chromatograph is the Pye series 104 model 15 with a dual flame ionisation detector with recorder. A 5 ft glass column of 10% polyethylene glycol 6000 on 100–120 mesh diatomite treated with KOH is suitable. Special glass-stoppered tubes ('amphetamine tubes') of 15 ml capacity are required; these have a finely tapered base and are used for concentrating the ether extract to a volume of about 50 μl. The dimethylaniline solution used as internal standard is made by adding 0·1 ml of the base to 24·9 ml of $CHCl_3$. It is kept in the refrigerator and renewed every 2 weeks. The stock standards of amphetamine and methylamphetamine in ethanol contain 1 mg of base/ml and are kept cold. The working standard is prepared by mixing 0·25 ml of the dimethylaniline solution, 1 ml of stock amphetamine and 1 ml of stock methylamphetamine solution.

Procedure. Approximately 10 ml of urine is poured into a glass plastic-capped Universal container, 1 ml of 4 M-NaOH is added followed by approximately 10 ml of diethyl ether and the lid made tight. The container is shaken vigorously in a mechanical shaker for a time period of 10 min. The ether layer is then allowed to separate, or centrifuged if necessary, and transferred to a glass-stoppered tube which is placed in the water bath at 37°C and the ether extract evaporated to approximately 0·2 ml under gentle suction with a stream of air, using a filter pump. The tube is removed from the bath, a small quantity of anhydrous sodium sulphate is added and the water-free, concentrated ether extract is transferred to an 'amphetamine' tube. This is placed in a 37°C bath and the contents allowed to evaporate to approximately 50 μl (halfway down the narrow part of the tube); evaporation must never go to dryness or amphetamine will be lost. Dimethylaniline (1 μl) is added to the final concentrated ether extract and the sample is then ready for gas chromatography. If this procedure is not to be carried out immediately, the tube is tightly stoppered and stored in the refrigerator until wanted. A modification of the method of Beckett et al., (1967) is used with a 1 μl injection. The retention times of amphetamine and methylamphetamine are determined at the beginning of each day's run using standard solutions. Optimum operating conditions vary a little from column to column; the minimum interval at which it is possible to inject consecutive specimens is about 12 min. Chromatographic conditions are: flow of nitrogen, 5 ml in 5 s; flow of hydrogen; 10 ml in 5 s; pressure of air; 13 lb/in^2; temperature; 140°C; attenuation × 500; chart speed, 10 in./h.

Table 13.3 Some drugs for which urine tests are available

Drug	Proprietary name	Source of test and remarks
Amphetamine, methylamphetamine		Beckett et al., 1967
Antihistamines		Davidow et al., 1965; some only
Barbiturates		Sunshine, 1963; group test
Caffeine		Davidow et al., 1966
Chloralhydrate		Fujiwara, 1916; group test
Chlordiazepoxide	Librium	Squirrell & Buckell, unpublished; common metabolite with diazepam and oxazepam
Cocaine		Davidow et al., 1966
Codeine		Davidow et al., 1966
Dextromoramide	Palfium	Davidow et al., 1966
Dextropropoxyphene	Distalgesic	Davidow et al., 1966
Diazepam	Valium	See chlordiazepoxide
Diethylpropion	Tenuate	Beckett et al., 1967
Dihydrocodeine	DF 118	Davidow et al., 1966
Dipipanone	Diconal	Davidow et al., 1966
Ethanol		Curry et al., 1966; fluoride container needed
Fenfluramine	Ponderax	Beckett et al., 1967
Heroin		Davidow et al., 1966
Methadone	Physeptone	Davidow et al., 1966
Methaqualone	Mandrax	Streck & Buckell, unpublished
Methylphenidate	Ritalin	Beckett et al., 1967
Morphine		Davidow et al., 1966
Nitrazepam	Mogadon	Squirrell & Buckell, unpublished;
Oxazepam	Serenid	See chlordiazepoxide
Paracetamol		Tompsett, 1969
Pentazocine	Fortral	Davidow et al., 1966
Pethidine		Davidow et al., 1966
Phenacetin		Tompsett, 1969; metabolised to paracetamol
Phenmetrazine	Preludin	Beckett et al., 1967
Phenothiazines		Forrest & Forrest, 1960; group test
Phenytoin	Epanutin	Sunshine, 1963
Propylhexadrine	Benzedrex	Beckett et al., 1967
Salicylates		Keller, 1947
Tricyclic antidepressants		Forrest et al., 1960; Davidow et al., 1966; group test

The tests in this Table are in routine use in the Drug Division of the Chemical Pathology Laboratory of the Bethlem Royal & Maudsley Hospitals. We are grateful to Dr M. Buckell, M.R.C.Path., and Mr J. E. Squirrell, F.I.M.L.T., for permission to publish this information and for help in its compilation. Many of the tests are modified versions of the published procedures referred to in the last column.

Comment. The procedure readily detects 0·5 µg of amphetamine/ml of urine; the lower limit is about 0·3 µg/ml using a 10 ml urine sample.

The extract prepared for analysis by gas chromatography may also be examined by thin-layer chromatography on silica gel G plates developed in methanol–ammonia (100:2 v/v). For details of typical thin layers methods for detecting amphetamine see Davidow et al. (1966), Cartoni & Cavalli (1968) and Marks & Fry (1968).

REFERENCES

Allan, J. D. & Raine, D. N. Eds (1969) *Some Inherited Disorders of Brain and Muscle.* Edinburgh: E. & S. Livingstone.
Armstrong, M. D. & Shaw, K. N. F. (1956) *J. biol. Chem.* **225,** 921.
Armstrong, M. D., Yates, N. K. & Stemmerman, M. G. (1964) *Pediatrics* **33,** 280.
Atfield, G. H. & Morris, C. J. O. R. (1961) *Biochem. J.* **81,** 606.
Baldridge, R. C. & Greenberg, N. (1963) *J. Lab. clin. Med.* **61,** 700.
Beckett, A. H., Tucker, G. T. & Moffat, A. C. (1967) *J. Pharm. Pharmac.* **19,** 273.
Beecher, H. K. (1958) *Experimentation in Man.* Springfield, Illinois: Thomas.
Benson, J. V. Jun., Gordon, M. J. & Patterson, J. A. (1967) *Analyt. Biochem.* **18,** 228.
Berlin, R. (1957) *Acta med. scand.* **158,** 113.
Berry, D. J. & Grove, J. (1971) *J. Chromat.* **61,** 111.
Berry, D. J. & Grove, J. (1973) *J. Chromat.* **80,** 205.
Bessman, S. P., Koppanyi, Z. H. & Wapnir, R. A. (1967) *Analyt. Biochem.* **18,** 213.
Beutler, E. (1969) In *Biochemical Methods in Red Cell Genetics,* ed. Yunis, J. J. New York: Academic Press.
Bollard, B. M., Culpan, R. H., Marks, N., McIlwain, H. & Shepherd, M. (1960) *J. ment. Sci.* **106,** 1250.
Bowers, M. B. Jun. & Gerbode, F. A. (1968) *Nature, Lond.* **219,** 1286.
Brenton, D. P., Cusworth, D. C. & Gaull, G. E. (1965) *Pediatrics* **35,** 50.
Brigham, M. P., Stein, W. H. & Moore, S. (1960) *J. clin. Invest.* **39,** 1633.
Cain, A. R. R. & Holton, J. B. (1968) *Archs Dis. Childn.* **43,** 62.
Carson, N. A. J., Cusworth, D. C., Dent, C. E., Field, C. M. B., Neill, D. W. & Westall, R. G. (1963) *Archs Dis. Childh.* **38,** 425.
Cartoni, G. P. & Cavalli, A. (1968) *J. Chromat.* **37,** 158.
Chandler, A. B. (1960) *Nature, Lond.* **185,** 697.
Chattaway, F. W., Hullin, R. P. & Odds, F. C. (1969) *Clin. chim. Acta* **26,** 567.
Clarke, E. G. C., ed. (1969) *Isolation and Identification of Drugs.* London: Pharmaceutical Press.
Cochin, J. (1966) *Psychopharmacology Bull.* **3,** 53.
Connell, P. H. (1957) *Amphetamine Psychosis.* Maudsley Monograph No. 5. London: Chapman & Hall.
Coppen, A. J. & Mezey, A. G. (1960) *J. psychosom. Res.* **5,** 56.
Crome, L. C. & Stern, J. (1967) *The Pathology of Mental Retardation,* p. 338. London: J. & A. Churchill,.
Curry, A. (1969) *Poison Detection in Human Organs,* 2nd edn. Springfield, Illinois: Thomas.
Curry, A. S., Walker, G. W. & Simpson, G. S. (1966) *Analyst, Lond.* **91,** 742.
Cusworth, D. C. (1969) *Biochem. J.* **111,** 1P.
Cusworth, D. C. & Westall, R. G. (1961) *Nature, Lond.* **192,** 555.
Davidow, B. & Fastlich, M. S. (1974) *Prog. clin. Path.* **5,** 85.
Davidow, B., Li Petri, N., Quame, B., Searle, B., Fastlich, E. & Savitzky, J. (1966) *Am. J. clin. Path.* **46,** 58.
Davis, J. M. (1970) *Int. Rev. Neurobiol.* **12,** 145.
DeJong, R. N. (1958) *The Neurologic Examination,* 2nd edn. London- Pitman.
De Laey, P., Hooft, C., Timmermans, J. & Snoek, J. (1964) *Ann. Paediat.* **202,** 321.

Efron, M. L. (1965) *New Eng. J. Med.* **272**, 1058, 1107.
Efron, M. L. (1966a) In *Proceedings of the Technicon Symposium on Automation in Analytical Chemistry*, p. 637. New York: Mediad Inc.
Efron, M. L. (1966b) In *The Metabolic Basis of Inherited Disease*, ed. Stanbury, Wyngaarden & Frederickson. 2nd edn, p. 393. New York: McGraw-Hill.
Ellis, G. & Goldberg, D. M. (1969) *Ann. clin. Chem.* **6**, 70.
Ertingshausen, G. & Adler, H. J. (1970) *Am. J. clin. Path.* **53**, 680.
Folin, O. (1904) *Am. J. Insanity* **61**, 299.
Forrest, I. S. & Forrest, F. M. (1960) *Clin. Chem.* **6**, 11.
Forrest, I. S., Forrest, F. M. & Mason, A. S. (1960) *Am. J. Psychiat.* **116**, 1021.
Fotherby, K., Ashcroft, G., Affleck, J. W. & Forrest, A. D. (1963) *J. Neurol. Neurosurg. Psychiat.* **26**, 71.
Fujiwara, K. (1916) *Sber. Abh. naturf. Ges. Rostock* **6**, 33.
Fuller, R. N. & Shuman, J. B. (1969) *Nature, Lond.* **221**, 693.
Garelis, E. & Sourkes, T. C. (1973) *J. Neurol. Neurosurg. Psychiat.* **36**, 625.
Gerritsen, T. & Waisman, H. A. (1964) *Pediatrics* **33**, 413.
Gerritsen, T. & Waisman, H. A. (1966) In *The Metabolic Basis of Inherited Disease*, ed. Stanbury, Wyngaarden & Frederickson. 2nd edn., p. 420. New York: McGraw-Hill.
Goodwin, F. K., Post, R. M., Dunner, D. L. & Gordon, E. K. (1973) *Am. J. Psychiat.* **130**, 73.
Gottfries, C. G., Gottfries, I., Johansson, B., Olsson, R. Persson, T., Roos, B. E. & Sjostrom, R. (1971) *Neuropharmacology*, **10**, 665.
Greer, M. & Williams, C. M. (1967) *Analyt. Biochem.* **19**, 40.
Groth, U. & Rosenberg, L. E. (1972) *J. clin. Invest.* **51**, 2130.
Guthrie, R. & Susi, A. (1963) *Pediatrics* **32**, 338.
Hamilton, P. B. (1963) *Analyt. Chem.* **35**, 2055.
Hamilton, P. B. (1968) Selected data for molecular biology, p. 343. In *Handbook of Biochemistry*, ed. Sobers, H. A. Cleveland: Chem. Rubber Publ. Co.
Hammar, C.-G., Holmstedt, B., Lindgren, J. E. & Tham, R. (1969) *Adv. Pharmacol. Chemotherap.* **7**, 53.
Hansen, I. L. & Crawford, M. A. (1966) *J. Chromat.* **22**, 330.
Harvey, D. (1968) M.Sc. Thesis. London University.
Hersov, L. A. & Rodnight, R. (1960) *J. Neurol. Neurosurg. Psychiat.* **23**, 40.
Himwich, H. E. (ed.) (1970) *Biochemistry, Schizophrenias and Affective Illnesses*. Baltimore: Williams & Wilkins.
Himwich, H. E., Kety, S. S. & Smythies, J. R. (eds.) (1967) *Amines and Schizophrenia*. Oxford: Pergamon Press.
Holton, J. B. (1972) *Ann. clin. Biochem.* **9**, 118.
Horwitt, M. K. (1963) *Recent Advances in Biological Psychiatry* **5**, 257.
Ireland, J. T. & Read, R. A. (1972) *Ann. clin. Chem.* **9**, 129.
Isselbacher, K. J. (1966) In *The Metabolic Basis of Inherited Disease*, ed. Stanbury, Wyngaarden & Frederickson. 2nd edn., p. 178. New York: McGraw-Hill.
Jepson, J. B. (1966) In *The Metabolic Basis of Inherited Diseases*, ed. Stanbury, Wyngaarden & Frederickson. 2nd edn., p. 1283. New York: McGraw-Hill.
Jepson, J. B. (1968) *Adv. Pharmacol.* **6B**, 171.
Kaeser, A. C., Rodnight, R. & Ellis, B. A. (1969) *J. Neurol. Neurosurg. Psychiat.* **32**, 88.
Kalckar, H. M., Kinoshita, J. H. & Donnell, G. N. (1973) In *Biology of Brain Dysfunction*, ed. G. E. Gaull. Vol. 1, p. 31. New York: Plenum Press.
Kaldewey, H. (1969) In *Thin-Layer Chromatography: A Laboratory Handbook*, ed. E. Stahl. 2nd edn., p. 471. London: Allen & Unwin.
Kedenburg, C. P. (1971) *Analyt. Biochem.* **40**, 35.
Keller, W. J. (1947) *Am. J. clin. Path.* **17**, 415.
Knox, W. E. (1966) In *The Metabolic Basis of Inherited Disease*, ed. Stanbury, Wyngaarden & Frederickson. 2nd Edn, p. 258. New York: McGraw-Hill.

La Du, B. N. (1966) In *The Metabolic Basis of Inherited Disease*, ed. Stanubry, Wyngaarden & Frederickson. 2nd edn., p. 366. New York: McGraw-Hill.
La Du, B. N. & Michael, P. J. (1960) *J. Lab. clin. Med.* **55**, 491.
Lajtha, A. (ed.) (1972) *Handbook of Neurochemistry* **7**.
Lups, S. & Haan, A. M. F. H. (1954) *The Cerebrospinal Fluid*. Amsterdam: Elsevier.
Mann, J. G. & La Brosse, E. H. (1959) *Archs gen. Psychiat.* **1**, 547.
Marks, V. & Fry, D. (1968) *Proc. Assoc. Clin. Biochem.* **5**, 95.
Maxwell, A. E. (1958) *Experimental Design in Psychology and the Medical Sciences*. London: Methuen.
McBean, M. S. & Stephenson, J. B. P. (1968) *Archs Dis. Childh.* **43**, 1.
McCaman, M. W. & Robins, E. (1962) *J. lab. clin. Med.* **59**, 885.
Medical Research Council Working Party on Phenylketonuria (1968) *Br. med. J.* iv, 7.
Milne, M. D. (1969) *Biochem. J.* **111**, 3P.
Moir, A. T. B., Ashcroft, G. W., Crawford, T. B. B., Eccleston, D. & Guldeberg, H. C. (1970) *Brain* **93**, 357.
Moser, H. W., Efron, M. L., Brown, H., Diamond, R. & Neumann, C. G. (1967) *Am. J. Med.* **42**, 9.
Mozziconacci, P., Boisse, J., Lemonnier, A. & Charpentier, C. (1968) *Les Maladies Métaboliques des Acides Aminés avec Arriération Mentale*. Paris: l'Expansion Scientifique.
Mudd, S. H., Edwards, W. A., Loeb, P. M., Brown, M. S. & Laster, L. (1970) *J. clin. Invest.* **49**, 1762.
Mudd, S. H., Laster, L., Finkelstein, J. D. & Irrevere, F. (1967) In *Amines and Schizophrenia*, ed. Himwich, Kety & Smythies, p. 247. Oxford: Pergamon Press.
Nyhan, W. L. (ed.) (1967) *Amino Acid Metabolism and Genetic Variation*. New York: McGraw-Hill.
Perry, T. L. (1967) In *Amino Acid Metabolism and Genetic Variation*, ed. W. L. Nyhan, p. 279. New York: McGraw-Hill.
Perry, T. L., Hansen, S., Tischler, B., Bunting, R. & Berry, K. (1967) *New Eng. J. Med.* **277**, 1219.
Perry, T. L., Hansen, S., Love, D. L., Crawford, L. E. & Tischler, B. (1968) *Lancet* ii, 474.
Peters, J. H., Berridge, B. J. Jun., Cummings, J. G. & Lin, S. C. (1968) *Analyt. Biochem.* **23**, 459.
Post, R. M., Goodwin, F. K., Gordon, E. & Watkin D. M. (1973) *Science, N.Y.* **179**, 897.
Post, R. M. Kotin, J., Goodwin, F. K. & Gordon, E. (1973b) *Am. J. Psychiat.* **130**, 67.
Prockop, L. D. (1973) In *Biology of Brain Dysfunction*, ed. G. E. Gaull. Vol. 1, p. 229. New York: Plenum Press.
Price, J. (1971) *J. nerv. ment. Dis.* **153**, 280.
Pscheidt, G. R., Berlet, H. H., Spaide, J. & Himwich, H. E. (1966) *Clin. chim. Acta* **13**, 228.
Psychopharmacology Bulletin (1970), Vol. 6, No. 1. (National Institute of Mental Health, Bethesda, Maryland.)
Raine, D. N. (1969) *Ann, clin. Chem.* **6**, 29.
Rodnight, R. (1961) *Int. Rev. Neurobiol.* **3**, 251.
Rodnight, R. (1968) In *Applied Neurochemistry*, ed. Davison, A. N. & Dobbing, J. p. 377. Oxford: Blackwell.
Rodnight, R. & McIlwain, H. (1955) *J. ment. Sci.* **101**, 884.
Rosenthal, A. F. & Yasen, A. (1969) *Clin. chim. Acta* **26**, 363.
Rupe, C. O. & Free, A. H. (1959) *Clin. Chem.* **5**, 405.
Samuels, S. & Ward, S. S. (1966) *J. Lab. clin. Med.* **67**, 669.

Samyn, W., Carton, D. & Hooft, C. (1970) *Clin. chim. Acts* **28**, 83.
Scott, C. D. (1972) *Adv. Clin. Chem.* **15**, 1.
Seakins, J. W. T. & Holton, D. C. (1969) *Biochem. J.* **111**, 4p.
Seakins, J. W. T., Haktan, M., Andrew, B. C. & Erssen, R. S. (1972) *Ann. clin. Chem.* **9**, 103.
Sharlit, H. (1933) *J. biol. Chem.* **99**, 537.
Shih, V. E. & Efron, M. L. (1970) *New Eng. J. Med.* **283**, 1206.
Sjostrom, R. & Roos, B. E. (1972) *Eur. J. clin. Pharmacol.* **4**, 170.
Smith, I (Editor) (1960) *Chromatographic and Electrophoretic Techniques*, 2nd edn., Vol.1. London: Heinemann.
Smith, D. M., Paul, R. M., McGeer, E. G. & McGeer, P. C. (1959) *Can. J. Biochem. Physiol.* **37**, 1493.
Snedecor, G. W. (1956) *Statistical Methods*, 5th edn. Iowa: Iowa State College Press.
Sprinkle, T., Greer, M. & Williams, C. M. (1969) *Clin. chim. Acta* **23**, 27.
Sunshine, I. (1963) *Am. J. clin. Path.* **40**, 576.
Tomlinson, S. & Westall, R. G. (1964) *Clin. Sci.* **26**, 261.
Tompsett, S. L. (1969) *Ann. clin. Biochem.* **6**, 81.
Troughton, W. D., St. Clair Brown, R. & Turner, N. A. (1966) *Am. J. Clin. Path.* **46**, 139.
Udenfriend, S., Lovenberg, W. & Sjoerdsma, A. (1959) *Archs Biochem. Biophys.* **85**, 487.
Uhendorf, B. W. & Mudd, S. H. (1968) *Science, N.Y.* **160**, 1007.
Varley, H. (1968) *Practical Clinical Biochemistry*, 3rd edn. London: Heinemann.
Vestergaard, P. & Leverett, R. L. (1958) *J. Lab. clin. Med.* **51**, 211.
Walker, W. H. & Bark, M. (1966) *Clin. chim. Acta* **13**, 241.
Weil-Malherbe, H. & Szara, S. I. (1971) *The Biochemistry of Functional and Experimental Psychoses*. Springfield, Illinois: Thomas.
Westall, R. G. (1960) *Biochem. J.* **77**, 135.
Whitfield, A. E. & Shepherd, J. (1970) *Clin. chim. Acta* **29**, 181.
Wong, D. W. K. & Pillai, P. M. (1966) *Archs Dis. Childh.* **41**, 383.
Woody, N. C., Snyder, C. H. & Harris, J. A. (1965) *Am. J. Dis. Childh.* **110**, 606.
Woolf, L. I. (1967) In *Phenylketonuria and Allied Metabolic Diseases*, ed. Anderson, J. A., & Swaiman, K. F., p. 50. Department of Health, Education and Welfare, U.S.A.
Woolf, L. I. (1972) In *Biochemical Aspects of Nervous Diseases*, ed. Cumings, J. N., p. 214. New York: Plenum Press.
Yunis, J. J. (ed.) (1969) *Biochemical Methods in Red Cell Genetics*. New York: Academic Press.

Index

Acetate into lipids, cortex, 277
Acetic thiokinase, extracted, 249
Acetoacetate, adenosine triphosphate and, 138
  oxidative utilization, 137
Acetone in enzyme extraction, 249
Acetone-dried brain, 247
Acetone extraction: brain tissue, 14, 89
Acetone powders: extraction of, 248
  preparation, 247
Acetylaspartic acid: 28
Acetylaspartylglutamate, 28
Acetylated hydroxy fatty acid esters, 81
Acetylcholine: bound transmitter, 234
  fractions, 231
  gel filtration, 235
  output, 186
Acetylgalactosamine, 39
Acetylneuraminic acid, by g.l.c., 84
Acid formation: anaerobically, 141
Acyl ester: determination, 93
Adenine: separation, 46
Adenine derivatives: output, 186, 188, 284
Adenine nucleotides: 44, 52
Adenosine: separation, 46
Adenosine cyclic monophosphate: noradrenaline, 139
Adenosine diphosphate: methods, 52
Adenosine nucleotides, 51, 139
Adenosine triphosphatases, calcium and magnesium, 277
  and see ATPase
Adenosine triphosphate: lost, 138
Adenylic acid: method, 52
Affinity chromatography: tyrosine hydroxylase, 250
After-discharges: fluid composition, 203, 205
  recording, 205
Air encephalograph, c.s.f., 299
Alanine: determination, 28
Alcohol dehydrogenase to measure NAD, 53
Aldehydes: g.l.c., 77
Aldolase isoenzymes, 251
Aldopentoses: determination, 42
Alkylating agents and tissue, 187
Alternating pulses, stimulation, 174
Alumina chromatography: phospholipid, 70
Amacrine cell processes, 262
Amine oxidase isoenzymes, 252

Amino acids: and acetoacetate, 138
  from the amygdala, 156
  analysis, 25, 302, 304
  chromatography, 13, 26
  content of tissue, 139
  disorders, 303–313
  electrophoresis in urine, 302
  in neural tissues, 25
  retina, 269, 272
  and tissue response, 186
Amino acid oxidase procedure, 313
Aminoaciduria in Hartnup, 309–311
  tubular, 308
Aminoacidurias, 302 seq.
Aminobutyric acid: determination, 27
  retina, 269
Ammonia: determination, 30
  increases postmortem, 30
  intoxication, 313
Amperometric method: chloride, 24
  titration, 25
Amphetamine, gas chromatography, 316
  in urine, 317
Amplifier: electrical, 176, 199
Amyloglucosidase: 38
Anaerobic conditions, 21
  experiments: manometric, 141
  glycolysis, 137, 277
Anaesthesia: cerebral changes, 34
Anaesthetics in gas phase, 196
Analgesics: and tissue, 187
Anterior commisure, 285
Anthrone: determination, 37
Anticoagulant: use of, 298
Anticonvulsants: and tissue, 187
Antidepressants in urine, 317
Antimony trichloride for retinol, 273
Apical dentrites, cortex, 278
Apiezon L for g.l.c., 77
Arachidic acid, g.l.c., 80
Araldite embedding, 239
Arginine and fumaric acid, 313
Argininosuccinate synthetase, 257
Argininosuccinic aciduria, 313
Ascorbate: noradrenaline oxidation, 139
Assimilation of tissue constituents, 138
Astrocytes: lipids, 61
  perikarya, 130
Ataxia: cerebellar, 308
ATP cellular, 233
  conversion to ADP, 177

x   323

ATPase activity: enriched, 250
  membrane-bound, 250
  sodium iodide reagent, 251
Atropine, inhibition, 187
Autoradiography, retina, 272
Axon, giant, 210
Axoplasm extruded, 210

Bacterial growth, urine, 294
Bacterial metabolism, mental disorder, 301
Barbiturates, urine assay, 317
Barium salt fractionation, 45
Base-ratio analysis, nerve cells, 46
Basic proteins: action, 187
Bicarbonate: removal, 140
Bicarbonate-CO_2 system, buffer, 134
Bicarbonate saline: composition, 133
Biochemical markers, subcellular organelles, 214, 216, 223
Bipolar electrodes, 196
Bipolar terminals, retina, 262
Blade: bow cutters, 119
Blade and guide, cutting, 107, 111, 118, 279
Blenders in dispersing, 245
Blocks, regional, brain, 284–289
Blood analyses, 297–304, 308
  apprehension, 297
  collection, 298, 302
  constituents, 136, 293, 297, 304
  enzymic activities, 297, 298
  tissue metabolism, 297
Blood plasma: composition, 136, 297, 304
Blood-retinal barrier, 267
Blood supply, retina, 271
Body fluids and antibiotics, 294, 301
  investigation, 293, 297, 301
  nervous disorders, 294
Bolting cloth: for tissues, 130
Bow cutter, blades, 117, 119
  and coverslip support, 118
Brain: anaesthetized animals, 5
  fixation *in situ*, 4–10
  newly-born guinea pigs, rats, 123
  obtaining, 1–8
  regional analysis, 7, 284–289
Brain subsystems: excitation, 188, 275, 281–284
Breis: prepared by rubbing, 128
Bromothymol blue, to detect lipids, 68
Burst response to stimulation, 202, 204–205
Butobarbitone on isolates, 204, 275

Calcein: calcium reagent, 22
Calcium: brain, 21–2
  fluid media 134
  frog retinae, 270
  ion movement, 186
Cannulae: for superfusion, 156

Capillary electrodes, 197
Carbohydrate: abnormality 257, 300, 314
  enzymes, stability, 244
  intermediates, 35–44, 257
  trimethylsilyl derivatives, 75, 77, 82
Carboxymethyl amino acids, 307
Carbowax: 20M polyglycol, 77
Cardiolipin: of brain, 61, 101
  R_F, 66
  mitochondria, 101
Catecholamine: output 184–186, 188
  uptake, 183
  urine, 300
Cation movements: inhibition, stimulation, 178, 186–188
Cations: brain, 20–23, 139
Caudate nucleus: action potentials, 275
  synaptic structures, 284
  tyrosine hydroxylase, 250
Cell bodies: collected, 129
  from fresh tissues, 130
Cell contents: release, 210
Cell discharge: extracellular Cl, 202
  extracellular observation, 196, 202
  recording, 196
  spontaneous, 205
Cell-firing, observed intracellularly, 201
Cell-fractions by centrifuging, 217 *seq*.
Cell-free preparations, 208 *seq*.
Cell-membrane potentials, 199
Cell nuclei in preparations, 129, 219, 226
Cell processes: lost, 128
Cell regions of negative potential, 199–203
Cells, rupturing, 210–214
Cell structure, broken, 128, 208
  disrupting, 213
Cell suspensions: from neural tissues, 128
Cellular phases: fluid, 17, 287
Centrifugal forces: values, 217
Centrifugation density-gradient procedure, 131, 222
  primary fractions, 217–220
Centrifuge: heads, 217, 223
Centrifuge-tube, 217
  layers, 221
  plastic, 221, 223
Cephalin: fraction from brain, 89
Ceramide of gangliosides, 101
Cerebellar block, 285, 287
  preparations: response, 203
Cerebellum: synaptosomes, 227
Cerebral cortex: creatine, 117
  heterogeneity, 276
  lactic acid, 117, 142
  metabolic properties, 104 *seq*., 277, 285
Cerebral hemisphere: cutting, 110, 123, 285
Cerebral peduncle, 283

Cerebral phosphates: barium salts, 43, 45
Cerebral subsystems, stimulation, 188, 275
Cerebral systems: metabolic responses, 159, 185
Cerebral tissues, electrical measurements, 191, 202
  isolated, 105, 133, 159, 191
Cerebroside: determination, 73, 95
  preparation, 94
  R_F, 66
Cerebroside sulphatases: defects, 253
Cerebrospinal fluid abnormal, 299
  analysis, 298
  chloride, 24, 134
  composition, 134
  and cortex, 138
  psychosis, 299
Cetylpyridinium chloride: 39
Charcoal: adsorption, 46
  removal: responses, 205
Chlordiazepoxide, urine assay, 317
Chloride, cellular, 21, 233
  determination, 23
  exchange, 24
Chloroform-methanol: extraction, 37, 62
  in chromatography 90, 91
Chlorpromazine: concentration-action curves, 182
  electrical pulses, 179
  inhibition, 187
  on isolates, 275
Cholesterol: assay after thin-layer chromatography, 66
  from brain, 64
  brain content, 61
  chromatography, 91, 92
  crystallization, 93
  determination, 71–3
  g.l.c., 75, 77
  Hanel and Dam's method, 72
  preparation, 92
  R_F, 66
Cholesterol esters: chromatography, 91
  determination, 71, 72
  by g.l.c. 75, 77
Choline: detection, 68
  determination, 99
  uptake, 234
Choline kinase extraction, 248
Cholinergic synapses: vesicles, 231
Cholinergic vesicles: ganglion, 232
Cholinesterases liberated, 246
Chopped tissue: dispersing, 126, 289
  extracting, 127, 289
  preparation, 121, 125, 288
  sampling, 123
  specified regions, 288
  transference, 127
Chromatography: adsorbents, 64, 88
  amino acids, 26
  argininosuccinic acid, 313

basic plates, 100
blood extracts, 302
choline, 236
developing solvents, 65
drugs, 315
elution, 90
gas–liquid, 74
hexose phosphates, 47
high pressure liquid, 46
isoenzyme, 251
lipids, column, 88
plasmalogen, 69
plates, 64
in reversed-phase, 67
Sephadex, 251
sodium borate plates, 67
thin-layer, 64
urine, 302, 310, 315
Chromatoplates: preparations, 64
Ciliary blood vessels, 262
Citrate anticoagulant, 298
Citric acid cycle: intermediates, 43
Citrullinaemia, enzymic abnormality, 257, 313
Clinical chemistry: abnormalities, 303 seq.
Clinician and chemist, 294
Clinical conditions: screening, 301
Cocaine: inhibition by, 150
  urine assay, 317
Coenzymes: saturating, 254
Collagenase: on tissue, 227
Collection of specimens, clinical, 294
Collicular subsystem, 282
Collodion specimen grid, 238
Concentric-electrode vessels, 165
Condenser pulses: alternating, 166, 172
  generators, 173–5
  stimulating, 166
Conduction velocity: tissue structure, 202–5
Cones, retina, 262
Control subjects: sex, weight, 299–301
Convulsants: and tissue, 187
Copper: brain, 22
  determination, 23
Cord blood: galactosaemia, 314
Corpus callosum impulse, 281
  response, 188
Corpus striatum: tetrodotoxin, 276
Cortex cellular make-up, 277
  sheet, 278
Coverslip: support for tissue, 117
Creatine: measurement, 51
Creatine phosphate: barium precipitation, 50
  enzymic methods, 51
Creatinine urinary, 296
Crushed tissues, 128
  suspension, 129
Cutting guide: for tissue, 108, 110
Cutting table: for tissue, 109–111
Cyanide: plating solution, 162

Cyclic AMP: estimation, 50
  neocortical tissues, 150
  response, 187
Cystathionine synthetase, 305
Cystathioninuria, enzymic dysfunction, 258
Cysteic acid, urine, 306
Cytochrome-c spectrophotometrically, 52
Cytoplasm, preterminal, 228
Cytoplasm from disrupted cells, 225

DEAE-cellulose chromatography, 90, 249
DEAE-cellulose column, 91, 251
Decapitation: chemical change, 10
Density gradient centrifuging, 218 seq.
  fractionation, 218
  hypertonic, 220
  subfractions, 224, 225
Dentate gyrus and hippocampus, 276, 284
  spike response, 205, 285
Deoxyribonucleic acid: phenol extraction, 55
Depolarization: in adverse conditions, 199
  Na content, 201, 202
  and permeability to K, 177
Depressive illnesses: isoenzymes, 252
  laboratory study, 293 seq.
Diazepam, urine assay, 317
Dibenamine: action, 187
Diet galactose-free, 314
  in phenylketonuria, 303
Dietary variation and urine, 296, 300
Diethyldithiocarbamate: copper, 22
Diethylene glycol succinate, 77
Differential centrifugation neural tissue, 217
Diffusion through vitreous, 265
Digitonin: cholesteral assay, 71
Diglyceride acetates: 67
Dimethylaminobenzaldehyde: 40, 310
Dimethylaminocinnamaldehyde: indoles, 310
Dinitrophenol: in paraffin, 143
Diphenylcarbazone: 24
Direct cortical response, 276
Disialogangliosides: fractionation, 102
Dopamine on action potentials, 276
  retinal, 264
Dragendorf reagent: to detect choline, 68
Drug abuse suspected, 315
Drug dependence, 315
  in body fluids, 314
Drugs: and excitability, 178, 187
  detection tests, 317
  nervous system, 294
Eating habits, disturbed, 300

Efflux, retinal, 269
  tissues, 182
Ehrlich's reagent, indoles, 310
  total hexosamines, 40
Electric organ: extraction, 11
  organelles, 209
Electrical characteristics: and metabolic response, 169
Electrical environment of tissue, 159
Electrical events: in fluid, 166 seq.
Electrical excitation: chemical changes, 172, 177
Electrical increase in respiratory rate, 180, 186
Electrical measurements in cerebral tissues, 191 seq., 275, 282
  resting potentials, 198
  transmitted response, 202–205, 284
Electrical pulses: sources, 173
Electrical responses to stimulation, 191, 202
  of retina, 268
  of subsystems, 276
Electrical stimulation: of metabolism, 159
  peak potential, 169
  of synaptosomes, 236
  of tissue metabolism, 177
Electrically-coupled synapse, hippocampus, 276
earthing, 194, 199
Electrode arms: flexibility, 169
  capillary, 197
  exposing limited portions, 171
  impedance, 173, 197–200
  micropipette, 197, 282–283
  penetration, 196, 199
  rapid-transfer, 145, 152
  recording extracellularly, 196
  resistance, 161, 197
  silver, 160
  stimulating and extracellular recording, 196
  superfusion arrangements, 151–156
  in test-tube, 168
Electrode materials: for tissue, 160 seq.
Electrode-puller, 197
Electrode systems: appraisal, 172
  in neurochemistry, 165
Electrode tubing, 197
Electrode vessels, 165, 169
  plating solution, 162
Electrodeposition: of gold, 162
Electrolytes: brain, 20–25
Electron microscopy: particles, 238
  piriform study, 281
  vesicles, 239
Electrophoresis, in clinical work, 252–3, 301
  high voltage, 302
  isoenzymes, 251
  starch-gel, 251
  of urine, 302
Electrophorus electric organ, 230

INDEX 327

Electroretinogram, excited, 261
  measured, 268
Emotional stress, blood, 297
Endoplasmic reticulum: cytochrome-c reductase, 219
  fragments, 229
  purification, 229
Entry of Na: electrically-stimulated, 150
Environmental stress, adrenaline, 297
Enzyme: acetone powders for, 247
  assay systems, 249
  cortex, 277
  fixing, 5, 14, 244
  inactivation, 11
  isolation, 243 seq.
  kinetic studies, 253
  masked, 246
  membrane-bound, 247
  occulded, 234, 239
  purification, 250
  stable, 244, 246
  in tissue dispersions, 216, 225, 245
Enzyme abnormality: kinetics, 253
Enzymic constituents, markers, 216, 224, 287
Enzymic cycling: glutathione, 29
  of NAD, 52
Enzyme kinetics: dysfunction and disease, 253
Enzyme stability during purification, 250
Enzyme treatment: of tissue, 130
Enzymology: brain disorders, 244, 252
Erythrocytes: argininosuccinase, 313
  blood collection, 297
  on centrifuging, 218
Ergot derivatives, inhibition, 187
Esterified phosphate: in extracts 45, 49
Ethanol: and tissue, 187
  in urine, 317
Ethanol extraction, 14
Ethanolamine phospholipids: column chromatography, 90
  components, 95
  preparation, 95
Ethanolamine plasmalogen: from white matter, 96
Ethyleneglycol succinate silicone, 77
Ethylmercurithiosalicylate: bacteriostatic, 295
Excitation: efflux or influx rates, 184, 186
  intracellular Na content, 177
Excitatory activity, retina, 268
Exclusion chromatography, esterases, 231
Excretion: patterns, diet, 300
Excretion rates urinary, 296
Exocytosis of vesicle, 230
Extracellular fluid: sampling, 156
Extracellular recordings, 196
  l.o.t., 202, 204

Extracellular response: barbiturates, ether, chlorpromazine, 202
Extracting agents: 11–15
  neural tissues for analysis, 10
Extraction from acetone powders, 248
  of enzymes, 246
Eye: dissection block, 266
  enucleated, 264
  species, 263
  structure, 262
Eye-cup buffer flow-rates, 271
  preparation, retina, 262, 266

Fatty acids: by g.l.c., 75
  of lecithin, 99
  separation, 67
Fatty-acid ester analysis, g.l.c., 79
Fatty acid esters, by g.l.c., 80
Fatty acid methyl esters: g.l.c., 79, 82
Fenfluramine in urine, 317
Ferric chloride-sulfosalicylic acid: to detect phosphate, 68
Fibroblasts cultured, 257, 308
  cystathionine synthetase, 305
Fibre-tracts white-matter, 275, 279, 281
Ficoll homogenizing medium, 228
  polydextran, 220
Field potentials: conditioning stimuli; Mg, Ca, 202
Fiske and Subbarow method, 47
Fixation: brain, 7
Fixation methods, validity, 9
Fixing and extracting neural tissues, 1
Flame ionization detector: for g.l.c., 76
Flame photometry: 20–22
Florosil: chromatography, 90
Fluid spaces: 19, 264
Fluids: brain, 17 seq., 134
Fluoride blood glucose, 298
Folch extraction, for lipids, 62
Fovea, retinae, 262
Fractionation schemes organelles, 209
Freeze blowing: brain, 6, 34
Freezing and thawing: ATPases, 246
Frozen-dried tissues: dissection, 129
Fucose, 40
Fumarate of brain, 41
  respiratory rate, 139
Funnel freezing: brain, 34

Galactolipid: eluates, 69, 90
  g.l.c., 75
Galactolipids, minor: separation, 95
Galactosaemia: defect, 314
Galactosamine: 39
Galactose: determination, 73, 75
Galactose phosphate uridyl transferase, 314
Galactosidase deficiency, 253
  lysosomal, 253

Ganglia: extraction, 129, 156
nutrient fluids, 156
preparation, 129, 227
superfusion, 156
superior cervical, 209, 227
Ganglion cells, retina, 262
Gangliosides: cerebral content, 102
dialysis, 63
extraction, 63, 102
galactosidase, 253
g.l.c., 82
glycosphingolipids, 101
separation, 63, 66
tissue excitability, 139
t.l.c., 66
trimethylsilyl derivaties for g.l.c., 77
Gargoylism: 39
Gas: supply and measurement, 140, 143
Gas change in manometric experiment, 180
Gas chromatography argininosuccinic acid, 314
retention data, 84
Gas mixture: for tissue, 153
Gaseous exchange: tissue, 140
Gas-liquid chromatography, 74 *seq.*
carrier gas, 76
column packings, 77
injection of sample, 83
lipids, 74, 77
preparation of samples, 80
Gasometric: experiments, 140, 142, 180
Genetic disorders: 94, 253, 303 *seq.*
Geniculate body, lateral, 283
Glass capillary microelectrodes, 197
Glial cells nuclei, 226
Glial clumps: preparation, 129
Glial fractions: density gradient separation, 131
Glomeruli of the cerebellum, 131
Glucosamine: phosphorylated, 40
Glucose: determination, 35
energy-yielding substrate, 137
in oils, 143
omitted: response lost, 186
radioactive, 36
Glucose oxidase: 37
Glucose 6-phosphate dehydrogenase: 36
estimating NADP, 53
Glucose salines: cerebral cortex, 137 *seq.*
Glutamate determination, 27
depolarizes, 201, 202
increased tissue K, 139
induces firing, 276
oxidizable substrate, 138
substrate, 186
Glutamate decarboxylase: acetone powders 249
regional, 289
Glutamate dehydrogenase, 28
and NAD, 53
Glutamate solutions: penetration, 201

Glutamine: determination, 27
hydrolysis, 28
Glutamine-synthesizing enzymes, 249
Glutaraldehyde, prefixation, 239
Glutathione: determination, 29
specific co-enzyme, 29
Glutathione reductase, 30
extraction, 249
Glycerides: chromatography, 90, 91
Glycerol: determination, 93
Glycine: output, 184
Glycogen: debranching complex, 38
cortical layers, 277
enzymic determination, 38
extraction, 37
Glycogen phosphorylase, 38, 49
cyclic AMP, 50
Glycolysis: brain, 135
inhibitors, 35
maximal excitation, 155, 179
retina, 268
Glycolytic enzymes, 243
cerebral tissues, 135, 155
intermediates, 35
Glycoproteins: 39
Glycosaminoglycans, 39
Glycylglycine, buffer, 136
Glyoxalase: assay, 29
Gold: electrodes, 162, 167
in metabolic experiments, 163, 172
sol, 163
and tissue slices, 163
Gold-electrode vessels: cleaning, 163, 165
Gold-plated platinum electrodes, 162
Gradients continuous, 220
discontinuous, 220, 221–225
Gradient displacement: tube sectioning, 222
Granules supraoptic nucleus, 230
Graphical: analysis, enzymic, 256
Grey-matter regions, 277
Grid electrodes, 168, 178
construction, 169
in electrode vessel, 169, 180
Grid electrode wires template for, 172
Grinding: neural tissues, 128, 210
Guanidinobutyric acid, 14
Guanine, chromatography, 46
Guanosine, 47
Guthrie method, phenylketonuria, 304
Gyrus, dentate, 284

Haemolysis, centrifugal, 298
prevention, 298
Half-cell, silver, 199
Hanel and Dam: reagent, 72
Hartnup disease abnormalities, 308
alanine, asparagine, 309
indicanuria, 310, 311
stature, 308
urine, 309
Heat precipitation: in analysis, 15

Heparin anticoagulant, 298
Heterozygote carriers diseases, 303
 histidine loading tests, 311
 phenylketonuria, 304
Hexokinase activity, extracting, 246
 occluded, 246
Hexosamines: determination, 39, 40
Hexose monophosphate, retina, 268
Hexose phosphates, 44
 in tissues, 49
Hexuronic acids: 39
Hexylene glycol: microtubules in, 230
Hippocampal block, 285, 287
Hippocampal neurones, responses, 284
Hippocampus and dentate gyrus, 284
 intracellular electrodes, 203
 negative spike, 276
 response, 205
 sections, 284
Histamine and cyclic AMP, 150
Histidinaemia, enzyme defect, 311
 $FeCl_3$ test, 311
Histidine, plasma, 311
Histidine ammonia-lyase, 312
Histidine deaminase, skin, 312
Histidine metabolism disorder, 311
Histochemistry: retina, 272
Homocysteic acid urinary, 306
Homocysteine urine, plasma, 307, 308
Homocystine, urine, 306
Homocystinuria, defects in, 305, 308
 enzymes, 305
Homocystinuric subjects, 305
 plasma from, 307
Homogenates: primary fractions, 219
 sucrose, 212–214
Homogenized in acetone, 247
Homogenizer: pestle and motar, 211
 piston-press, 212
 test-tube, 8, 211
 tube, ground, 211
 wear, 211
Homogenizing: tissues, 128, 213
 death and, 213
 procedure, 8, 213
Homovanillic acid, cerebrospinal, 299
Horizontal and bipolar cells, 262
Human serum: components, 134, 297
Hurler's syndrome: 39
Hyaluronidase on retina, 273
Hyamine 10-X hydroxide, 37
Hydrocarbons: g.l.c., 76
Hydrochloric acid: extracts, 12
Hydrogen peroxide in reactions catalysed by platinum, 164
Hydroxy fatty acid esters: g.l.c., 77
Hydroxybutyrate: cerebral substrate, 137
Hydroxyphenylacetic acid, urine, 303, 304
Hyperammonaemia, symptoms, 313
Hypoglycaemia: tissue modified, 186
Hypo-osmotic treatment, 239
Hypothalamic block, 285, 287
Hypothalamic nuclei in samples, 288

Hypothalamus: chlorpromazine, 276
 dissection, 285
 posterior, 283
 serotonin, 188
Hypoxanthine: 46
Hypoxia: incubation, 155
 output of amino acids, 156
Hypoxic excitation conditions, 186

Imidazole acids detected, 311
Imidazole derivatives, excreted, 303, 311
Imidazole propionic acid, 312
Imidazole pyruvic and lactic acids, urine, 311, 313
Immunological cross-reactivity, 253
Impedance change on penetration, 200
 of electrode vessels, 174
 of electrodes, 168, 198
Impulse-conducting tracts, 204
Impulse propagation: phenobarbitone, 202
Inborn errors, nervous system, 294, 301
Incubation: of neural tissues, 133, 140, 151, 156
Incubation chamber, 194
Incubation fluids: sampling, 144, 151
Incubation media, cellular environment, 233
 for tissues, 134, 139
Indican: Hartnup disease, 310
Indoles: mentally ill, 301
Indoles, urinary, 310
Indolic products, tryptophan, 311
Indoxylglucuronide, Hartnup, 310
Inorganic constituents: brain, 20, 135
Inorganic phosphate: enzymic determination, 48
 buffer, 135
Inosine: 46
Inositol: determination, periodate oxidation, 98
 phospholipids, 61, 65
Inositol phosphatides: three groups, 97
 turnover, extraction, 63
Intestinal flora, body fluids, 310
Intestinal function, 301
Intracellular: fluids, 17, 233
 recordings, 197–202
 substrate, 257
Intracellular micropipette measurement, 199
Intra-terminal constituents: 227, 234
Inulin: determination, 19
 isotopic, 20
Inulin space: measurements, 19
Iodine vapour, to detect lipids, 68
Iodoacetate, retina, 272
Iodotyrosine linked to Sepharose, 250
Ion content: cerebral cortex, 20, 134
Ion determinations: multiple, 24
Ion exchange, amino acids, 302, 304
 hexose phosphates, 50
 resins, 39, 40, 67, 302

Ion movement: normal tissue, 186
Ionic fluxes retina, 269
Isocitrate, 41
Isoenzymes defective, 257
   separation, 251
Isolated subsystems of the brain, 275 *seq.*

$KHCO_3$ extractant, 249
Kidney damage aminoaciduria, 314
Kinetic constants, enzymic, 256
Kinetics, neural enzymes, 253
Krebs-Ringer: salines, 134
   solutions, 133, 139
Kynurenine from tryptophan, 308

Labile intermediates: brain, 33
Lactate: estimating, 41
   extraction 14, 40
   output stimulated, 183
   in superfusion, 155
Lactate accumulation: in fluid, 178
Lactate dehydrogenase and NAD, 53
   occluded, 228
   of synaptosomes, 225, 228
   tissue contact, 225
Lactic acid: from glucose, 142
   on superfusion, 155
Lateral geniculate body, 283
Lateral olfactory tract: conduction, 204, 275
   piriform cortex, 275, 281
   respiration, 188
Layers I, II, III, cortex, 277
Lecithin: from brain, 89, 91
   in mammalian brain, 61
   measurement, 70, 99
   phosphatidyl choline, 99
   R_F values, 66
Lecithins: of lipoproteins, 99
Leucine: output, 155
Liebermann-Burchard reagent, 71
Light on retinal metabolism, 268
   stimulus, retina, 273
Lignoceric acid: column g.l.c., 80
Lipid amine: determination, 96
Lipid extractions: acetone, 89
   total, 62
Lipid content, brain, 61, 89
   slices, 277
   preparative and analytical methods, 88
Lipids: analytical techniques, 70, 88
   chloroform-methanol extraction, 62
   chromatography, 64, 74, 88
   classes in the brain, 61, 67, 89
   degree of unsaturation, 67
   detection reagents, 68
   highly polar, 67
   iodine staining, 68
   phosphorus determinations, 68, 71
   quantitative analysis, 64
   quantitative recovery, 69
   relative band speeds, 66
   subclasses, polar, 67
   thin-layer chromatography, 64
   t.l.c. subfractionation, 67
Lipoprotein matrix, membrane, 250
Liquid-liquid partition, 62, 64
Liquid nitrogen: fixation, 4–10
Liquid paraffin: cerebral cortex, 143
Liver biopsy cystathionine, 305
Liver nicotinamide, 308
Local addition made by micrometer syringe, 195
Localization of substance intracellular 225, 233
Localized: application of pulses, 171
Luciferase system, estimating ATP, 52
Lumbar region, c.s.f., 299
Lymphocytes homocystinuria, 305
Lysergic acid diethylamide on colliculus 276, 284
Lysocompounds: of lecithin and ethanolamine, 70
Lysolecithin: haemolytic, 246
Lysolecithin: acetone powder, 249
   R_F, 66
   use in extracting, 249
Lysophosphatidyl serine: separation, 101
Lysophospholipids: preparation, 100
Lysosomes: centrifuging, 209, 229
   components, 228
   glucuronidase, 216, 228
   isolation procedure, 228
   marker components, 216

Macerators for enzymes, 245
Magnesium: brain, 21, 22
Malate: brain, 41
Man: body fluid investigations, 293
Manic activity simulated, 299
Mannose: 40
Manometric apparatus, 140
   non-aqueous solvents, 143
   reading, 141
   synaptosomes, 238
Martin and Doty: phosphate, 48
Meat extracts creatinine output, 296
Medulla metabolic responses, 276
Medulla-pons dissection, 285
   region, 286, 287
Membrane ATPase, 225
   increase in permeability, 177
   resealing, 209
Membrane-bound enzymes, 246
Membrane potentials. cerebral cortex, 197–200
   and chemical measurements, 200, 202
   glutamate, 139
   observation, 198
Mental illness body fluids, 293–314
   diet, 301
   drug abuse, 314

nutrition, 300
venipuncture, 297
Mental retardation: galactose, 314
Mentally disturbed subjects creatinine, 296
urine, 294
Mescaline, actions, 187
Metabolic activities integrated, 294
Metabolic change: cerebral subsystems, 188, 275
stimulating tissue, 159, 186–8
Metabolic conditions: phosphocreatine 139, 150, 201
Metabolic disorders: screening, 301
Metabolic effects: of electrical stimulation, 177, 186, 276
Metabolic experiments: cerebral cortex, 141
quick-transfer, 149
with neural tissues, 133, 188
Metabolic inadequacies, 201
Metabolic mixtures: sampling and analysis, 144, 178, 180
Metabolic responses to electrical pulses, 177, 180, 186, 276
Metabolic status: abnormal, 299
Metabolic studies: isolated tissues, 105, 133
Metabolism, errors, 294
Metabolites bodily, 293
cerebrospinal fluid, 298
in urine, 294
Methanolysis: for g.l.c., 81
reagents, 81, 82
Metachromatic leucodystrophy: enzymes, 253
mucopolysaccharides, 39
Methadone, urine, 317
Methionine, raised, 305
test, 306
Methylamphetamine assay, 316
on caudate, 276
Methylphenidate in urine, 317
Michaelis constants, 254 *seq.*
Microdiffusion: method, 24
Microelectrode artefacts, 200
carrier, 198
filling, 197
penetrated cortex 197, 199, 201, 282
Micro-electrophoresis to tissue, 196
nucleotides, 46
Micromanipulator, electrode, 196
Micropipette electrodes and glutamate, 202
and intracellular recording, 196, 197, 199
responses, 202, 205
Microscopy: cell bodies, nuclei, 238
phase contrast, 238
Microsomal contamination: eliminating, 210
Microsomal fraction, cerebral tissues, 229
Microsomes: centrifuging, 209, 229

gangliosides, 102
lipids, 61
Microtubules isolating, 230
Micro-wave: fixation, 6, 34
Midbrain block, 285–287
Mitochondria amine oxidases, 252
enzymes, 224, 225, 227
hexokinase: solubilization, 246
lipids, 61
purified, 228
respiratory control, 228
respiratory quotient, 238
succinic dehydrogenase, 225
Mitochondrial spin, 209
Mitochondrial fraction, crude, 219
washed, 218
Molybdenum: electrodes, 164, 168
Molybdic acid: to detect phosphatides, 68
Monoamine metabolites, cerebrospinal, 299
Monoamine oxidase: occluded, 246
isoenzymes, 252
Monosaccharides by g.l.c., 75, 77
Monosialogangliosides: separation, 101
Morphology: myelin fragments, 225
of subfractions, 225, 238
Mossy fibre endings: synaptosomes, 277
Mossy fibres, path of, 284
Mucopolysaccharides, 39
Multiple forms of enzymes, 251
Muller cells retina, 262
Myelin: carbonic anhydrase, 216
centrifuging, 131
cholinesterase, 216
contamination, 227
fractions 97, 227
fragments, 224, 226
lamellae prepared, 131
lecithin levels, 99
phosphoinositides, 97
purified, density gradient, 226
Myelinated fibres, 99
Myoinositol: phosphate diester, 97

Narcotics in urine, 315
Negative potential: compounds modifying, 200, 202
Negative wave response, 204
Neocortical samples electrical activity, 276
cyclic AMP, 188
Neonatal screening procedures, 302
tissues: superfusion, 155
Neoplasma: metabolic response, 187
Nerve terminals: preparations, 219, 224, 227
re-sealed, 227
synaptosomes, 216, 227
Neural lipids: separation, 88
Neuraminidase and tissue, 187
Neurohumoral agents, cortex, 276

332 PRACTICAL NEUROCHEMISTRY

Neuronal: perikarya 129, 131
  nuclei, 226
Neurophysin granules, 231
Neurosecretory granules, 231
  markers, 216
  sedimenting, 209
Neurotransmitter metabolism, 234, 243, 252
  metabolic effects, 150
  regions, 287
  translocation, 184, 233
Neutron activation: determinations, 23
Nicotinamide nucleotides: 52, 138
  enzymic methods, 53
  estimation, 52
  tissue content, 44, 45, 138
Ninhydrin: to detect aminophosphatides, 68
Nitric acid: digestion, 21
  extraction, 21, 24
Non-aqueous fluids: for tissue, 143
  metabolic experiments, 144
Noradrenaline: and cyclic AMP, 150
  nialamide, 183
  output stimulated, 184, 186
Noradrenaline vesicles, 231
Noradrenergic terminals, varicosities, 230
Nuclear fraction, centrifugal, 218, 219
Nuclei: DNA marker, 216
  hypothalamic, 288
  markers for, 216
  neuronal, 226, 238
  recognized anatomically, 288
Nucleic acids: denatured, 54
  DNA, 216
  phenol extraction, 55
  Schmidt and Thannhauser, 54
  sugar and phosphate moieties, 55, 56
Nucleoplasm: microscopy, 226
Nucleosides, 46, 47
Nucleotidase of myelin, 216, 243
Nucleotides: 46, 52
  paper electrophoresis, 55
Nutritional status, body fluids, 300
  mentally ill, 301
Nylon cloth: sieving, 130

Occluded enzyme activity, 239
  lactate dehydrogenase, 234
Ocular abnormalities, 305
Olfactory bulbs, 281
Olfactory tract: electrical response, 204, 282
Oligodendrocytic nuclei, 226
Oligodendroglia: lipid, 61
Opiate-binding brain, 286
Optic nerve, 265
Optic tract, chiasma, 285
  conduction, 275, 283
  electrodes, 196
Optic tract-superior colliculus,
  electrical response, 276

  preparation, 283
  serotonin, 188
Ora serrata attachment, 265
  eye, 262
Orcinol: to detect galactolipids, 68
Organelles: biochemical markers, 216
  density in sucrose, 209
  fluid environment, 233
  isolation and properties, 223, 237
  marker substances, 215, 216
  microscopical examination, 214, 238
  release from, 236
Organic constituents: of fluid media, 137–139
Organic solvents: as tissue extractants, 13, 60, 89
Oscilloscope: monitor, 176
  camera, 198, 200
Osmotic shock deleterious, 220, 233
Osmotically ruptured fractions, 223
Oxaloacetate: 41
Oxidation: of glutathione, 29
  of noradrenaline, 183
Oxoglutarate, 41
Oxygen absence: diminish response, 186
  diffusion, 105
  high pressures, 139
  membrane potentials, 199, 202
  solubility, 143
Oxygen-sensitive: electrodes, 156, 237
Oxygen uptake: 180, 188
Oxygenation: tissues slices, 105, 140, 142, 145, 153, 178, 198

Pecten, retina, 264
Peduncle cerebral, 283
Pellagra symptoms, 308
Pentazocine in urine, 317
Pentoses: acid-soluble, 42
Pentose phosphates: 42
Peptides: 28
Perchloric acid: extractant, 12, 44, 49
Perfusion: by blood vessels, 156
  of the brain *in situ*, 156
  evaluation, 154
  retina, 269
Perikarya membrane removed, 131
Peripheral nerve: extraction, 11, 245
Peripheral nervous tissues vesicles, 230
Peristaltic pump, 151–3
Pestles plastic; clearance, 211
Phenacetin metabolised, 317
Phenistix test, 303
Phenmetrazine in urine, 317
Phenobarbital: response, 202
Phenolic derivatives, urine, 300
Phenothiazines in urine, 317
Phenylalanine administration, 303
  blood, 304
  reagent, procedure, 305
  tolerance tests, 303
  urinary, 303

Phenylalaninemia assay, 304
Phenylalanine-ninhydrin product, 304
Phenylketonuria: chemical tests, 303
  fluorometric procedure, 305
  metabolism, 303
  treatment, 303
Phenylpyruvic acid, urinary, 303
Phenytoin in urine, 317
Phosphate esters of choline, 99
Phosphate esters of ethanolamine, 95
Phosphates: acid-insoluble, 54
  acid-soluble, 33, 43 seq.
  fractionation by chromatography, 45
  incorporation, 135
  of nucleic acids, 54
  precipitation, 45
  in tissues, 44, 47
  use as a buffer, 135
Phosphatidic acid: myelin, 101
Phosphatidyl inositol: separated, 65, 66
Phosphatidyl serine: column chromatography, 66, 90, 91
  preparation, 96
Phosphocreatine: breakdown, 149, 177, 188
  extraction, 45
  glucose and oxygen supply, 138, 139
  tissue content, 44
Phosphofructokinase: labile, 244
Phosphogluconate: 37
Phosphoinositides: with brain proteins, 97
  fatty acid pattern, 97
  in solvent fractionation, 89
  by t.l.c., 65
Phospholipase D: and chromatography, 101
  splitting, 67
Phospholipid, by g.l.c., 75
  minor, 100
  precipitate, 44
  protein ratio, enzyme, 251
  separation: chromatography, 90, 91, 95
  turnover, 187
Phosphomolybate: extraction, 48
  method, 47
Phosphoprotein-phosphorus: turnover, 186
Phosphorus, total, 49
  to maintain anaerobiosis, 141, 185
Phosphorylase kinase: cyclic AMP, 50
Phosphotungstate: electron-dense, 239
Photochemical elements, 262
Photoreceptor cells degenerate, 272
Photoreceptor outer limbs, 263, 273
  population, species, 263
Photosensitive systems, retina, 261, 267, 272
Picric acid: extracts, 13
Pigment epithelium, 262
  retina relationship, 267
Piriform cortex K content, 188
  preparations, 204, 281

stimulation, 188, 282
Piriform lobe: anatomical, 281
Pituitary homogenization, 227, 231
  subcellular fractions, 209, 216
  synaptosomes, 227
Plasma argininosuccinic acid, 313
  citrulline, 313
  histidine, 312
  in homocystinuria, 305
  phenylalanine, 304
  S-amino acids, 307
  serine, asparagine, 302–2
Plasma membrane: nucleotidase, 216
  preparation, 229
  sacs, 227
Plasmalogen: by aldehyde, 73
Plasmalogen: determination, 73
Platinum: catalysis, 163
  caution, 163
  electrode material, 163–4, 196
  recording electrodes, 164, 196
Plating: gold, 162
Plexiform layer, 262
P/O ratio calculated, 238
Polarization: at silver, 160
Polarography for oxygen, 237
Polyphosphoinositides: calcium and magnesium salts, 97
  extraction, 63
  purification, 97
  separation, 62, 65, 91, 98
Post-mitochondrial fraction, 218, 229
Postsynaptic membrane: detergents to solubilize, 229
Postsynaptic potentials: compounds modifying, 202
Postsynaptic response: adenosine triphosphate of tissue, 202
  cortex, 282
Post-tetanic potentiation, 275
Potassium: brain, 21
Potential gradients: metabolic response, 167, 169
  in saline, 166, 169
  stimulating, 167, 179
Potentials after displacement, 202, 276
  intracellular 197–200
  resting, 201
Power amplifier, 175
Pre-synaptic responses, colliculus, 276
Preterminal organelles, 230
Primer: for superfusion, 153
Probenecid cerebrospinal fluid, 299
Procaine: inhibition, 187
Processes: entangled, 128, 210
Protein: acid insoluble, 54, 56
  chloroform phase, 63
  nucleoproteins, 56
  separations, 251
  in subcellular fractions, 234, 239
Proteolipid protein: 63
Proteolytic activity inhibiting, 246
Proteolytic digestion: 39
Proteolytic enzymes: releasing, 246

Protoveratrines: action, 187
Psychiatric complaints, 308
 hospitals, 294
Psychosis, homocystinuria, 305
Pulse frequency: response, 169, 175
Pulses, electrical: changes anaerobically, 185
 polarity, stimulating, 205
 separated by intervals, 177
Pulse generator, 174
Purification ratio, 215
Purine and pyrimidine bases, 46
Putamen, tissue, 285
Pyramidal neurons cortex, 277
Pyridoxine-deficient regions, 289
Pyridoxine feeding, 305
Pyruvate assay, 41
Pyruvate: hydrazone, 42
Pyruvate: in oils, 143
 oxidizable substrate, 137, 139
Pyruvate kinase: in assay, 50

Quick-transfer: apparatus, 145, 152
Quick-transfer holder, tissue, 150, 172
 in superfusion, 151
 stimulation, 150, 178, 183

Raffinose: space, 20
Receptor tips, pigment epithelium, 267
Recruitment of response, 204
Rectangular: pulses, 175
Refractory period, colliculus, 276
Refrigerated centrifuge, 217
Regional analysis: brain, 6, 7
Regional blocks: cell integrity, 288
 chemical constituents, 287
 components, 285
 preparing, 285
 rat brain, 284
 reproducibility, 286
 samples from, 289
Regions, brain, 261, 284
Regression analysis weighted, 254
Reineckate: choline, 99
Relative detector response: g.l.c., 83
Relative specific activity, marker, 215
Resistance: method of substitution, 175
Resorcinol: to detect gangliosides, 68
Respiration in bicarbonate saline, 134
 maintenace, 139
 of mammalian cerebral cortex, 139, 141
 of synaptosomes, 237
 tissue, 140, 177, 181
Respiratory experiments: minimal fluid, 142
 non-aqueous fluids, 143
Respiratory measurements; aqueous media, 140, 186
Respiratory response cerebral cortex, 180
 chlorpromazine, 180
 parts of the brain, 188
 requirements, 169, 182, 186
Responsibility in clinician experiment, 294
Retinae amphibians, birds, 263, 267
 anoxia, 272
 cell numbers, 264
 dark-adapted, 267
 electrical activity, 271
 frozen, 272
 glutamate, 272
 handling, 264
 infrared illumination, 268
 in situ, 271
 in vivo, 271
 ischaemia, 272
 isolated, 264
 metabolic studies, 268
 respiration, 268
 species, 263, 267, 271
 specific layers, 262
 structure, 262, 272
 subcellular fractionation, 273
 superfusion, 269
 vascularized, 272
 weight, 263
Retinal extracts, assay, 273–4
Retinal isomers, 273
Retinitis pigmentosa, 272
Retinol dehydrogenase assay, 274
Retinol, vitamin A, 273
Rhinal fissure cut, 281
Rhodanine reagent, 23
Rhodopsin lipoprotein, 273
 solubilizing, 273
Ribosomes desoxycholate, 229
 from membranes, 229
 neuronal, 216
 separation, centrifugal, 209, 219
Ribonuclease: degradation by, 55
Ribonucleic acid: column chromatography, 54
 enzyme, occurrence, 216
Ribose phosphate, 43
Ringer's physiological salines, 133
Rod cells, retina, 263
Rod outer segments, 273
Rotors centrifugal, 217

Saccules artificially-formed, 227
Salicylates in urine, 317
Salines: aqueous suspending media, 134
 for metabolic work, 109, 133
Schizophrenic subjects urine: 296
Schiff's reagent, p-rosaniline, 73
Schuster tube cutter, 221
Scintillation spectrometry of ATP, 52
Sedimenting conditions: nuclei *et al.*, 219, 226
Seizure discharges, dentate, 284
 media, 276

Sephadex to manipulate synaptosome, 221, 234
Serine phospholipids: preparation, 96
Serotonin: on field potential, 283
  output, 186, 188, 284
Sialic acid: 39, 101
Sialoaminoglycans, 39
Sieve: for tissues, 129
  steel, 131
Silica gel: silver-impregnated, 68
Silicic acid columns: lipids, 69
Silicone-coated needle, blood, 298
Silver: electrode material, 160
Silver electrodes: 'chloriding', 161, 197
  cleaned mechanically, 161, 170, 178
  half cell, 198, 199
  metabolic experiments, 169, 178, 180
  stimulating, 169, 172, 183, 192, 196
Silver-grid: electrodes, 168, 172, 178
Silver wire: enamelled, 169
Sine-wave: current, 169, 172, 175
  generators, 178
Slice chamber: electrodes and pipettes, 192, 196
  gas mixture, 193, 195
  procedure, 198, 203
Slice-grid, glass fibre, 193
Slices cellular elements, 277
  cutting, 106-126, 278
  guinea pig neocortex, 106, 277
  maximal yield, 278
  support, during metabolism, 118, 144
Slicing: with blade and guide, 110, 118
Skin, sampling, 312
  histidine deaminase, 311, 312
Skull: cutter, 3, 281
Sodium: brain, 21
Solvents: cerebral white matter, 62
  extraction with, 62, 89, 92
  fractionation of lipids, 89, 91
Specimens collection, 294
Spike, chloride-deficient media, 205
Spike discharges recorded, 201
Spinal cord preparation, 280
  regions, 299
Sphingolipids: hydrolysis, 100
  white or grey matter, 61, 100
Sphingomyelin: from brain, 89, 100
  cerebral content, 61, 100
  chromatography, 70, 89, 90, 91
  distribution, 61
  R_F, 66
  separated from lecithin, 100
Square-wave: pulses, 175
Staining technique: negative, 239
Stainless steel: electrode material, 164
Steel electrodes, 164, 196
Steroids: g.l.c., 75, 77
  and see Cholesterol
Stimulating electrodes tissue, 165, 196
  position, 200, 204
Stimulation: cation movements, 178
  optic tract, 282

respiratory and glycolytic response, 180
retina, 268, 271
of tissue, 177
Stimulator: with superfusion, 183
Stimulus artefacts, 163, 172, 205
Striatal block, 285
  tissue, 287
Striatum: chopped, 288-289
Subcellular-fractionation, neural tissues, 208
  apparatus, 217
Subcellular fractions metabolic properties, 232
  from regional block, 289
Subcellular organelles: isolation, 223
Subcellular particles from neural tissues, 208
Subiculum and presubiculum, 284
Substrates: oxidizable, 137, 139, 185
  utilization, normal tissue, 186
Subsystems, mammalian brain, 275
  components in, 287
  response to excitation, 188
Subthalamic tissue, 285
Succinate, assay, 41
  oxygen uptake, 138
  respiration increased, 186
Sucrose flotation mitochondria, 209, 273
Sucrose gradients, centrifuging, 220
  photoreceptors, 273
  space, 20
Sucrose solutions subcellular particles, 209
Sugar phosphates, chromatographic procedures, 47
Sulphated mucopolysaccharides, 39
Sulphatide: column chromatography, 90
  determination, 95
  preparation, 94
Sulphosalicylic acid: extracts, 13
Superfusate: amino acids, 154
Superfusing fluid, tissue, 153, 206
Superfusion: apparatus, 151, 156
  of chopped fragments, 156
  electrical conditions, 151
  flow-rate, 154
  of isolated tissues, 151, 203
  procedure, 154
  of retina, 271
Superfusion system: output of K+, 156
Superior colliculus, optic tract, 282
Surface electrodes: responses, 196, 203
Surface response piriform cortex, 204
Surface stimulation slices, 278
Surface-negative waves, cortex, 278
Surgical specimens, handling, 123, 213
Swing-out heads, 221
Synapses in cerebellum 227
Synaptic contact, retina, 263
Synaptic vesicles aggregated, 230
  bound transmitter, 231

from cerebral cortex, 209, 230
isolated, 231
morphology, 230, 239
negative staining, 239
separation, 231
Synaptosomal fractions of regions, 289
Synaptosome beds, 221, 236, 237
Synaptosome contaminants, 227
Synaptosome density layers, 233
Synaptosome: formation, 210
Synaptosomes: acetylcholine, 224, 225, 235
artificial particles, 210, 233
centrifuging, 224, 225, 227
electrical stimulation, 237
homogenizer, 210–211
isolation, 227
lactate dehydrogenase, 216, 225, 228, 234
leakage, 234, 236
morphology, 230
release experiments, 236
respiration, 237
ruptured, 232
transport experiments, 233
uptake, choline, 235

Tays-Sachs disease: enzymes, 253
TEAE-cellulose columns, 88, 92
Template, cerebral cortex, 277
Tetrodotoxin and stimulation, 179, 187
inhibition, 150
specific action, 178
Thin-layer chromatography, body fluids, 315
for lipids, 74, 88, 95
Thiobarbituric acid: retinal, 274
Thiolesterase: extraction, 249
Time-constant: pulses, 172, 174
Time-voltage: exponential, 173
Tissue: accumulation of metabolites, 117
chopped: dispersing, 126
chopped specimens, 122, 124
cold media, 187
crushing, 128
without fluid, 117
grinding in isotonic media, 128, 210
incubation: procedure, 144, 154, 178, 199
oxygenation, 105, 186, 199
preincubation bath, 121, 153
prompt transfer, 145
radioactivity released, 184
superfusion, 151, 182
surviving, 275
traumatic removal, 144
Tissue analysis: preparation for, 33
Tissue block: chopping, 122, 288
cutting, 107, 119, 288
regional, 285–289
Tissue chamber and electrodes, 192
assembly, 194, 199

and recording apparatus, 198
recording cell-discharge, 203
Tissue chopper: apparatus, 124
plane chosen, 288
preparation method, 121
when advantageous, 122
Tissue cutting: guide; preparation, 113
table, 109, 111
without added fluid, 115
Tissue disintegration: for different cell-types, 129
aided by enzymes, 130
Tissues electrically stimulated, 160, 177, 180, 188, 203
electrodes, 178, 194, 204
Tissue extraction: 10, 35, 37, 44, 60
agents, 11–15, 62, 89
Tissue fixation: 4, 44
Tissue fluid: 17
absorption, 115
extracellular, 19
minimal aqueous fluid, 116
Tissue-holding electrodes, 146, 152, 169, 172
Tissue-homogenizer, 211
Tissue integrity: blocks, 288
Tissue metabolism: procedure, 140 seq.
experiments, 133, 143
study in isolation, 105
Tissue normality: phosphocreatine, K, 135
Tissue preparation 107, 142 seq.
bow cutter, 117, 119
Tissue sampling: after incubation, 149
Tissue slice: drained of fluid, 115
in absence of fluid, 117
protein content, 116
template, 107
transfer and weighing, 112, 115
Tissue slicing: apparatus, 107
with blade and guide, 110
with bow cutter, 117
Tissue support, 118, 120, 142, 144
Tissue weight: determination, 112, 119
fresh weight, 115, 122
Toluene-ethanol, extraction, 62
Torpedo electric organ, 232
Transfer-stick: superfusion, 154
Transmission: within tissue, 171, 187, 204, 281
mossy fibres, 284
Transmitter, bound, 228, 231
quanta, 230
Transport, c.s.f and blood, 299
Trichloroacetic acid: extracts, 12, 44
Trimethylsilyl derivatives for g.l.c., 75, 82
Trinitrobenzene sulphonic acid: and sphingosine, 74
Triolein: R_F, 66
Triose phosphates: 44, 45
Tris buffer: solutions, 136
Trisialogangliosides: structures, 101
Trypsinizing medium: for tissues, 130

Tryptophan metabolism, 308, 310
  oral loading, 311
  transport, 308
Tungsten electrodes, 164, 196
  welding of electrode arms, 164, 165
Tungstic acid: extractant, 12
Tyrosine content, plasma, 309

Ultracentrifuges, preparative, 217, 223
Ultrasonication for enzyme, 239, 246
Unit discharge: Ca, Mg, barbiturates, 202
Urea hereditary disorders, 313
Urine aminoaciduria, 302
  argininosuccinic acid, 313
  collection, 294
  composition, timing, 295–296
  drug overdose, 315
  drugs, 317
  excretion rate, 296
  indolic products, 310, 311
  labile constituents, 294
  phenylpyruvic acid, 303
  preservation, 294
  stimulants in, 316
  variation, dietary, 300
Urocanic acid content of skin, 312
  plasma histidine, 311

Varicosities isolated from vas deferens, 227
  yielding vesicles, 230
Velocity of transmission, 202, 204, 205
Venipuncture ethical aspects, 297
  technique, 297–8
Ventricular c.s.f., 299
  perfusion, 156
Vesicles electron-lucid, 230
  isolation, 225, 231, 232
  large granular, 230
  negative staining, 239
  small granular, 230

Vessels: tested for leak, 165
Vestibular nucleus, lateral, 129
Vinyl ethers: determined, 73
Vision, photopic, 263
Visual cells, retina, 262
Visual pigments, eye, 273
Vitamin A aldehyde: illumination, 273
Vitreous body, eye, 262
  injection, 272
  removed, 266, 267
Voltage gradient: tissue, 166, 169
Voltage-response: curve, 169, 176

Warburg: manometric vessels, 140, 166
Ward specimen container, 295
  metabolic, 297, 300
Waring blender, 211
Washing: recentrifugation, 218
Weight and sex: purines, 301
White matter: composition, 60, 61, 96, 102
  chopping, 123
  conducted impulses, 275
  fibre tracts, 279, 280
  subcortical, 280, 281
Wilson's disease: copper, 22
  proline, 309

Xanthine, chromatography, 47

Yohimbine: inhibition, 187

Zinc: determination, 23
Zinc hydroxide: protein precipitation, 14
Zonal centrifugation, neural tissue, 212, 220
Zonal rotor: bowl, 220